The Anatidae Family

Biology, Ecology, and Conservation

Editors

Ashish Kumar Arya
Department of Environmental Science
Graphic Era (Deemed to be University)
Dehradun, India

Archana Bachheti
Department of Environmental Science
Graphic Era (Deemed to be University)
Dehradun, India

Vinaya Kumar Sethi
Faculty of Modern Studies
Uttarakhand Sanskrit University
Haridwar, India

CRC Press
Taylor & Francis Group
Boca Raton London New York

CRC Press is an imprint of the
Taylor & Francis Group, an **informa** business
A SCIENCE PUBLISHERS BOOK

First edition published 2027
by CRC Press
2385 NW Executive Center Drive, Suite 320, Boca Raton FL 33431

and by CRC Press
4 Park Square, Milton Park, Abingdon, Oxon, OX14 4RN

CRC Press is an imprint of Taylor & Francis Group, LLC

Library of Congress Cataloging-in-Publication Data (applied for)

ISBN: 978-1-041-09995-6 (hbk)
ISBN: 978-1-041-13429-9 (pbk)
ISBN: 978-1-003-66978-4 (ebk)

DOI: 10.1201/9781003669784

Typeset in Times New Roman
by Prime Publishing Services

Preface

The avian family *Anatidae*, encompassing ducks, geese, and swans, constitutes one of the most diverse, widespread, and ecologically influential groups within the order *Anseriformes*. Species belonging to this family are distributed across all continents except Antarctica and inhabit an extensive range of aquatic and semi-aquatic ecosystems, including freshwater wetlands, rivers, lakes, estuaries, coastal zones, and marine environments. Through their roles as herbivores, omnivores, and, in some cases, predators, anatids significantly influence primary productivity, nutrient cycling, seed dispersal, and energy flow within wetland and terrestrial ecosystems. Their close association with water bodies and their sensitivity to environmental change further establish them as important bioindicators of ecosystem health.

The present volume, *The Anatidae Family: Biology, Ecology, and Conservation*, is designed to provide a comprehensive and scientifically rigorous synthesis of current knowledge on the structural, functional, ecological, and evolutionary dimensions of the family. This volume consists of 11 chapters, which have been thematically structured to address diverse aspects of family Anatidae. Chapter 1 introduces the Anatidae family within a taxonomic and evolutionary framework, outlining its systematic position, historical classification, and diversification patterns. This foundational overview sets the stage for a deeper understanding of the morphological and physiological attributes discussed in Chapter 2, which examines skeletal structure, musculature, integumentary features, respiratory and digestive systems, and physiological adaptations related to flight, thermoregulation, buoyancy, and aquatic locomotion. Chapters 3 through 5 focus on the ecological dimensions of Anatidae family, emphasizing habitat selection, niche differentiation, and resource utilization. These chapters explore how species partition habitats across spatial and temporal scales, adapt to fluctuating environmental conditions, and employ diverse foraging strategies such as grazing, dabbling, diving, and filter feeding. Behavioral and social aspects, including mating systems, parental care, flocking behavior, communication, and dominance hierarchies, have also been addressed in these chapters. Furthermore, attempts have been made to address their adaptive significance in relation to reproduction, predator avoidance, and resource acquisition. Migration, one of the most distinctive and intensively studied phenomena in the Anatidae family, is examined in Chapter 6. This chapter integrates ecological, physiological, and behavioral perspectives to elucidate migratory routes, stopover ecology, energetic demands, and navigational mechanisms. The role of migration in shaping population connectivity and its implications for international conservation and disease transmission have also been evaluated. Chapter 7 introduces molecular approaches, particularly DNA barcoding and phylogenetic analyses, to clarify species boundaries, resolve taxonomic ambiguities, and reconstruct evolutionary relationships, thereby contributing to a more robust understanding of genetic diversity and lineage divergence within the family. Chapter 8 provides an assessment of population trends, distribution patterns, and conservation status at regional and global scales, drawing on long-term monitoring and international databases. Chapter 9 examines the multifaceted interactions between anatids and humans, including their roles in agriculture, hunting, domestication, cultural traditions, and ecotourism, as well as the conflicts and management challenges that arise from these interactions. Chapter 10 discusses diseases associated with the Anatidae family, with particular emphasis on avian influenza, parasitic infections, and other pathogens of ecological, economic, and public health significance. Finally, Chapter 11 synthesizes the major anthropogenic and natural threats facing the Anatidae family, including habitat degradation and loss, climate change, pollution, overharvesting, and the spread of invasive species. It also outlines contemporary conservation strategies, policy frameworks, and management approaches, emphasizing habitat restoration, international cooperation along migratory flyways, and the integration of scientific research into conservation planning. By integrating classical biological knowledge with modern ecological and molecular perspectives, this volume aims to serve as a reliable reference for ornithologists, ecologists, wildlife biologists, conservation practitioners, and graduate students along with contributing to informed decision-making for the sustainable management and long-term conservation of the Anatidae family.

Ashish Kumar Arya
Archana Bachheti
Vinaya Kumar Sethi

Contents

Preface *iii*

1. Introduction to Family Anatidae 1

2. Anatomy and Physiology of the Anatidae Family 36

3. Ecological Niches and Habitats of Anatidae Family 62

4. Anatidae Family: Behavioral and Social Structure 93

5. Foraging Ecology of Anatidae Family 130

6. Migration of the Members of Anatidae Family 157

7. DNA Barcoding and Phylogenetic Relationships in Anatidae 200

8. Status of Anatidae Family 224

9. Human–Anatidae Interactions: Historical, Cultural, and Conservation Perspectives 249

10. Diseases Associated with the Anatidae Family 275

11. An Assessment of the Threats to Anatidae Birds on a Global Level and Potential Conservation Strategies 306

Index 333

About the Editors 337

CHAPTER
1

Introduction to Family Anatidae

Deepak Rai[1], Ashish Kumar Arya[2], Harsh Gulati[3*], Gunica Senjam[3], Piyush Goyal[1], Archana Bachheti[2] and Vinaya Kumar Sethi[4]

[1] Department of Zoology, Kurukshetra University, Kurukshetra, Haryana, India
[2] Department of Environmental Science, Graphic Era (Deemed to be University), Dehradun, Uttarakhand, India
[3] Department of Zoology, School of Bioengineering and Biosciences, Lovely Professional University, Phagwara, Punjab, India
[4] Uttarakhand Sanskrit University, Haridwar, Uttarakhand, India

1. Introduction and Taxonomic Overview

The Anatidae family, including ducks, geese, and swans, has been identified as one of the largest and most widespread families of waterfowl in the world. It is classified in the order Anseriformes, which includes three separate families, two of which being Anhimidae (screamer) and Anseranatidae (magpie goose), and the third Anatidae that includes ducks, geese, and swans. Out of these, Anatidae includes the highest number of species and with approximately 170–180 extant species, as classified recently (Carboneras et al., 2020). There exist various species living today that exhibit great differences regarding body size, feeding methods, habitat use, as well as lifespan. On the other hand, similarities between anatids exist concerning their physical characteristics that help them live as aquatic as well as semi-aquatic animals. These may include: the lamellate bill adapted for filtering and catching food, the webbed foot facilitating swimming in water effectively, as well as the thick feathered bodies that are water-repellent. Dabblers are mostly engaged in searching in surface water while positioned head first. Diving ducks are engaged in underwater searching through powerful

*Corresponding author: harsh12@kuk.ac.in

DOI: 10.1201/9781003669784-1

feats of leg propulsion. Geese and swans are primarily herbivores feeding on herbaceous vegetation.

The radial (sunburst) diagram of family Anatidae as shown in Figure 1 gives a concise and current taxonomic perspective of waterfowl as classified under the order Anseriformes. The levels of taxonomic hierarchy are placed concentrically from order and family at the center to subfamilies and genera toward the outside edges of the diagram to offer a quick glimpse at its structure and not at phylogenetic detail. The demarcations made for major subfamilies such as Anatinae, Anserinae, Aythyinae, Tadorninae, Merginae, Oxyurinae, and Dendrocygninae tend to acknowledge large-scale adaptations and specializations based on ecologies such as grazing and surface-feeding to diving and seabed foraging (Kear, 2005).

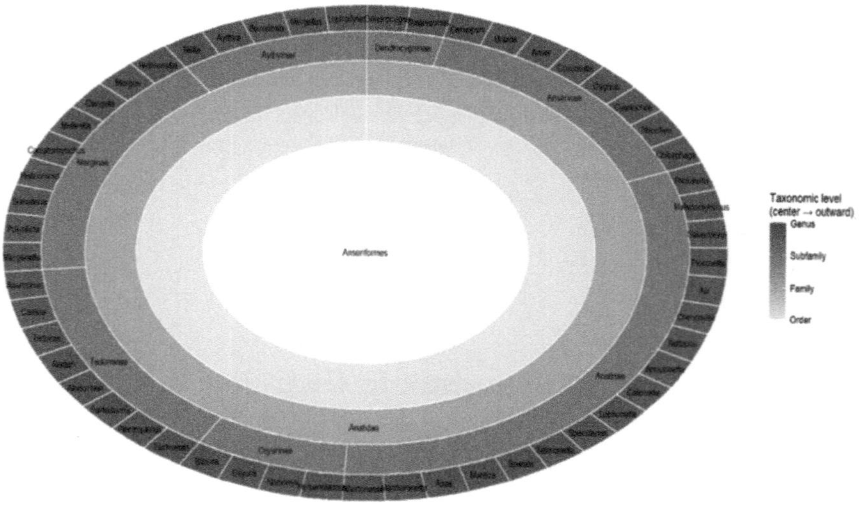

Figure 1. Radial (sunburst) representation of the taxonomic hierarchy of the family *Anatidae* (order: *Anseriformes*). Concentric rings depict successive taxonomic levels from the center outward (order, family, subfamilies, and genera). Graded shading corresponds to hierarchical depth, with lighter tones indicating higher taxonomic ranks and progressively darker tones representing lower ranks.

This classification conforms to the current systematic conventions of classification, as adopted by the International Ornithologists' Union and BirdLife International, taking into consideration the morphology of the species as well as genetic data (BirdLife International, 2023; Gill et al., 2024). Currently, research is underway regarding the waterfowl phylogeny, indicating the monophyly of the highest levels of the subfamily in the Anatidae family, but a respective improvement has been made in the topological relationships of the various genera (Sun et al., 2020; Zhang et al., 2022). Thus, Figure 1 serves as a basis of classification regarding various aspects of the evolution history, ecologic diversification, as well as the conservation pattern of the Anatidae species. The trait–taxonomy integration graph shows how the large taxa of Anatidae have large diagnostic differences based on prominence for different foraging behaviors, which reflect profound ecological specializations that have evolved over time along distinct lineages. The grouping of grazing, dabbling, diving, and piscivorous behaviors, for instance, for individual taxonomic groups demonstrates significant functional consistency in traditional systematic classification, although key emphasis has been placed on adaptive divergence between aquatic, wetland, or terrestrial ecological settings (Livezey, 1991; Green & Elmberg, 2014). The correlation between ecological traits, therefore, provides a clear ecological perspective that helps evaluate lineage relationships within Anatidae. Although ecological patterns derived from traits have significant validity for evaluating principles of adaptive divergence, the study findings raise concerns that lineage relationships, particularly when affected by rapid adaptation, convergence, or hybridization, cannot be fully resolved using ecological traits alone (Gonzalez et al., 2009; Ottenburghs et al., 2020).

2. Evolutionary History and Paleontology

The order Anseriformes, which includes anatids, might be one of the oldest surviving lineages among the Neornithes, the group comprising modern birds. The order Anseriformes might have crossed the great extinction event at the boundary of the Cretaceous and Paleogene eras because the fossil and genomic data indicate that its ancestors diverged from other bird groups during Late Cretaceous times, more than 65 million years ago (Clarke et al., 2005). The fossil record of Anatidae also varies greatly through time, with a clear escalation of fossil occurrences from the Paleogene to the Quaternary.

The fossil record of Anseriformes during the Late Cretaceous and Paleocene epochs is very sparse, and this can be attributed to the fact that birds of this group were of early evolutionary origin and that avian fossils did not conserve well at that time (Olson, 1985; Mayr, 2016). These fossils that date back to a very long time ago are highly significant to understanding the origin of early Anseriformes that lived before modern waterfowl originated. A considerable increase in fossil abundance is observed in the Eocene and Miocene ages, during global warming events, and development of wetland and freshwater biomes. Such climatic factors are considered as suitable for both evolution and fossilization of waterbird species, including the early ancestry of Anatidae (Mayr, 2016). The highest index of fossil abundance in the Miocene age is attributed to a rapid evolution of the family along with deposition in suitable fossilization environments. The high fossil occurrence reported for the Pleistocene epoch emphasizes the growing proportional representation of anatomically modern ducks in the fossil record. Further, the fossil occurrence is also affected by high sampling intensity and better resolution in the geological strata of successively younger geological ages (Olson,

1985). The fossil occurrence pattern reported supports the kind of progressive diversification that occurred in Anatidae, and which has been largely influenced by the paleoclimatic changes in the geological past.

The contrast between minimum fossil ages and divergence times based on molecular data reveals that there is a systematic discrepancy, where divergence times precede the oldest fossil records for large clades of Anatidae (Figure 2). This discrepancy is a common phenomenon in studies of avian evolutionary biology and arises because fossil ages are minimum divergence times rather than divergence times and are inherently incomplete, as ages refer to divergence times rather than origins of lineages (Clarke et al., 2005; Mayr, 2016). Nevertheless, small discrepancies between divergences and fossil records imply that there is considerable agreement between fossil and genomic data.

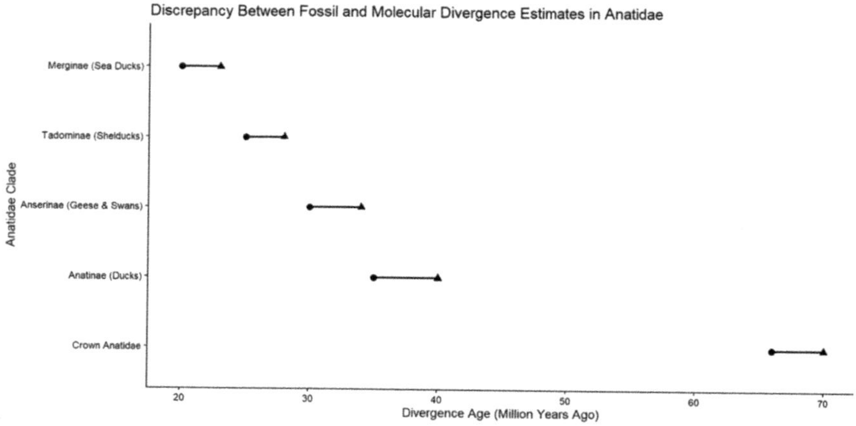

Figure 2. Fossil and molecular divergence estimates for major Anatidae clades showing systematic age offsets (Clarke et al., 2005; Jarvis et al., 2014; Mayr, 2016). Filled circles indicate fossil minimum ages; whereas triangles indicate molecular divergence estimates; connecting lines show divergence magnitude.

The greatest age disparities occur in Anatinae and Anatidae, indicating that initial periods of diversification for each likely predate the emergence of fossils that could be classified on the basis of morphology. The initial rapid evolution, followed by habitat range extensions into new Paleogene environments of freshwater, could have diminished the chances for fossilization or recognition (Olson, 1985; Worthy and Lee, 2008). Such observations have been recorded for other bird lineages that have undergone rapid evolution.

By contrast, smaller discrepancies between fossil and molecular divergence time estimates for Anserinae, Tadorninae, and Merginae suggest better fossil representation for these lineages, which could be due to a possible late emergence, larger body size, or habitat conditions conducive to fossilization. Lastly, molecular phylogenomic analyses confirm the quasi-simultaneous emergence of these three lineages around the transition from the Paleogene to the Neogene epoch (Prum et al., 2015; Jarvis et al., 2014).

Overall, the combined information provided by fossils and molecules gives strong support for understanding the evolutionary history of Anatidae species. On one hand, fossils provide all the required calibration data, and on the other hand, molecular data enhances estimates of divergence times and ultimately focuses on combined roles of paleoclimatic patterns and morphological innovation for the success of modern Anatidae (Mayr, 2016).

3. Morphological Adaptations and Functional Anatomy

The wide diversity of morphological adaptations present in the family Anatidae has enabled ducks, geese, and swans to thrive in aquatic and semi-aquatic habitats around the globe. These adaptations offer great diversity of species and also reflect specialization in swimming, feeding, flight, and thermoregulation.

One of the most distinctive characteristics of anatids is in the form of their bills, which may differ in dimensions depending on their diet. The majority of anatid ducks have flattened bills with small lamellae in the form of a comb that helps filter food debris or water. Although geese and swans have bigger bills that help the species feed on the grasses and water vegetation in ponds, others in their group called dabbling ducks apply a similar technique in consuming seeds as well as small crustaceans within ponds through filtration by their bills. River mergansers are small bill serrates that are fish eaters or predators (Kear, 2005).

Both the structure of the limbs and anatid locomotion are quite similar to each other. The position of the legs will be behind the body to enhance propulsion while swimming. Webbed feet will enhance swimming capabilities as the area of the foot will be greater. Further specialization, according to observation, will be there in divers, which will include greater bone density and more developed leg muscles to aid them in foraging in water (Livezey, 1996).

Anatid feathers are dense and very water-resistant as their uropygial gland produces oils. This waterproofing is necessary to maintain the insulation and buoyancy particularly in cold climates. Seasonal change also favors the development of structurally distinct breeding plumage, and feathers which are replaced.

3.1. Bill Morphology and Feeding Adaptations

The morphology of the bill of the Anatidae family is highly variable in structure which can be considered as a strong indication of ecological specialization and dietary adaptation. The manner in which different species acquire and digest food is highly influenced by the size, shape, and the internal frame of the bills. Since feeding ecology between ducks, geese, and swans has major differences, the shape of a bill has evolved as an important adaptive trait that enables the usage and exploitation of the dietary resources (Kear, 2005).

Food Association and Adaptation: Anamorphic patterns of bill structure indicate a high degree of association between the ecology and functional adaptation of the digestion score. Its form–function interactions can be depicted through the diagnostics features such as lamellae development, bill robustness, and serrations, which are the relationships that are used in the trophic specialization among the waterfowl (Johnsgard, 2010). Anatids possess many distinctive features among which lamellae—small, comb-like structures—line the margins of the bill. In most of the duck species, these lamellae play a role of filtering as the birds can select the plant matter, seeds, and small aquatic life in the mud and water. To feed on the surface and forage in the shallow water, dabbling ducks normally have slightly flattened bills which are well developed in lamellae. This bill design supports feeding behaviors such as tipping and filtering, which are common in swamps and flooded environments (Carboneras et al., 2020).

The morphology of the bill in the family Anatidae has an intense functional resilience of adaptive coupling with the feeding ecology and habitat utilization (Kear, 2005). Dietary specialization and foraging efficiency also depend primarily on variation in bill length, width, curvature, and in lamellae development between different ecological guilds (Livezey, 1996; Baldassarre, 2018). It is the dabbling and filter-feeding ducks that usually have broad, dorsoventrally flattened bills, the lamellae of which are well developed and assist in the efficient sieving of the vegetation, seeds, and small invertebrates of the shallow water masses (Kear, 2005).

3.2. Locomotor Adaptations

Other locomotor characteristics of the Anatidae family enable the animal to move effectively in the traditional land, water, and air environments. Although all anatids can fly, their limb structure and body composition show distinctly how they have been specialized in terms of their major locomotion and ecological tasks. Most anatids have their legs oriented toward the rear of the body, which is better in water propulsion, though it is poorly suited to terrestrial movement and posture. The rearward placement of diving ducks is more severe, whereas the leg placement of dabbling ducks is more balanced, allowing them to be more flexible between the aquatic and terrestrial activities (Livezey, 1996). All anatids have entirely webbed feet providing increased surface area. The feet of species such as diving ducks and stiff-tailed ducks mostly relying on underwater locomotion are larger and stronger with more thrust. On the contrary, the feet of swans and goose are adapted for walking as well as to swim and browse on land.

To facilitate controlled movement in an underwater environment, animal species that dwell in water normally tend to have stronger leg muscles and increased bone density. Swans and geese on the other hand have longer legs and more erect posture of the body, which befits land locomotion. In spite of

Table 1. Digestive system specialization in Anatidae in relation to feeding guilds

Digestive Component	Grazers (Geese, Swans)	Dabblers (Surface-feeding Ducks; *Anas* spp.)	Divers (Diving Ducks, Mergansers)	Functional Significance	Key References
Overall gut strategy	Adapted for processing fibrous, low-quality plant material	Generalist digestive strategy	Adapted for rapid digestion of animal prey	Reflects trophic niche and energy demands	Barnes & Thomas, 1987
Esophagus	Moderately muscular; facilitates bulk intake of vegetation	Moderately developed	Relatively narrow	Enables efficient food transport prior to processing	Barnes & Thomas, 1987
Crop	Small or poorly developed	Poorly developed or absent	Poorly developed or absent	Limited food storage; reliance on rapid ingestion and processing	Svihus, 2014
Proventriculus	Well developed; high gastric secretion	Moderately developed	Moderately developed	Chemical digestion prior to mechanical breakdown	Barnes & Thomas, 1987
Gizzard (ventriculus)	Large, highly muscular; thick koilin layer	Intermediate size and musculature	Reduced; less muscular	Mechanical grinding of fibrous plants and seeds	Kehoe &Ankney, 1985; Durant, 2013
Grit use	Frequent and essential	Common	Occasional	Enhances mechanical digestion in gizzard	Kehoe & Ankney, 1985
Small intestine length	Relatively long	Intermediate	Shorter	Balances absorptive capacity and digesta passage rate	Duque-Correa et al., 2022
Intestinal absorptive surface	High (long villi, increased surface area)	Moderate	Lower relative surface area	Optimizes nutrient extraction from diet	Durant, 2013

(*Contd.*)

(*Contd.*)

Digestive Component	Grazers (Geese, Swans)	Dabblers (Surface-feeding Ducks; *Anas* spp.)	Divers (Diving Ducks, Mergansers)	Functional Significance	Key References
Digesta passage rate	Slow to moderate	Moderate	Rapid	Matches diet digestibility and foraging mode	Durant, 2013
Paired ceca	Large and well developed	Moderately developed	Reduced or poorly developed	Microbial fermentation, water and electrolyte resorption	Clench & Mathias, 1995
Fiber digestion	Limited but ecologically significant; hemicellulose partially utilized	Low to moderate	Minimal	Supports herbivory and grazing strategies	Durant, 2013
Microbial fermentation	Active cecal microflora	Moderate microbial activity	Minimal fermentation	Enhances utilization of plant material	Clench & Mathias, 1995
Nitrogen recycling	Present via cecal processes	Limited	Negligible	Supports protein economy on low-quality diets	Clench & Mathias, 1995
Digestive plasticity	High; seasonal enlargement of gizzard and intestine	Moderate	Low to moderate	Allows rapid adjustment to dietary shifts	Barnes & Thomas, 1987; Durant, 2013
Migration-related adaptation	Pronounced digestive remodeling between seasons	Moderate remodeling	Limited remodeling	Supports long-distance migration and staging diets	Durant, 2013

these differences, all anatids are capable of maintaining highly developed flying locomotion because of strong pectoral muscles. All these adaptations allow Anatidae to inhabit environments of diverse types, and remain one of the most successful bird families.

3.3. Digestive System Specialization

Adaptation to diverse feeding habits which include grazing, dabbling, diving, and omnivory is reflected in the digestive system specialization of the Anatidae. Although the overall avian gastrointestinal model is also preserved, predictable variation exists in the relative development of the proventriculus, gizzard, intestine, and ceca, which reflects the disparities in dietary constituents and mechanical specifications (Barnes and Thomas, 1987; Kehoe and Ankney, 1985). The gizzard in herbivorous and omnivorous taxa, especially geese and swans, is extremely muscular and thus makes it easier to grind the fibrous plant debris and the seeds, which makes digestion more effective (Durant, 2013). Enzymes primarily carry out digestion and the small intestine is the primary site of nutrient absorption. Interspecific difference in small intestine length and absorptive surface depends on the food quality, habitat use, and energy constraints (Duque-Correa et al., 2022). Many of the anatids that feed on plants possess a well-developed paired ceca that aid in microbial fermentation, absorption of water and electrolytes, and the digestion of fiber (Clench & Mathias, 1995; Durant, 2013). Digestive organ plasticity seasonally enables the anatids to adapt quickly to the altered dietary environment, especially during migration (Barnes & Thomas, 1987).

This represents diet-based diversity in the organization of the digestive system in Anatidae, with grazing taxa and dabblers having prominent investment in the gizzard, small intestine, and ceca, and diving species having less investment in mechanical processing capacity, short intestinal length, and cecal development (summarized in Table 1).

4. Ecological Diversity and Behavioral Adaptations

The duckling diving nature tends to have a narrower, thinner-determined bill, which is designed to grab the benthic food such as mollusks and crabs. These species do not rely on filtration as much as they understand or probe food items in water. A change in feeding behavior where filtration is replaced by direct food capture is reflected in the loss of interest in lamellae of diving ducks. These physiological variations in the structure of the bills minimize the competition between sympatric organisms because they enable them to exploit different food resources in the same aquatic environment (Kear, 2005).

Forging behavior represents another of the great lines of ecological division among anatids. To reduce interspecific competition and enhance the niche munching, species have shaped rare feeding methodologies. Dabbling ducks typically forage on or near the water surface (or in low-lying marshes), and they have techniques such as tipping and filter feeding to consume aquatic vegetation and invertebrates (Johnson et al., 2014). However, diving ducks plunge into deeper waters and feed on submerged vegetation, crustaceans, and mollusks. As a result of behavioral changes in the utilization of land resources, geese and swans became primarily herbivores and occasionally feeding on grasses, sedges, and grown crops (Weller, 1999).

There is behavioral plasticity in the patterns of habitat use of anatids, as well as specialization. Many animals exhibit strong habitat selectivity during the mating season, and the broader use of the habitat outside the mating season, in response to seasonal changes in food availability and environmental conditions. They require behavioral flexibility in order to exploit various habitats on large spatial scales because migratory species, especially, rely on a network of breeding, stopover, and wintering locations that are geographically dispersed (Baldassarre & Bolen, 2006). Due to their adaptability, the waterfowl have a short adjustment period to their changing ecological conditions such as climatic and water-level changes. Social behavior is also used to enhance ecological success in Anatidae. The breeding season alters the social organization to territoriality and pair bonding as a sign of a shift in the selective forces around the fields of reproduction and offspring survival of the oppressed (Afton and Paulus, 1992). Some examples of the anti-predation behavior of Anatidae are vigilance, hiding of nesting habits, and responding to possible threats as a group. Although adults may perform the displays of diversion or even actively defend broods when predators approach, nesting females often conceal and camouflage their broods to reduce detection. Escape behavior and flock cohesiveness could reduce the threat of predation in open habitats significantly (Weller, 1999).

4.1. Foraging Guilds and Feeding Ecology

Based on similarities in feeding methods, habitat uses, and nutritional demands, the species of the family Anatidae may be separated into various foraging guilds. Besides representing ecological specialization, these guilds are also necessary in reduction of competition among coexisting species. Waterfowl can coexist in large numbers particularly in bountiful wetlands habitat as a result of exploiting multiple food resources or feeding areas in the same habitat (Baldassarre & Bolen, 2006). Another foraging guild even better adapted by a more aquatic habitat consists of diving ducks. Examples of benthic prey which these species preys using a diving strategy include mollusks, crustaceans, and submerged macrophytes. Underwater foraging requires specialization for both buoyancy regulation and submerged locomotion, and it demands high energy inputs. Due to this, diving ducks often display high habitat specialization and are therefore more vulnerable to changing prey levels and water quality (Johnson et al., 2014). Swans and geese form a largely terrestrial guild that forages rapidly through grubbing and grazing. Their diet is mostly made up of grass, sedges, and agricultural crops, and they often feed in the open terrestrial habitats alongside water bodies where they nest. The feeding ecology is assisted by social foraging behavior in which individuals tend to feed in organized groups, which enhances predator detection and efficiency in foraging (Baldassarre and Bolen, 2006). Overall, the foraging guilds into which Anatidae are subdivided lend support to the importance of feeding ecology in shaping waterfowl populations. The significance of such guilds to ecological research and conservation planning is that they influence the migration patterns, habitat selection, and vulnerability to environmental change.

Anatidae has very clear taxonomic segments, each of which is associated with a set of feeding strategies that is optimal to use the resources in freshwater, marine, and on the land (Kear, 2005; Baldassarre, 2018). Surface feeding and lamellate bill forms are mainly the reasons behind the dabbling and filter-feeding ducks to exploit aquatic vegetation, seeds, plankton, and small invertebrates, especially in shallow wetlands and marshes (Kear, 2005). Conversely, diving ducks and sea ducks are more commonly found in deep freshwater lakes and the coastal seas where they use foot-propelled diving to reach benthic invertebrates and mollusks to create less competition with surface feeding taxa (Livezey, 1996; Delacour & Mayr, 1945). The grazing of geese is an evident ecological transition in the context of Anatidae because they have a high dependency on the land habitat, including grasslands and fields, where their feeding economy is dominated by herbivorous feeding (Kear, 2005). Swans also prey on aquatic macrophytes but can forage in deeper water due to their long necks and larger bodies (Baldassarre, 2018). Piscivorous anatids, especially the mergansers, have specialized ecological niche designed by the presence of clear rivers and lakes wherein fish are hunted using slick bodies and narrow-serrated bills (Delacour & Mayr, 1945). Figure 3 shows that the allocation of foraging guilds has facilitated niche partitioning and co-occurrence among Anatidae in a wide range of aquatic and terrestrial habitats through ecological specialization in feeding behavior (Kear, 2005; Livezey, 1996).

4.2. Habitat Preferences and Distribution Patterns

Seasonal environmental fluctuation, behavioral flexibility, and biological requirements all influence habitat preferences and dispersal patterns within the Anatidae family. Ducks, geese, and swans inhabit a great variety of habitats in fresh wetlands, rivers, lakes, and estuaries, as well as coastal marine ecosystems and the surrounding terrestrial habitat. Intensively, species preferences of habitat are widely influential in giving way to spatial differentiation and reduce on competition within waterfowl populations (Baldassarre & Bolen, 2006). The Anatidae species are pro-shallow in that many of them prefer habitats with shallow freshwater that have enough nesting cover, food store, as well as protection from predators during mating time. Wetlands having emergent plants are critical because they facilitate emergent feeding of adults as well as raising brood. Conversely, the habitat decisions become broadened during periods out of the mating season where large lakes, reservoirs, estuaries, and agricultural scenery are resorted to where food supply might be more predictable (Weller,

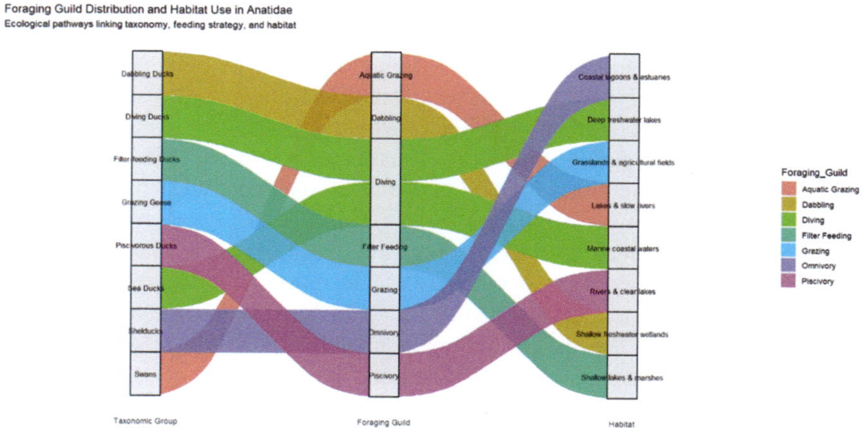

Figure 3. The Sankey diagram illustrates major ecological pathways linking taxonomic groups of Anatidae with dominant foraging guilds and habitat use, highlighting the diversity of feeding strategies across freshwater, marine, and terrestrial environments.

1999). This is a change in habitat utilization, which is an adaptation of reaction to changing energy and reproduction demands.

Patterns of distribution and migratory behavior in anatid evolved together. Long-distance migrations between breeding and wintering areas entail the use of established flyways by many species between the continents. Consequently, population distribution is highly dynamic in space, and the regional abundance changes depending on habitat availability and seasonal migration (Johnson et al., 2014). Small-scale habitat selection is influenced by environmental factors such as depth of water, vegetation cover, prey density, and the level of disturbance. Whereas individual species are more opportunistic and faster to colonize new environments, others are highly site-faithful, moving to and away across multiple years onto the same breeding or wintering sites. In spite of variations in reliance on such habitat, numerous waterfowl species are increasingly using human-made artificial ecosystems including the irrigation system, rice fields, and urban wetlands as part of their habitat networks (Baldassarre & Bolen, 2006). Overall, the patterns of distribution of the anatids and their habitat preferences are balanced in terms of ecological specialization and behavior adaptability. Understanding these patterns is required to predict population responses to the change in the environment and the development of effective conservation efforts.

4.3. Migration Pattern and Flyaway Systems

Seasonal environmental fluctuation, behavioral flexibility, and biological requirements all influence habitat preferences and dispersal patterns within the Anatidae family.

For this family, one of the biggest characteristics is migration, which allows the species to get the optimum level of resources throughout any season. Anatids make long-distance migrations, linking breeding and non-breeding regions of high latitude and low latitude, becoming patterned in globally defined migratory flyway networks (Kear, 2005; Baldassarre, 2018). Population connectivity within Eurasia, Africa, East Asia, and Australasia via major important flyways such as the Central Asian Flyway, East Asian–Australasian Flyway, and African–Eurasian Flyway is a way of connecting breeding, stopover, and wintering places (Wetlands International, 2021). Such flyway regimes facilitate the rich variety of ecological plans, reaching both short-range movements at a regional scale up to trans-continental scaled ones, and they are essential to keep population feasible among the migratory waterfowl (Livezey, 1996; Kear, 2005). The sketch map of flyways in Figure 4 highlights the migration route patterns on large scales and puts significant stress on the interdependence of Anatidae on the wetlands and staging networks, which are present in several countries and biogeographic regions. This reliance on spatially discontinuous habitats causes migratory anatids to be highly susceptible to habitat degradation, phenological injuries caused by climate changes, and distraction of important stopover locations on flyways (Baldassarre, 2018; Wetlands International, 2021).

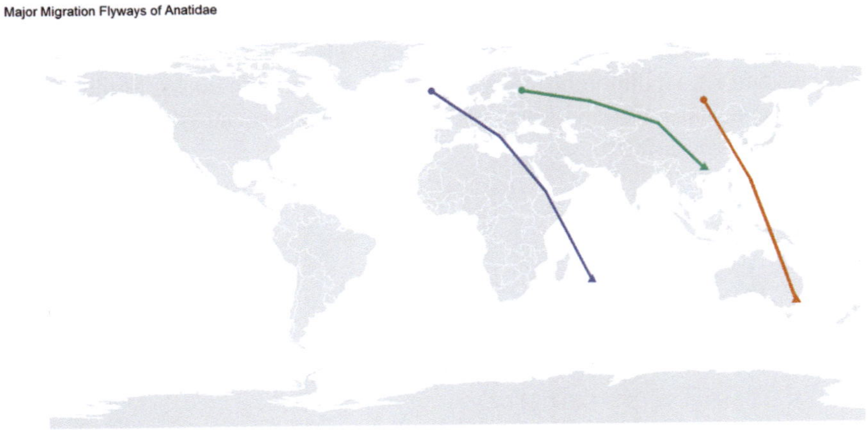

Figure 4. Major migration flyways of the family Anatidae illustrating broad-scale movement corridors between breeding and wintering regions across Eurasia, Africa, East Asia, and Australasia.

5. Reproductive Biology and Social System

Reproductive biology in Anatidae is characterized by predominantly seasonal monogamy, diverse nesting strategies, and marked variation in parental care, all closely linked to ecological conditions and life-history traits. In most duck species, pair bonds are formed during the non-breeding period or early migration and typically persist for a single breeding season, facilitating synchronized migration and reproductive timing (Baldassarre & Bolen, 2006). Nesting strategies range from concealed ground nests to cavity or cliff nesting, reflecting habitat use and predation pressure. Females generally undertake nest construction, incubation, and early brood care, whereas male involvement after egg-laying is limited. In contrast, geese and swans commonly exhibit long-term pair bonds with substantial male participation in territory defense and offspring protection (Afton & Paulus, 1992; Black, 2001).

Clutch size and incubation duration vary among species in relation to latitude, food availability, and predation risk, with larger clutches often occurring at higher latitudes to compensate for shorter breeding seasons (Weller, 1999). Social organization in Anatidae is seasonally dynamic, shifting from territoriality during breeding to gregarious flocking outside the breeding period. These guild-level patterns in pair bonding, parental roles, mating competition, nesting strategies, and territoriality are synthesized in Figure 5. In this figure, dot size represents the relative level of expression of each reproductive or social trait (low, moderate, high), derived from comparative synthesis of published studies, while color distinguishes major functional groups (dabbling ducks, diving ducks, sea ducks, mergansers, geese, and swans). This relative, nonquantitative framework is intended to emphasize broad functional contrasts among groups rather than species-specific variation (Afton, 1981; Black, 2001; Johnsgard, 2010). Courtship displays and sexual dimorphism, particularly pronounced in ducks, further reflect the role of sexual selection in shaping reproductive and social systems within the family (Johnsgard, 2010).

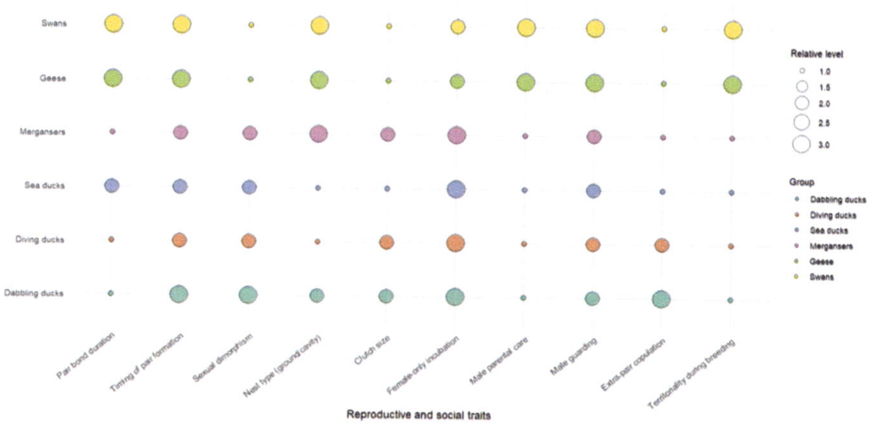

Figure 5. Comparative visualization of reproductive biology and social system traits across major functional groups of Anatidae.

5.1. Mating System and Sexual Selection

The Anatidae family includes a variety of mating strategies that balance the demands of sexual selection, parental investment, and ecological limitations. Seasonal monogamy, in which pair bonds are formed before or during the breeding season and usually disintegrate once reproduction is finished, is the most common mating pattern among ducks, geese, and swans. According to Baldassarre and Bolen (2006), this system maintains flexibility over subsequent reproductive cycles while enabling synchronized migration and breeding. Even though monogamy is common, Anatidae species differ greatly in their mating habits. Males in many duck species compete fiercely for access to females because they offer little parental care outside of courtship and pair formation. Strong sexual selection has been fuelled by this disparity in parental investment, especially preferring characteristics that improve male appeal or competitiveness. Because of this, ducks frequently exhibit strong sexual dimorphism, with males displaying sophisticated vocalizations, elaborate courtship displays, and vividly colored plumage during the breeding season (Johnsgard, 2010).

In Anatidae, courtship behavior frequently includes verbal cues, postures, and ritualized gestures that serve as markers of individual quality. Before establishing pair bonds, females typically assess multiple potential mates, and social dominance, plumage condition, and display intensity can all have an impact on mate choice. These behaviors can reinforce reproductive isolation among closely related species and promote assortative mating (Emlen & Oring, 1977).

Swans and geese, on the other hand, typically exhibit higher levels of biparental care and longer-term pair bonds. Sexual dimorphism is less noticeable in these species, and compatibility and the capacity to defend one's territory seem to be more important factors in mate selection than ostentatious decoration. Long-term monogamy promotes selection for characteristics linked to cooperation and parental investment, while decreasing opportunities for repeated mate competition (Baldassarre & Bolen, 2006).

Anatidae's mating strategy is further complicated by reports of forced copulation and extra-pair mating in a number of duck species. The evolution of reproductive architecture and female choice mechanisms are affected by these behaviors, which also affect the success of reproduction. These exchanges demonstrate how sexual conflict functions as an extra factor influencing family mating choices (Johnsgard, 2010).

In general, interactions between environment, behavior, and life-history strategies influence Anatidae mating systems and sexual selection. Interpreting patterns of reproductive success, population structure, and evolutionary diversification in ducks requires an understanding of these dynamics.

5.2. Nesting Ecology and Habitat Requirements

Anatidae exhibit strong habitat-specific nesting preferences, with a pronounced bias toward freshwater wetlands that provide shallow water, dense emergent vegetation, and reduced predation risk. Synthesis of published studies indicates that freshwater marshes support the highest proportion of nesting activity, reflecting their structural complexity and availability of concealment cover, which enhances nest survival (Kaminski & Gluesing, 1987; Baldassarre, 2018). Habitats such as lakes and ponds also contribute substantially to nesting distribution, particularly for dabbling ducks and geese that utilize shoreline vegetation and islands for nest placement (Kear, 2005). Riverine floodplains represent an important secondary habitat, especially in dynamic fluvial systems where seasonal flooding creates temporary nesting opportunities, though nest success in these areas may be constrained by water-level fluctuations (Elmberg et al., 2005). Agricultural wetlands account for a smaller but notable proportion of nesting sites, reflecting adaptive use of human-modified landscapes where natural wetlands are limited; however, nests in these habitats are often exposed to higher disturbance and predation pressures. Coastal wetlands contribute the lowest proportion of nesting activity, largely restricted to specific taxa adapted to saline or brackish conditions, highlighting the overall freshwater dependency of most Anatidae species during the breeding season (Baldassarre, 2018).

5.3. Parental Care and Brood Ecology

Parental care in Anatidae varies widely among species in relation to mating systems, ecological conditions, and predation pressure, and is a critical determinant of reproductive success. In most ducks, males typically disengage after copulation or early incubation, leaving females to provide the majority of parental care, a pattern reflecting sex-specific investment strategies (Afton & Paulus, 1992). Although Anatidae produce precocial young capable of early movement and feeding, broods remain highly dependent on maternal guidance to food-rich, shallow wetlands offering adequate cover. In contrast, swans and geese frequently exhibit biparental care, with joint defense, vigilance, and brood guidance substantially enhancing offspring survival, particularly in open habitats (Weller, 1999; Baldassarre & Bolen, 2006).

Brood vulnerability in Anatidae declines progressively from hatching to the pre-fledging stage as chicks undergo rapid physiological and behavioral development. Newly hatched broods are characterized by limited mobility and poor thermoregulatory capacity, making them highly dependent on maternal brooding and concealment to minimize exposure to predators and adverse environmental conditions (Afton & Paulus, 1992; Kear, 2005). During early brood stages, increasing locomotor ability and partial thermoregulation allow broods to exploit shallow, food-rich habitats; however, vulnerability remains elevated due to limited vigilance and escape performance (Baldassarre, 2018). As development proceeds, improved swimming ability, enhanced vigilance, and more effective habitat selection substantially reduce predation risk, resulting in near-independent movement prior to fledging (Afton & Paulus, 1992; Elmberg et al., 2005).

6. Biography and Global Distribution

The family Anatidae exhibits one of the broadest global distributions among avian groups, occupying freshwater, brackish, and coastal marine habitats across all continents except Antarctica. Members of this family are ecologically diverse, ranging from sedentary tropical species to highly migratory taxa that traverse hemispheric distances, thereby linking distant biogeographic regions through seasonal movements (Newton, 2017). As a result, the global distribution of Anatidae reflects a combination of deep evolutionary history, climatic gradients, wetland availability, and large-scale dispersal processes.

At evolutionary timescales, continental drift, historical land connections, and Pleistocene climatic oscillations have played a major role in shaping patterns of diversification and distribution within Anatidae (Livezey, 1991). Repeated cycles of glaciation and deglaciation in the Northern Hemisphere promoted range fragmentation, secondary contact, and the evolution of migratory strategies, particularly among temperate and boreal lineages (Kear, 2005; Newton, 2017). In contrast, long-term climatic stability in some tropical regions has favored persistence and ecological specialization rather than extensive diversification. Understanding the biogeography of Anatidae therefore requires analysis at multiple spatial scales, beginning with broad continental patterns before considering island endemism and recent range dynamics.

6.1. Continental Distribution Pattern

Continental patterns of Anatidae diversity are strongly structured by latitude and climatic regime, with species richness peaking in regions characterized by extensive temperate and boreal wetland systems. As illustrated in Figure 6, the Palearctic realm represents a major center of diversity, reflecting the combined effects of historical glaciation, large-scale wetland availability, and the evolution of long-distance migratory strategies supported by pronounced seasonal productivity (Carboneras, 2020). Comparatively high richness in the Neotropical realm further highlights the importance of large tropical river basins and floodplains in promoting ecological differentiation and year-round resource stability (Livezey, 1991; Kear, 2005).

In contrast, lower diversity in the Afrotropical and Australasian realms may be attributed to historical aridity, reduced glacial-driven diversification, and a higher prevalence of habitat-specialist taxa (Carboneras, 2020; Newton, 2017). Together, these patterns emphasize the dominant role of large-scale biogeographic processes in shaping the global distribution of Anatidae.

6.2. Island Biogeography and Endemism

Island systems exert a strong influence on the distribution, diversification, and persistence of Anatidae, producing biogeographic patterns that differ markedly from those observed in continental regions. Although islands generally support

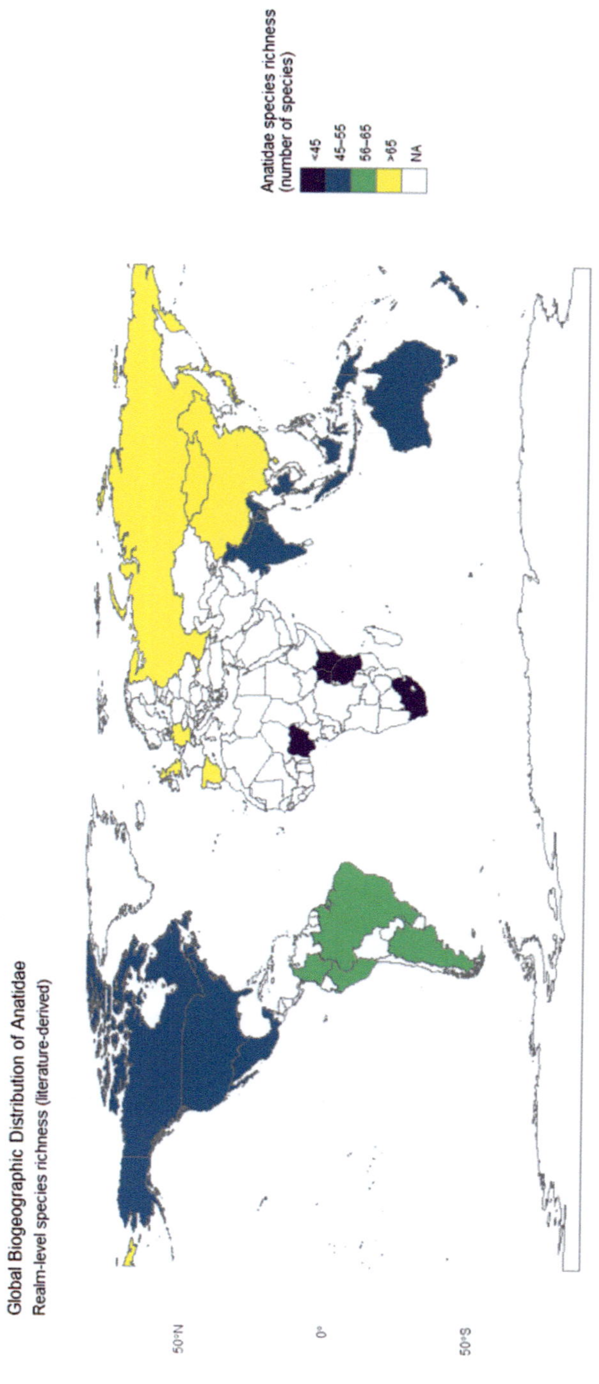

Figure 6. Global biogeographic patterns of Anatidae species richness across major realms.

fewer waterfowl species than mainland systems, they frequently exhibit disproportionately high levels of endemism and distinct evolutionary trajectories (Carboneras et al., 2020). Most island-dwelling anatids are derived from migratory or highly dispersive continental ancestors.

Island systems generate distinctive biogeographic outcomes within the family Anatidae by combining reduced colonization with prolonged geographic isolation. Although islands typically support fewer waterfowl species than continental regions, they frequently harbor a disproportionately high level of endemism driven by in situ divergence and niche specialization. Highly isolated archipelagos such as Hawaii and New Zealand are dominated by endemic taxa, whereas island systems experiencing frequent colonization retain a greater proportion of non-endemic species. Endemism in island anatids is closely linked to elevated extinction risk, as restricted ranges and small populations increase sensitivity to habitat loss, invasive species, and climate-driven environmental change (Runge et al., 2017; Carboneras et al., 2020).

6.3. Range Dynamics and Climate Change

Climate change has emerged as a dominant driver reshaping the global distribution, migration strategies, and seasonal ranges of Anatidae. Rising temperatures, altered precipitation regimes, and changes in wetland hydrology have collectively modified the suitability of breeding, stopover, and wintering habitats, leading to pronounced spatial redistribution across latitudinal gradients. Long-term studies consistently document poleward shifts in breeding and wintering ranges, particularly in temperate and high-latitude regions, alongside changes in migration timing and reduced dependence on traditional wintering areas (Lehikoinen et al., 2013; Amano et al., 2019).

The global synthesis reveals pronounced heterogeneity in climate-driven range-change intensity among major Anatidae groups. Swans (Cygninae) and true geese (Anserinae) exhibit the strongest responses, particularly across the Northern Hemisphere, reflecting high migratory propensity and dependence on climate-sensitive tundra and boreal breeding systems. These taxa show rapid redistribution of wintering ranges and increased overwintering at higher latitudes under warming conditions (Fox et al., 2019; Lehikoinen et al., 2013). In contrast, shelducks (Tadorninae) display more regionally constrained responses, while ducks (Anatinae, Aythyini, and Mergini combined) show widespread but comparatively moderate shifts linked to greater ecological plasticity. Overall, the literature-synthesized patterns indicate that climate change is amplifying lineage-specific ecological differences within Anatidae, with important implications for future wetland conservation planning (Amano et al., 2019).

7. Conservation Biology and Population Dynamics

Conservation biology in Anatidae integrates population ecology, threat assessment, and adaptive management to ensure the long-term persistence of ducks, geese, and swans across their global ranges. Population dynamics in this family are strongly structured by wetland availability, habitat quality, and connectivity among breeding, staging, and wintering areas, particularly for migratory species operating at flyway scales (Newton, 2017; Fox et al., 2019; Wetlands International, 2021). Anthropogenic pressures such as wetland drainage, agricultural intensification, water regulation, hunting, and infrastructure development continue to drive population changes, often interacting with climatic variability to amplify demographic instability (Kirby et al., 2021; IUCN, 2023).

Recent global assessments highlight increasing divergence in population trajectories within Anatidae, with generalist species showing relative resilience while habitat specialists and sea ducks exhibit sustained declines (Carboneras et al., 2020; Lehikoinen et al., 2013). Consequently, effective conservation depends on integrated monitoring, habitat-based management, and international cooperation across flyways (Davidson et al., 2020; Wetlands International, 2021).

7.1. Population Trends and Threat Assessment

Population dynamics within Anatidae exhibit pronounced spatial and taxonomic heterogeneity, reflecting differences in life-history strategies, migratory behavior, and degrees of habitat specialization. Recent flyway-scale assessments indicate that while several generalist taxa—particularly dabbling ducks and some goose species—have maintained stable or increasing populations, many specialist groups, including diving ducks, sea ducks, island endemics, and long-distance migrants, are experiencing sustained declines (Wetlands International, 2021; IUCN, 2023). Long-term monitoring data derived from standardized international waterbird counts identify habitat loss and degradation, hydrological alteration, overexploitation, pollution, and emerging pressures such as disease outbreaks and infrastructure development as the principal drivers of population declines across multiple flyways (Fox et al., 2019; Kirby et al., 2021).

These contrasting long-term trajectories are synthesized in Figure 7, which shows increasing trends in geese and generally stable or slightly increasing patterns in dabbling ducks, reflecting their ecological flexibility and adaptation to human-modified landscapes (Fox et al., 2019; Wetlands International, 2021). In contrast, diving ducks exhibit persistent declines, highlighting their sensitivity to wetland degradation and eutrophication, whereas swans display comparatively stable or modestly increasing trends following long-term protection measures (Kirby et al., 2021; IUCN, 2023). Overall, the figure underscores strong guild-dependent population responses within Anatidae, shaped by ecological specialization, anthropogenic pressures, and climate-driven variability (Lehikoinen et al., 2013).

7.2. Habitat Conservation Strategies

Effective conservation of Anatidae populations depends on the protection and coordinated management of wetland networks across entire migratory flyways

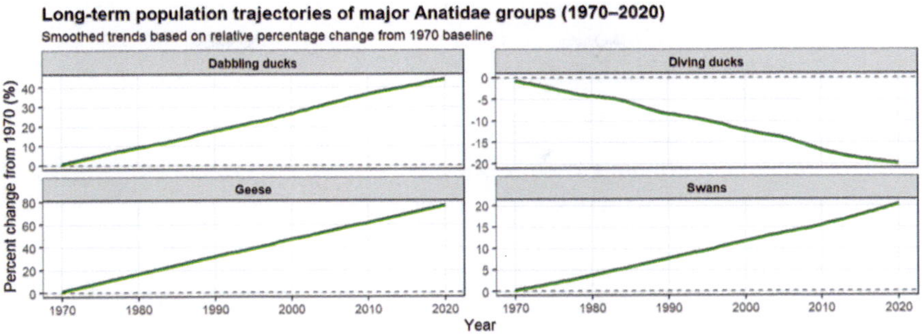

Figure 7. Long-term population trajectories of major Anatidae groups (1970–2020) based on relative population indices.

rather than focusing on isolated sites. Contemporary conservation strategies therefore prioritize the safeguarding of key breeding, stopover, and wintering habitats through protected area designation, sustainable water management, and large-scale habitat restoration initiatives (Davidson et al., 2020; Wetlands International, 2021). Maintaining shallow-water mosaics, floodplain connectivity, and natural seasonal inundation regimes has been shown to enhance foraging efficiency, nesting success, and juvenile recruitment in multiple Anatidae species, particularly within highly dynamic floodplain systems (Kingsford et al., 2022). Increasingly, adaptive and community-based management approaches that integrate local livelihoods with wetland conservation are recognized as essential for reducing anthropogenic pressures and improving compliance at landscape scales (Reis et al., 2023).

Consistent with these principles, population responses across different management contexts indicate that protected and restored wetlands support more stable and positive Anatidae population trends than community-managed, artificial, or degraded systems, underscoring the critical role of habitat quality, hydrological integrity, and management effectiveness in driving long-term population stability, as illustrated in Figure 8 (Fox et al., 2019; Wetlands International, 2021).

7.3. Climate Change Adaptation

Climate change is reshaping the distribution, phenology, and demographic structure of Anatidae populations through altered temperature regimes, precipitation patterns, and wetland hydrology. Recent modelling studies predict poleward and altitudinal range shifts, changes in migration timing, and increasing dependence on climate-resilient refugia, such as groundwater-fed wetlands (Haest et al., 2020). Consequently, adaptation strategies increasingly emphasize maintaining landscape connectivity, protecting hydrologically stable wetlands, and integrating climate projections into site prioritization and management planning, supported by long-term population monitoring and flexible, adaptive management frameworks.

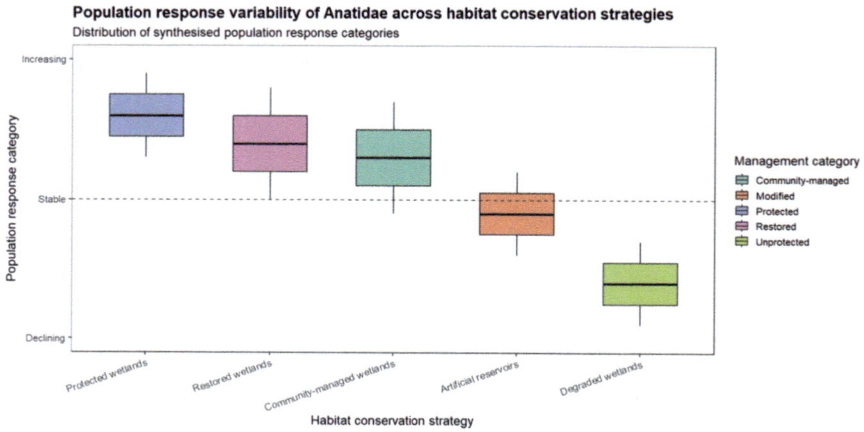

Figure 8. Population responses of Anatidae under different wetland conservation and management strategies, highlighting the importance of habitat quality and protection for long-term population stability.

Differences in adaptive capacity among Anatidae groups help explain their contrasting responses to climate change. In Figure 9, adaptive capacity is expressed using a relative, literature-based scale ranging from 1 (very low) to 5 (very high), reflecting variation in migration flexibility, diet breadth, habitat plasticity, and breeding phenology. Generalist groups such as dabbling ducks

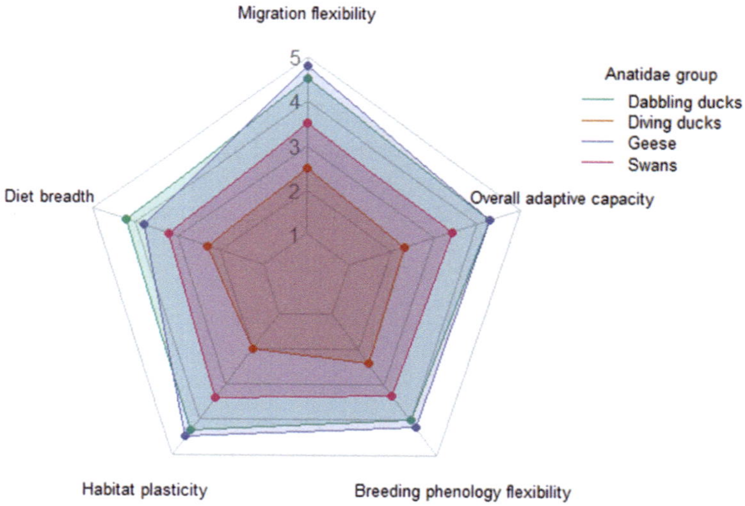

Figure 9. Trait-based adaptive capacity of major Anatidae groups in response to climate change.

and geese exhibit higher adaptive capacity, consistent with greater ecological flexibility and resilience to climatic variability, whereas diving ducks show lower scores, indicating heightened vulnerability to climate-driven changes in wetland structure and resource availability (Fox et al., 2019; IUCN, 2023). Swans display intermediate adaptive capacity, suggesting partial resilience constrained by habitat specificity and slower life-history strategies. These trait-based differences provide a mechanistic basis for observed population trends under climate change and align with global assessments of climate vulnerability in waterbirds (Wetlands International, 2021; Lehikoinen et al., 2023).

8. Economic and Cultural Significance

Members of the family Anatidae (ducks, geese, and swans) have long been intertwined with human societies through domestication, subsistence and commercial exploitation, recreation, and symbolic representation. Their economic importance is reflected in global poultry production systems, while their cultural significance is evident in traditional cuisines, rituals, folklore, and recreational practices across continents (Guo et al., 2021; El Sabry et al., 2023). Furthermore, waterfowl-based recreation and hunting industries generate substantial economic returns and contribute to conservation financing through licensing and habitat management programs (USFWS, 2023; Southwick Associates, 2024).

8.1. Domestication and Agricultural Value

Ducks and geese are among the earliest domesticated avian taxa, providing meat, eggs, feathers, and down across diverse agro-ecological systems. Recent genomic studies indicate that duck domestication followed a complex evolutionary trajectory involving multiple wild populations and historical introgression, rather than a single linear domestication event (Guo et al., 2021).

Globally, duck and goose production is strongly concentrated in Asia, particularly China and Southeast Asia, which together account for the majority of the world's duck population and nearly all reported goose meat production (FAO, 2021; El Sabry et al., 2023). Ducks are especially valued in small-scale and integrated farming systems because of their adaptability to wetlands, rice paddies, and low-input production environments (El Sabry et al., 2023). Improvements in stocking density, welfare-oriented housing, and nutrition have been shown to enhance productivity and economic returns, highlighting the link between animal welfare and agricultural sustainability in Anatidae farming systems (El Sabry et al., 2023).

8.2. Hunting and Recreation Industries

Wild Anatidae are central to hunting and recreational economies, particularly in North America and Europe. Contemporary economic assessments demonstrate that waterfowl hunting generates billions of dollars annually through equipment sales, travel, accommodation, and license fees, supporting rural livelihoods and employment (USFWS, 2023; Southwick Associates, 2024).

Beyond direct economic benefits, hunting revenues play a critical role in conservation funding, supporting wetland restoration, population monitoring, and species management under frameworks such as adaptive harvest management. Wildlife watching and bird-based ecotourism further enhance the recreational value of Anatidae, especially for charismatic species such as swans and migratory ducks, contributing to nonconsumptive use economies (USFWS, 2023). Changes in participation rates, habitat availability, and regulatory policies therefore have cascading socio-economic and conservation implications.

8.3. Cultural and Traditional Values

Anatidae possess deep cultural significance across societies, often extending beyond their direct economic value. Duck eggs and derived products such as salted eggs and balut form integral components of traditional diets in Southeast and East Asia and represent culturally embedded food economies (Mopera et al., 2021). The cultural and traditional roles of Anatidae vary regionally, reflecting differences in dietary practices, symbolism, and historical human–bird interactions (Table 2). Geese and ducks also feature prominently in rituals, seasonal festivals, and symbolic traditions.

Table 2. Cultural and traditional significance of Anatidae across major regions

Region	Cultural and Traditional Roles of Anatidae	Sources
East and Southeast Asia	Traditional foods (salted duck eggs, balut), symbolism in art and folklore, household food security	Alejandria (2019); Mopera et al. (2021)
Europe	Festive cuisine, agrarian folklore, use of feathers and down in traditional crafts	El Sabry et al. (2023)
North America	Hunting heritage, indigenous traditions, recreational and community-based practices	USFWS (2023)
South Asia	Backyard duck rearing, subsistence use, culturally influenced consumption patterns	FAO (2021)
Africa (selected regions)	Small-scale duck keeping, livelihood support in wetland areas, cultural norms shaping use	FAO (2021)

Note: "Selected regions" refers to illustrative examples reported in the literature and does not imply uniform cultural practices across entire continents.

In East Asian cultures, mandarin ducks symbolize marital fidelity, while in Europe, geese have historically been associated with agricultural calendars, festivities, and folklore (Alejandria, 2019). Feathers and down from Anatidae have long been used in bedding, clothing, and handicrafts, reinforcing their role in traditional livelihoods. Cultural perceptions, beliefs, and taboos can strongly influence patterns of domestication, consumption, and conservation, underscoring the importance of socio-cultural context in Anatidae management (Mopera et al., 2021).

9. Systematics and Molecular Phylogeny

9.1. Historical Classification Systems

The classification of the family Anatidae has undergone substantial revision over time, reflecting progressive shifts in the types of characters considered informative for phylogenetic inference. Early taxonomic frameworks were necessarily based on external morphological traits, including plumage characteristics, bill structure, and overall body form, which provided a practical basis for species recognition and higher-level grouping (Linnaeus, 1758). Despite being useful to descriptive taxonomy, these methods were also restricted in determining the difference between homologous characters and convergent ones, especially in waterfowl that is ecologically specialized.

Later improvements included anatomical and osteological features, including skeletal structure, syrinx design, and tracheal features that were deemed less prone to environmental plasticity, and thus more dependable in determining higher taxonomic relationships in Anatidae (Delacour & Mayr, 1945). Nevertheless, anatomical categories did not entirely solve deep evolutionary, but were often applied in conjunction with external morphology and not as independent systematic systems. In the middle of the twentieth century, as an additional taxonomic evidence, behavioral and ecological characters, such as courtship displays, vocalizations, and habitat associations, were presented. The behavioral characters were viewed as the evolutionary history in waterfowls especially in the lineages with the complex mating system (Johnsgard, 2010). However, their usefulness as pure measures of phylogenetic relatedness was limited by the possibility of behavioral convergence. It is the constrained nature of morphology-, anatomy-, and behavior-based systems which eventually drove the use of molecular techniques, starting with mitochondrial DNA markers and techniques of DNA–DNA hybridization in the late twentieth century (Sibley & Ahlquist, 1990). These molecular structures gave the original genetic view of relationships among Anatidae and showed conflicts with classical classification. The further methodological developments resulted in multilocus phylogenetic studies, which combine mitochondrial and nuclear loci to enhance resolution and strength (Livezey, 1996; Gonzalez et al., 2009).

9.2. Molecular Phylogenetic Insights

These successive classification systems, as Figure 10 shows, did not follow each other but rather overlapped in time as the old methodologies and systems of character continued to be pertinent together with new approaches. All of these integrative and cumulative developments eventually led to the current genome-

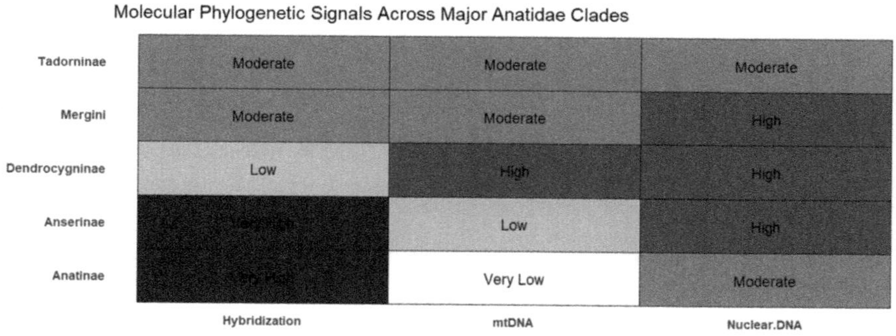

Figure 10. Molecular phylogenetic signals across major clades of Anatidae. The figure summarizes clade-specific variation in the extent of hybridization or introgression and the relative phylogenetic resolution provided by mitochondrial DNA (mtDNA) and nuclear datasets.

scale phylogenomic models of Anatidae, which now form the basis of the modern systematic knowledge of the group (Jarvis et al., 2014).

Molecular phylogenetic investigations have greatly advanced our understanding of evolutionary relationships in Anatidae, which frequently contradicts conventional morphology-based classifications. Mitochondrial DNA marker studies including cytochrome b, ND2, and control region were early evidences of the monophyly of Anatidae and large-scale deep divisions between major duck and goose clades (Donne-Goussé et al., 2002). Further taxon-dense studies of the mitochondria showed that many of the phenotypic appearances of waterfowl are due to ecological convergence and not high evolutionary affinity, which underscores the inefficacy of morphology-only methods in resolving anatid phylogeny (Gonzalez et al., 2009).

In Anserinae (geese and swans), molecular data consistently indicate rapid diversification accompanied by widespread hybridization. Initial mitochondrial analyses demonstrated poor performance in Anser and Branta speciation, and often non-monophyly could be explained by a recent split and a common lineage polymorphism (Ruokonen et al., 2000). Hundreds of nuclear loci phylogeny studies have since shown that introgression has occurred many times across the evolutionary history of true geese, with extensive gene tree discordance in spite of a recoverable species-tree signal (Ottenburghs et al., 2016; Ottenburghs et al., 2017).

Comprehensive molecular analysis of mallards and spot-billed ducks shows that introgressive hybridization has dominated the formation of the mitochondrial variation, and it should not be assumed that a single-locus dataset can be used to define species boundaries in this group (Wang et al., 2019). Although mitogenomic studies have enhanced the support of some more profound relationships, their interpretation is limited by the commonness of reticulate evolution (Buckner et al., 2018).

Table 3. Contemporary systematic framework of the family Anatidae integrating molecular phylogenetic evidence

Subfamily	Major Tribes/ Groups	Representative Genera	Molecular and Systematic Basis	Key References
Dendrocygninae	Dendrocygnini	*Dendrocygna*	Early-diverging, monophyletic lineage with stable mitochondrial and nuclear signals	Donne-Goussé et al. (2002); Gonzalez et al. (2009)
Tadorninae	Tadornini	*Tadorna, Alopochen*	Intermediate placement with variable support across datasets, suggesting early radiation	Gonzalez et al. (2009); Hackett et al. (2008)
Anserinae	Anserini, Cygnini	*Anser, Branta, Cygnus*	Extensive hybridization with coherent nuclear genomic structure supporting current genera	Ruokonen et al. (2000); Ottenburghs et al. (2016, 2017)
Anatinae	Anatini, Aythyini, Mergini, others	*Anas* (sensu lato), *Aythya*	High mitochondrial discordance and introgression; classification relies on multilocus data	Gonzalez et al. (2009); Buckner et al. (2018); Wang et al. (2019)
Anatinae (Mergini)	Sea ducks	*Somateria, Clangula, Mergus*	Phylogenomics resolves intergeneric relationships despite historical gene flow	Lavretsky et al. (2021)

Note: Classification reflects consensus from recent molecular and phylogenomic studies; alternative treatments exist for some genera, particularly within Anatinae.

Phylogenomic methods also defined relationships within ecologically speacialized groups such as sea ducks (Mergini), and have both well-supported lineages and evidence of historical gene flow with marine adaptation (Lavretsky et al., 2021). Basal and intermediate lineages such as Dendrocygninae and Tadorninae are always found as early-divergent members of Anatidae but again their position varies across datasets, possibly due to early rapid radiation and ancient introgression (Donne-Goussé et al., 2002; Gonzalez et al., 2009). Taken altogether, recent genomic works highlight the idea that evolutionary history of Anatidae can be characterized as a dynamic continuum, which is affected by rapid radiation, hybridization, and incomplete lineage sorting, and, therefore, requires integrative, coalescent-aware analytical approaches to the robust phylogenetic inference (Hackett et al., 2008; Jarvis et al., 2014).

9.3. Contemporary Systematic Framework

The modern systematic scheme of Anatidae is an embodiment of the integration of old-fashioned morphological features and more and more solid molecular and phylogenomic data (Table 3). The molecular analysis has continually shown that the early-divergent lineages such as Dendrocygninae have consistent mitochondrial and nuclear signatures, which justify their classification as discrete and consistent subfamilies (Donne-Goussé et al., 2002; Gonzalez et al., 2009). Conversely, they are found in intermediate groups such as Tadorninae, which have inconsistently been placed in datasets, presumably due to early radiation events and the absence of sufficient historical introgression, thus continuing to cause uncertainty in their exact phylogenetic placement (Gonzalez et al., 2009; Hackett et al., 2008).

In Anserinae, highly hybridized and deeply mitochondrial discordant relationships confound relationships at the species level, but phylogenomic studies indicate that there is a high signal of nuclear species-tree obstructing the erosion of existing generic boundaries between geese and swans (Ruokonen et al., 2000; Ottenburghs et al., 2016, 2017). The most taxonomically complicated part of Anatidae is Anatinae: in many cases of mitochondrial phylogenies, introgression and mitochondrial captures have led to non-monophyly, so the classification of Anatinae now requires the use of multiple locus and genome-scale nuclear data (Gonzalez et al., 2009; Buckner et al., 2018; Wang et al., 2019). At the tribal level, phylogenomic studies have played a key role in stabilizing such ecologically specialized groupings as the sea ducks (Mergini) wherein genome-scale studies have helped to clarify intergeneric relationships despite evidence of historic gene flow (Lavretsky et al., 2021).

10. Research Frontiers and Future Directions

Rapid advances in molecular biology, ecological modelling, and socio-ecological research are transforming the study of waterfowl biology. Contemporary research on Anatidae is increasingly interdisciplinary, integrating genomics, ecosystem science, and human dimensions to address pressing conservation challenges in the context of global environmental change. Future progress will depend on coupling high-resolution biological data with adaptive management frameworks capable of responding to ecological uncertainty.

10.1. Genomic Approaches to Waterfowl Biology

The use of genomic technologies is one of the greatest frontiers in waterfowl studies. Although early molecular work was primarily based on mitochondrial markers and small sets of nuclear loci, recent advances in next-generation sequencing have made it possible to map the process of evolution genome-wide which has never reached such a level of detail in Anatidae (Jarvis et al., 2014; Ottenburghs et al., 2016). Genomic data also allow loci associated with ecological important phenotypes, such as migratory behavior, salinity, and immune response, to be identified, and thus a connection is drawn between evolutionary history and functional adaptation (Lavretsky et al., 2021). In addition to phylogenetics, conservation genomics has emerged as a critical tool to waterfowl management. Large-scale genome-wide data can be used to precisely calculate the effective population size, inbreeding, and genetic connectivity of populations, which can be used to make evidence-based conservation decisions in threatened and geographically separated species (Frankham et al., 2017). Future studies would focus more on integrative genomic models that incorporate nuclear genomes, epigenetic information, and transcriptomics to obtain insights in phenotypic plasticity and fast adaptation in dynamic wetlands.

10.2. Ecosystem Services and Human Dimensions

The strength of Anatidae ecosystem service relationships with the various human dimensions varies considerably, reflecting differences as well as disparities in service delivery, and imbalance in research focus. There is a stronger evidence base in provisioning and cultural services than in regulating and supporting services, which suggests that direct and socially visible benefits are primarily concerned. Local livelihoods and subsistence use are most strongly correlated with provisioning services, whereas the linkage to human–wildlife conflict is supported by more moderate evidence implying that the documentation varies regionally (Green & Elmberg, 2014; Karanth et al., 2018). Good, intensive connections to cultural identity, education, and ecotourism have a great presence in cultural services provided by Anatidae, which makes them flagship species that contribute to the improvement of the population's exposure and economic use of nature (Ballantyne et al., 2011; Steven et al., 2016; Ramsar Convention Secretariat, 2018). Comparatively, the regulation and support services demonstrate relatively lower relationships with governance and climate-related human aspects, which suggests a lack of integration of the indirect ecological functions in socio-economic and policy systems (Millennium Ecosystem Assessment, 2005; Kingsford et al., 2017). Inequality of evidence distribution, in general, highlights the necessity of interdisciplinary research in the future that is more inclusive of regulating and supporting services integrated into human-centered conservation strategies.

10.3. Climate Change and Adaptive Management

Climate change is progressively altering the ecological environment in which Anatidae populations continue to exist so as to pose a challenge to any conservation plans formulated in relatively constant past climatic conditions. Wetlands used by Anatidae are inherently dynamic, and the growing variability of precipitation and water availability challenges management strategies that depend on a set of spatial or hydrological assumptions (Kingsford et al., 2017). In a similar manner, conservation targets based on past population base lines are generally ill adapted to circumstances where phenological shifts, redistribution of wintering and breeding ranges, and higher demographic heterogeneity occur (Crick, 2004; Newton, 2017). Conversely, the flyway-scale planning demonstrates a comparatively higher correspondence to various climate pressures in terms of the ability to support the redistribution of the population of Anatidae across time and space at a large scale. With climate change modifying the migration pathways and the relative significance of the stopover and wintering sites, conservation policies that are conducted within the framework of a network of wetlands and political boundaries are of growing concern (Runge et al., 2017). The adaptive management focuses on flexibility, learning, and iterative decision-making, which enables the management actions, including water-level regulation, disturbance control, and harvest policies to be modified as a result of continuous monitoring and the emergence of the new climatic trends (Williams, 2011). In general, the synthesis shows that climate change does not only have a direct impact on Anatidae but also reveals the vulnerability of the traditional conservation strategies. One of the research frontiers to improve adaptability of the Anatidae population to climatic conditions in the future is to address these mismatches using adaptive frameworks at flyway scales.

11. Conclusion

The family Anatidae is an example of how an evolutionary history of species, ecological diversification, and functional adaptation form the basis of the success of waterfowl in a large variety of aquatic and terrestrial habitats. This chapter contributes to the understanding of the evolution of avian diversity on global scales through the use of the Anatidae model, which uses evidence in taxonomy, paleontology, morphology, behavior, biogeography, and conservation biology. The development of fossil and molecular remains supports a profound evolutionary background of Anatidae, and the diversification was largely affected by the climatic changes and the growth of wetlands. Meanwhile, modern phylogeny indicates that hybridization, introgression, and incomplete sorting of lineages are all part of anatids' evolutionary history and are problematic to strictly bifurcating perspectives of the avian systematics. These results underscore the importance of integrative, genome-based solutions of taxonomic relationships in groups with complex and rapid lineage evolution. Ecological and functional differentiation in the Anatidae is manifested in striking variations in feeding behavior, locomotor patterns, digestive specialization, and social structure. The segregation of the species into individual foraging guilds and habitats types depicts the way morphological and behavioral characteristics minimize interspecific competition and promote coexistence in the heterogeneous wetland landscapes. Migration turns out to be a characteristic ecological phenomenon connecting far-off biogeographic areas by flyway networks and establishing the reliance of the waterfowl groups on spatially consistent wetland structures. In conservation perspective, Anatidae are examples of both strength and susceptibility to quickening global change. Though not all, some generalist species are enjoying the advantages of agriculturalization and human-altered wetlands; many specialists, especially diving ducks, sea ducks, island endemics, and long-distance migrants, are suffering long-term declines in population. Climate change is also altering distributional ranges, migration pattern, and habitat suitability, magnifying the pressure and providing new challenges to the conservation planning. The synthesis illustrates that adaptive capacity is quite different across ecological guilds, the implications of which are significant in prioritization and management in the future climate scenario. In addition to their ecological and evolutionary importance, Anatidae play a central role in human cultures in the forms of domestication, subsistence, hunting, recreation, and cultural symbolism. They are ecosystem service providers, bioindicators, and flagship species at the nexus of the conservation of biodiversity to socio-economic choices. It is important to identify these human aspects when developing biologically effective and socially sustainable conservation approaches. The Anatidae are a promising example of how to study the evolution of birds, their ecological role in the wetlands, and the threat of extinction in a fast-evolving world. The further development will rely on the interdisciplinary research that will involve the combination of genomics, long-term ecological observation, climate modelling, and socio-ecological views in adaptive management frameworks. With global environmental change still continuing to reshape the form and role of wetland systems, the lessons learned about ducks, geese, and swans will still be at the focus of not only the future advancement of avian biology but also the future of sound conservation practice.

References

Afton, A. D. (1981). Courtship and pair bonding of wild ducks (*Anatidae*). *Wildfowl*, 32, 83–88.

Afton, A. D., & Paulus, S. L. (1992). Incubation and brood care. *In:* B. D. J. Batt et al. (Eds.), *Ecology and Management of Breeding Waterfowl* (pp. 62–108). University of Minnesota Press.

Alejandria, M. C. (2019). Symbolism of waterfowl in East Asian and European cultural traditions. *Journal of Ethnic and Cultural Studies*, 6(2), 45–58. https://doi.org/10.29333/ejecs/256

Amano, T., Székely, T., Sandel, B., Nagy, S., Mundkur, T., Langendoen, T. et al. (2019). Successful conservation of global waterbird populations depends on effective governance. *Nature*, 553, 199–202. https://doi.org/10.1038/nature25139

Baldassarre, G. A., & Bolen, E. G. (2006). *Waterfowl Ecology and Management* (2nd ed.). Krieger Publishing Company.

Baldassarre, G. A. (2018). *Ducks, Geese, and Swans of North America*. Johns Hopkins University Press.

Ballantyne, R., Packer, J., & Falk, J. (2011). Visitors' learning for environmental sustainability: Testing short- and long-term impacts of wildlife tourism experiences. *Tourism Management*, 32, 1243–1253. https://doi.org/10.1016/j.tourman.2010.11.003

Barnes, G. G., & Thomas, V. G. (1987). Digestive organ morphology, diet, and guild structure of North American Anatidae. *Canadian Journal of Zoology*, 65, 1812–1817. https://doi.org/10.1139/z87-274

BirdLife International. (2023). *Species Factsheets and Global Distribution Data for the Family Anatidae*. BirdLife International, Cambridge. https://datazone.birdlife.org/species/taxonomy

Black, J. M. (2001). *Fitness Consequences of Long-Term Pair Bonds in Barnacle Geese*. *In:* J. M. Black (Ed.), *Partnerships in Birds: The Study of Monogamy* (pp. 91–112). Oxford University Press.

Buckner, J. C., Provost, K. L., Drovetski, S. V., & Peters, J. L. (2018). Mitogenomic and mitochondrial gene phylogeny improve estimates of phylogeny within the Anatinae (Aves: Anatidae). *Molecular Phylogenetics and Evolution*, 129, 180–191.

Carboneras, C., Boesman, P. F. D., Kirwan, G. M., & Sharpe, C. J. (2020). Northern Screamer (*Chauna chavaria*), version 1.0. *In:* J. del Hoyo, A. Elliott, J. Sargatal, D. A. Christie, and E. de Juana (Eds), *Birds of the World*. Cornell Lab of Ornithology, Ithaca, NY, USA. https://doi.org/10.2173/bow.norscr1.01

Clarke, J. A., Tambussi, C. P., Noriega, J. I., Erickson, G. M., & Ketcham, R. A. (2005). Definitive fossil evidence for the extant avian radiation in the Cretaceous. *Nature, 433*(7023), 305–308. https://doi.org/10.1038/nature03150

Clench, M. H., & Mathias, J. R. (1995). *The Avian Cecum: A Review*. *The Wilson Bulletin*, 107(1), 93–121.

Crick, H. Q. P. (2004). The impact of climate change on birds. *Ibis*, 146, 48–56.

Davidson, N. C., Fluet-Chouinard, E., Finlayson, C. M. et al. (2020). Global extent and distribution of wetlands: Trends and drivers. *Marine and Freshwater Research*, 71(6), 650–664.

Delacour, J., & Mayr, E. (1945). The family Anatidae. *The Wilson Bulletin*, 57, 3–55.

Donne-Goussé, C., Laudet, V., & Hänni, C. (2002). A molecular phylogeny of Anseriformes based on mitochondrial DNA analysis. *Molecular Phylogenetics and Evolution*, 23(3), 339–356. https://doi.org/10.1016/S1055-7903(02)00019-2

Duque-Correa, M. J., Clauss, M., Hoppe, M. I., Buyse, K., Codron, D., Meloro, C. et al. (2022). Diet, habitat and flight characteristics correlate with intestine length in birds. *Proceedings of the Royal Society B: Biological Sciences*, 289: 20220675. https://doi.org/10.1098/rspb.2022.0675

Durant, D. (2013). The digestion of fibre in herbivorous Anatidae: A review. *Wildfowl*, 54, 7–24.

El Sabry, M. I., Hassan, S. M., & Abd El-Hack, M. E. (2023). Cultural, nutritional, and socioeconomic importance of ducks and geese in traditional food systems. *Frontiers in Veterinary Science, 10*, 1189452.

Elmberg, J., Nummi, P., Pöysä, H., & Sjöberg, K. (2005). Breeding success of sympatric dabbling ducks in relation to habitat quality. *Oikos*, 108, 467–477.

Emlen, S. T., & Oring, L. W. (1977). Ecology, sexual selection, and the evolution of mating systems. *Science*, 197(4300), 215–223. https://doi.org/10.1126/science.327542

FAO. (2021). *FAOSTAT: Livestock Production.* Food and Agriculture Organization of the United Nations. https://www.fao.org/faostat/

Fox, A. D., Elmberg, J., Tombre, I. M., & Hessel, R. (2019). Agriculture and herbivorous waterfowl: A review of the scientific basis for improved management. *Biological Reviews*, 94(6), 2023–2044.

Frankham, R., Ballou, J. D., Ralls, K., Eldridge, M. D. B., Dudash, M. R., Fenster, C. B. et al. (2017). *Genetic Management of Fragmented Animal and Plant Populations*. Oxford University Press.

Gill, F., Donsker, D., & Rasmussen, P. (Eds.). (2024). *IOC World Bird List* (Version 14.1). International Ornithological Committee. https://www.worldbirdnames.org/

Gonzalez, J., Düttmann, H., & Wink, M. (2009). Phylogenetic relationships based on two mitochondrial genes and hybridization patterns in Anatidae. *Journal of Zoology*, 279(3), 310–318. https://doi.org/10.1111/j.1469-7998.2009.00622.x

Green, A. J., & Elmberg, J. (2014). Ecosystem services provided by waterbirds. *Biological Reviews*, 89(1), 105–122. https://doi.org/10.1111/brv.12045

Guo, Y., Huang, Y., Gong, Y., Li, H., & Wang, Y. (2021). Global development of duck and goose production systems and their economic significance. *Poultry Science*, 100(9), 101328.

Hackett, S. J., Kimball, R. T., Reddy, S., Bowie, R. C. K., Braun, E. L., Braun, M. J. et al. (2008). A phylogenomic study of birds reveals their evolutionary history. *Science*, 320(5884), 1763–1768. https://doi.org/10.1126/science.1157704

Haest, B., Hüppop, O., & Bairlein, F. (2020). Weather at the winter and stopover areas determines spring migration onset, progress, and advancements in Afro-Palearctic migrants. *Proceedings of the Royal Society B*, 287, 20193006.

IUCN. (2023). *The IUCN Red List of Threatened Species* (Version 2023-2). International Union for Conservation of Nature, Gland, Switzerland. https://www.iucnredlist.org/

Jarvis, E. D., Mirarab, S., Aberer, A. J., Li, B., Houde, P., Li, C. et al. (2014). Whole-genome analyses resolve early branches in the tree of life of modern birds. *Science*, 346(6215), 1320–1331. https://doi.org/10.1126/science.1253451

Johnsgard, P. A. (2010). *Ducks, Geese, and Swans of the World*. University of Nebraska, Lincoln Press.

Johnson, W. P., Schmidt, P. M., & Taylor, D. P. (2014). Foraging ecology of ducks and geese. *Journal of Avian Biology*, 45(5), 437–449.

Kaminski, R. M., & Gluesing, E. A. (1987). Density- and habitat-related recruitment in mallards. *Journal of Wildlife Management*, 51, 141–148.

Karanth, K. K., Gopalaswamy, A. M., DeFries, R., & Ballal, N. (2018). Assessing patterns of human–wildlife interactions and attitudes towards conservation in South Asia. *Conservation Biology*, 32, 1320–1331.

Kear, J. (2005). *Ducks, Geese and Swans: Species Accounts*. Oxford University Press.

Kehoe, F. P., & Ankney, C. D. (1985). Variation in digestive organ size among five species of diving ducks (*Aythya* spp.). *Canadian Journal of Zoology*, 63, 2339–2342. https://doi.org/10.1139/z85-346

Kingsford, R. T., Bino, G., & Finlayson, C. M. (2022). Wetlands and waterbirds under pressure: Climate change and water resource development. *Biological Conservation*, 266, 109454.

Kingsford, R. T., Bino, G., & Porter, J. L. (2017). Continental impacts of water development on waterbirds, wetlands and rivers. *Biological Conservation*, 210, 1–10.

Kirby, J. S., Stattersfield, A. J., Butchart, S. H. M., Evans, M. I., Grimmett, R. F. A., Jones, V. R. et al. (2021). Key conservation issues for migratory land and waterbird species on the world's major flyways. *Bird Conservation International*, 31, 1–22.

Lavretsky, P., McCracken, K. G., & Peters, J. L. (2021). Phylogenomics reveals ancient and contemporary gene flow in sea ducks. *Molecular Phylogenetics and Evolution*, 155, 106963.

Lehikoinen, A., Jaatinen, K., Vähätalo, A. V., Clausen, P., Crowe, O., Deceuninck, B. et al. (2013). Rapid climate-driven shifts in wintering distributions of three common waterbird species. *Global Change Biology*, 19(7), 2071–2081. https://doi.org/10.1111/gcb.12200

Lehikoinen, A., Rajasärkkä, A., Møller, A. P., & Laaksonen, T. (2023). Trait-based vulnerability of waterbirds to climate change across flyways. *Global Change Biology*, 29(5), 1354–1368.

Linnaeus, C. (1758). *Systema Naturae per Regna Tria Naturae* (10th ed.). Holmiae: Laurentii Salvii.

Livezey, B. C. (1991). A phylogenetic analysis and classification of recent dabbling ducks (Tribe Anatini) based on comparative morphology. *The Auk*, 108(3), 471–507. https://doi.org/10.2307/4088089

Livezey, B. C. (1996). A phylogenetic analysis of modern pochards (Anatidae: Aythyini). *The Auk*, 113(1), 74–93. https://digitalcommons.usf.edu/auk/vol113/iss1/8

Mayr, G. (2016). *Avian Evolution: The Fossil Record of Birds and Its Paleobiological Significance*. Wiley-Blackwell.

Millennium Ecosystem Assessment (MEA). (2005). *Ecosystems and Human Well-Being: Synthesis*. Island Press, Washington, DC.

Mopera, L. E., Punzalan, A. E., & Lantican, F. A. (2021). Duck eggs, salted eggs, and balut: Cultural significance and food system roles in Southeast and East Asia. *Journal of Ethnic Foods*, 8(1), 32.

Newton, I. (2017). *The Migration Ecology of Birds* (2nd ed.). Academic Press, London.

Olson, S. L. (1985). The fossil record of birds. *In:* D. S. Farner, J. R. King, & K. C. Parkes (Eds.), *Avian Biology* (Vol. 8, pp. 79–238). Academic Press.

Ottenburghs, J., Kraus, R. H., Hooft, P. V., & Wieren, S. V. (2020). Avian introgression in the genomic era. *Avian Research*, 11, 1–15.

Ottenburghs, J., Megens, H.-J., Kraus, R. H. S., Madsen, O., van Hooft, P., van Wieren, S. E. et al. (2016). A tree of geese: A phylogenomic perspective on the evolutionary history of true geese. *Molecular Phylogenetics and Evolution*, 101, 303–313. https://doi.org/10.1016/j.ympev.2016.05.021

Ottenburghs, J., Megens, H. J., Kraus, R. H. S., van Hooft, P., van Wieren, S. E., Crooijmans, R. P. M. A. et al. (2017). A history of hybrids? Genomic patterns of introgression in the true geese. *BMC Evolutionary Biology*, 17, 201.

Prum, R. O., Berv, J. S., Dornburg, A., Field, D. J., Townsend, J. P., Lemmon, E. M. et al. (2015). A comprehensive phylogeny of birds (Aves) using targeted next-generation DNA sequencing. *Nature*, 526(7574), 569–573. https://doi.org/10.1038/nature15697

Ramsar Convention Secretariat. (2018). *Global Wetland Outlook: State of the World's Wetlands and Their Services to People*. Ramsar Convention Secretariat, Gland, Switzerland.

Reis, V., Kingsford, R. T., & Porter, J. L. (2023). Community-based wetland management and waterbird conservation outcomes. *Conservation Science and Practice*, 5(6), e1289.

Runge, C. A., Watson, J. E. M., Butchart, S. H. M., Hanson, J. O., Possingham, H. P., & Fuller, R. A. (2017). Protected areas and global conservation of migratory birds. *Science*, 356(6344), 1255–1258.

Ruokonen, M., Kvist, L. & Lumme, J. (2000). Close relatedness between mitochondrial DNA from seven *Anser* goose species. *Journal of Evolutionary Biology*, 13, 532–540.

Sibley, C. G., & Ahlquist, J. E. (1990). *Phylogeny and Classification of Birds: A Study in Molecular Evolution*. New Haven, CT: Yale University Press.

Southwick Associates. (2024). Economic Return from NAWMP Investments in Canadian Waterfowl Habitat within the U.S. plus Selected States. Prepared for the Association of Fish and Wildlife Agencies. Southwick Associates, Fernandina Beach, FL, USA.

Steven, R., Morrison, C., & Castley, J. G. (2016). Birdwatching and avitourism: A global review of research into its participant markets, distribution and impacts. *Biological Conservation*, 191, 171–179.

Sun, Z., Pan, T., Hu, C., Sun, L., Ding, H., Wang, H. et al. (2020). Phylogenomic evidence supports a revised classification of Anseriformes. *Molecular Phylogenetics and Evolution*, 146, 106746.

Svihus, B. (2014). Function of the digestive system. *Journal of Applied Poultry Research*, 23(2), 306–314. https://doi.org/10.3382/japr.2014-00937

U.S. Fish and Wildlife Service (USFWS). (2023). *2022 National Survey of Fishing, Hunting, and Wildlife-Associated Recreation*. U.S. Department of the Interior, U.S. Fish and Wildlife Service.

Wang, W., Li, S., Pan, T., Wang, Y., Hu, C., Tian, X. et al. (2019). Incomplete lineage sorting and introgression in the diversification of spot-billed ducks (*Anas zonorhyncha*) and mallards (*Anas platyrhynchos*). *Current Zoology*, 65, 589–597.

Weller, M. W. (1999). *Wetland Birds: Habitat Resources and Conservation Implications*. Cambridge University Press, Cambridge.

Wetlands International (2021). *Waterbird Population Estimates* (5th ed.). Wetlands International, Wageningen, The Netherlands.

Williams, B. K. (2011). Adaptive management of natural resources. *Environmental Management*, 48, 1–12.

Worthy, T. H., & Lee, M. S. Y. (2008). Affinities of Miocene waterfowl fossils and the early radiation of Anseriformes. *Zoological Journal of the Linnean Society*, 153(2), 351–377.

Zhang, G., Li, C., Li, Q., Li, B., Larkin, D. M., Lee, C. et al. (2022). Comparative genomics reveals insights into avian evolution and adaptive radiation. *Nature Ecology & Evolution*, 6(8), 1051–1062.

CHAPTER

2

Anatomy and Physiology of the Anatidae Family

Sk Imran#, Divya Kumari#, Trisha Mondal# and Ram Pratap Singh*

Department of Life Science, Central University of South Bihar, Gaya – 824236, Bihar, India

1. Introduction

The Anatidae family is one of the most diverse groups of waterfowl, which comprises subfamilies Anatinae (ducks) and Anserinae (swans and geese). The members of this family have a wide range of unique morphological and physiological adaptations that set them apart from other bird families. These include webbed feet for efficient swimming, soft and sensitive bills designed for filter-feeding, relatively long necks with short legs, powerful wings adapted for sustained long-distance flight, and dense waterproof plumage underlain by down feathers that provide excellent thermal insulation (Lucy & Karthiayini, 2022). Body mass within Anatidae varies substantially, ranging from 275 g to nearly 12 kg, with swans being the heaviest birds capable of accurate and sustained flapping flight (Backues, 2015). Anatidae are also highly social, often seen roosting and foraging in mixed-species flocks. They are adapted to both aquatic and terrestrial environments and employ diverse foraging strategies such as diving, dabbling, and grazing (Grimmett et al., 2016). The comparative anatomy and physiology of Anatidae underline their evolutionary importance, distinguishing them as a unique avian lineage. A detailed account of their anatomical and physiological characteristics is presented in the following text.

*Corresponding author: rampratap@cusb.ac.in
# Contributed equally

DOI: 10.1201/9781003669784-2

2. Skeletal System

The avian skeleton is unique, and is well known for its light weight, due to the presence of pneumatic bones and the strength provided by the fused bones. A complete skeleton of common shelduck (*Tadorna tadorna*) shows 14 cervical vertebrae, which together form an S-shaped structure that aids in controlling impact during flight. It has a total of 7 thoracic vertebrae, which articulate via the costal fovea with each rib. All the lumbar and sacral vertebrae (around 11 to 14), along with the last thoracic and first coccygeal vertebrae, are attached to the ilium. There are approximately 8 caudal vertebrae, also called coccygeal vertebrae, the initial of which forms the sacrum. Six of them are free, which enables the tail to move, while further caudally, a final ending segment called the pygostyle originates from the coccygeal vertebrae. The pygostyle is unique to birds and provides support for the uropygial gland, the tail feathers, and the muscles that control these feathers, and aids in stability during flight (Sarott et al., 2024).

The duck's skull is made up of multiple fused pneumatic bones, making it lightweight, specialized, and different from all other bird taxa. The skull is majorly kinetic with a movable upper jaw and quadrate bone, and has a large orbit to hold the extensive eyeballs (Feduccia, 1975). Birds' bill consists of a bony core, followed by a thin vascular tissue layer with nerve endings and blood vessels, encased in a soft and flexible sheath of keratinized integument called the rhamphotheca (Lovette & Fitzpatrick, 2016). Numerous mechanoreceptors in the tissues of the bill are distinctive in Anseriformes.

In Anatidae, the bill tip organ consists of the mandible and maxilla. They are covered by the tactile epithelial formations, which are more numerous on the mandible. Sensillar and asensillar are two types of tactile epithelial formations present in the Anatidae family. Asensillar structures are typically located at the tip of the maxilla, whereas sensillar structures occur at the tip of the mandible in most Anatidae species. However, in grazing species, sensillar structures are predominantly concentrated at the tip of the maxilla. The tissue of the bill tip is composed of connective tissue tubules, which are enclosed with corneous sheaths and consist of two types of encapsulated mechanoreceptive endings. The Herbst corpuscles in the proximal part of the tubule are sensitive to vibrations, and the Grandry corpuscles present at the distal part are sensitive to a change in the pressure applied to the rhamphotheca membrane (Avilova, 2018).

Cervical skeletons showed a considerable resemblance between ducks and geese vertebral morphology, despite differences in their vertebral counts. Both species show the regional functional patterns similar to those of other birds. Gross anatomy reveals 14–15 cervical joints in ducks (Landolt et al., 1989; Wang et al., 2025), while geese have 17–18 cervical segments (Zarnik, 1926; Wang et al., 2025). In both female and male greylag geese (*Anser anser*), the vertebral column includes 17 cervical, 10 thoracic, 10 caudal vertebrae, and 10 pairs of ribs (Doğan & Takcı, 2020). Domestic geese have five thoracic vertebrae, and all the vertebrae are free. The vertebrae are freely movable, as in ducks. They have 9 pairs of ribs compared to 7 pairs in ducks. The thoracic process is absent in both geese and ducks. The sternum of ducks and geese is rectangular, and middle piece of the sternum is separated from the caudolateral process by an oval notch as shown in Figure 1.

The skeletal adaptations of three duck species, including green-winged teal (*Anas crecca*), the mallard (*Anas platyrhynchos*), and tufted duck (*Aythya fuligula*), highlight different features: mallards have robust skeletal elements, which indicate their ability for powerful flight and adaptability across terrestrial and aquatic habitats. In contrast, teals have pneumatic lightweight bones optimized for agility and rapid flight. Tufted ducks display intermediate traits suitable for both diving and flying (Osiak-Wicha et al., 2024).

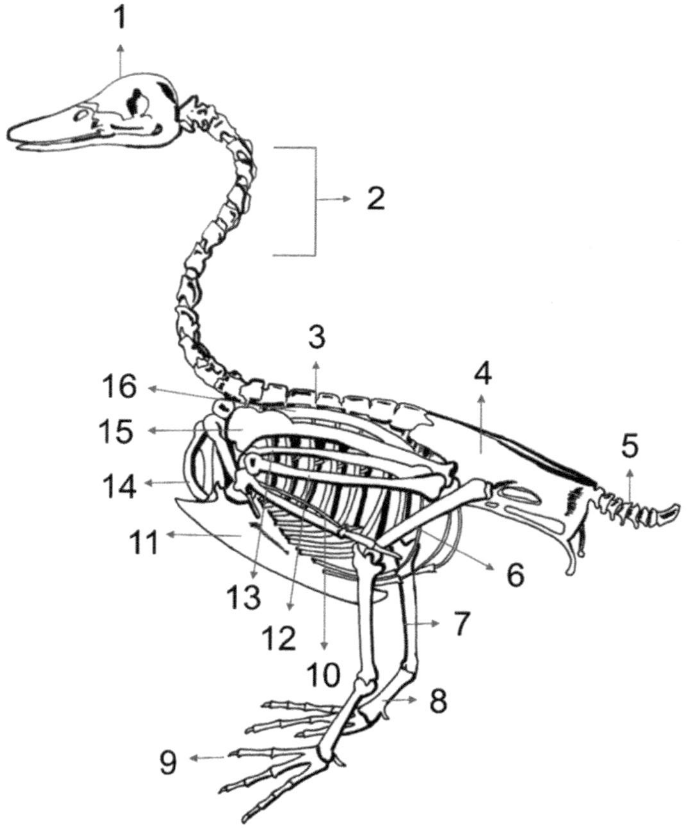

Figure 1. Drawing of the wild duck skeleton (1. skull, 2. cervical vertebrae, 3. thoracic vertebrae, 4. ilium, 5. coccygeal vertebrae, 6. femur, 7. tibiotarsus, 8. tarsometatarsus, 9. digits, 10. rib, 11. keel bone, 12. ulna, 13. radius, 14. furcula, 15. humerus, 16. scapula). Reproduced from a monograph on the Anatidae, or duck tribe (Longman, Orme, Brown, Green, Paternoster-Row, and Eddowes, Shrewsbury, London, 1838).

The appendicular skeleton of geese comprises the pectoral girdle with scapula, coracoid, and collar bone. The wing bones comprise the humerus, radius, ulna, carpus, carpometacarpus, and digits. The hand skeleton consists of the ossa carpi (wrist bones). In the ossa carpi, it was found that only the ossa carpi ulnare and the ossa carpi radiale were on the antebrachial line. The carpometacarpus (wrist-comb bone) was composed of 3 bones, formed by ossa carpi and metacarpus 2, 3, and 4. The ossa digitorum manus (finger bones) consist of three fingers of varying size and structure. Two were formed from 2 phalanges, and the third finger, the strongest, consists of a total of 3 phalanges. The fourth finger has a single phalanx. The pelvic girdle comprises the hip bone, the thigh bone, the knee cap, the foot, the wrist bone, and the fibula (calf bone). The foot bone consists of tarsals, tarsometatarsus (foot wrist-comb), toe bones, and phalanges. Two phalanxes on the first toe, three on the second, four on the third, and five on the fourth (Doğan & Takcı, 2020).

3. Respiratory System

The duck respiratory system comprises of a paired, non-inflatable lungs with several large air sacs as described in Figure 2. In lungs, the gaseous exchange with the blood takes places and the air sacs are organized as anterior and posterior sacs, which function as bellows to move gases across the exchange surfaces of the lungs but do not directly participate in the exchange of gases (Bretz & Schmidt-Nielsen, 1971). The exchange of gases usually occurs within fine air capillaries and branches of the bronchi. In contrast, gaseous exchange does not occur in the barely vascularized air sacs and alveoli. Generally, air flows in through the nostrils and mouth of a bird and then travels through the trachea and a series of increasingly smaller tubes into the lungs and air sacs (Lovette & Fitzpatrick, 2016). The larynx is situated above the trachea, which serves as the pathway for airflow. The respiratory system in the white pekin duck is characterized by an extensive network of 11 air sacs, including 4 paired (cervical, anterior thoracic, posterior thoracic, and abdominal) air sacs and 3 single air sacs (interclavicular, dorsal, and synsacral). It has complex bronchial branching leading to parabronchi and air capillaries for gas exchange, and a well-developed diaphragm (Mennega & Calhoun, 1968).

The airflow in duck lungs is organized into three different regions of the bronchi. The primary bronchus channels air from the trachea across the lungs to the posterior air sacs. The secondary bronchi arise from the primary bronchus, including the craniomedial secondary bronchi that connect both the anterior air sacs and the lung parenchyma. In contrast, the caudodorsal secondary bronchi connect exclusively with the lung parenchyma. In tertiary bronchi, gaseous exchange with blood occurs in the adjacent capillaries. Moreover, recurrent connections (recurrent bronchi) originate in the air sacs, forming part of the lung parenchyma and connecting to the tertiary bronchi. Gas exchange takes place exclusively in the region of the tertiary bronchi (Bretz & Schmidt-Nielsen, 1971).

The respiration process in ducks occurs in the following manner: during inspiration, air flows directly into the primary bronchus, and from there into the caudodorsal secondary bronchi. The anterior air sacs receive the air that comes from the lungs, after gaseous exchange has occurred, at the same time posterior air sacs get dead-space air. The airflow is channelized from the craniomedial secondary bronchi into anterior sacs through the tertiary bronchi. Then, the posterior sacs cannot fill a large quantity of air, as recurrent connections present in this region direct the air to exchange gases with the blood. The recurrent connections of the interclavicular and anterior thoracic sacs aid in filling the air sacs or bellows. Craniomedial secondary bronchi are presumed to be closed as no air flows into the anterior air sacs directly from the primary bronchus. Thus, during expiration, air moves from both the caudodorsal secondary bronchi to the tertiary bronchi, and from craniomedial secondary bronchi to the primary bronchus. In contrast, only minimal air flows cranially into the primary bronchus. At last, the expired air flows directly from posterior air sacs through the craniomedial secondary bronchi and tertiary bronchi to reach the primary bronchus (Bretz & Schmidt-Nielsen, 1971).

In geese, the larynx cartilage is made up of a single cricoid and a procricoid cartilage, and a paired arytenoid cartilage. Cricoid, the largest cartilage of larynx, forms the entire ventral and caudodorsal roof of the larynx, whereas a mature goose has ossified cricoid cartilage. The procricoid cartilage is the smallest and hammer-shaped. Trachea is made up of cartilage rings, which starts from caudal margin of the larynx, and extends ventromedially along the inguinal region before entering the thoracic cavity. The tracheal cartilage rings number between 137 and 140 in the domestic goose, depending on body size. In the mid portion of

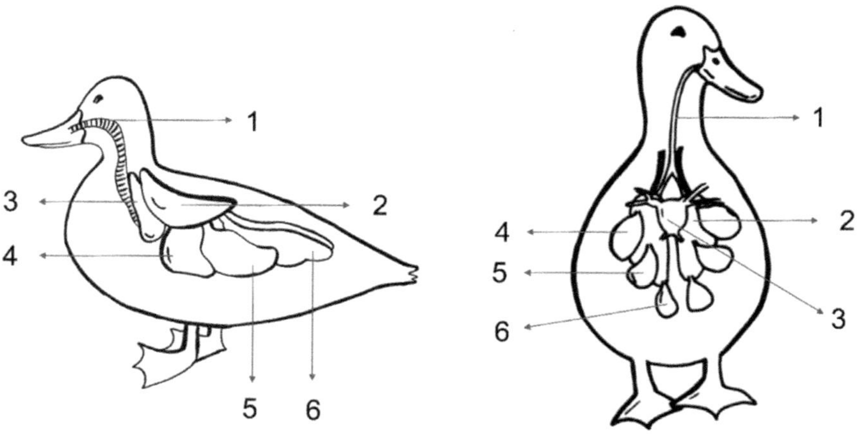

Figure 2. Respiratory system of ducks: lungs and air sacs (1. trachea, 2. lung, 3. interclavicular air sac, 4. anterior thoracic air sac, 5. posterior thoracic air sac, 6. abdominal air sacs). Reproduced from "The duck respiratory system: airways, lungs, and air sacs in detail (*Ducks of Providence*)," pet duck education blog, Melanie, E., 2025.

trachea, the cartilage rings are connected to the previous or following rings and are "H"-shaped. The muscles sternotrachealis, cleidotrachealis, and tracheolateralis facilitate tracheal movement (Onuk et al., 2010).

The syrinx, located at the base of the heart between the terminal end of trachea and the first part of right and left primary bronchi, is of tracheobronchial type. Goose syrinx is made up of the bronchosyringeal and tracheosyringeal cartilages. The "C"-shaped bronchosyringeal cartilages are formed by the first six cartilages in the goose. Anatomical studies in geese show that syrinx is observed within the thoracic cavity at the bifurcation of the trachea. It is composed of bronchosyringeal and tracheosyringeal cartilages, medial and lateral tympaniform membranes, pessulus, interbronchial ligament, foramen, and extrinsic syringeal muscles (Reda, 2017). In mallard, the syrinx is also a tracheobronchial type made from tympanum, bronchosyringeal, and tracheosyringeal cartilages. The tympanum and tracheosyringeal cartilages are made from four and six cartilage rings, whereas bronchosyringeal cartilages are made from C-shaped cartilage rings (Yilmaz et al., 2012).

4. Digestive System

In birds, the digestive system represents a remarkable facet of avian physiology, governing how food is ingested, processed, and absorbed through a series of specialized organs. In Anatidae, as in other avian families, modifications of the digestive tract mirror their diverse feeding strategies. The system comprises of oral cavity, tongue, pharynx, esophagus, crop, proventriculus, gizzard (ventriculus), liver, pancreas, small intestine, large intestine, and cloaca as shown in Figure 3. Ducks, which are primarily diving and filter-feeding species, display distinctive adaptations in their bills and tongues. The bill has a rounded tip with a keratinized nail, medium length, flattened at the center, and slightly raised at the base—a morphology typical of Anseriformes. The tongue is thick, fleshy, and spatula-shaped, lined with numerous papillae and horny spines whose lateral edges are interlocked with the lamellae of the bill, creating an effective filtering mechanism of ingestion that drains water laterally while retaining essential food particles.

The flow patterns of food in the digestive tract of ducks are as follows: at first from the oral cavity, the food enters into the crop, which serves as a temporary storage and processing organ and softens the food by the numerous chambers aided by mucus secretions. It then passes into the proventriculus, the glandular stomach, where digestion begins through the action of oxyntico-peptic cells secreting hydrochloric acid and pepsinogen, respectively. Then, the food subsequently reaches to the ventriculus (gizzard), which mechanically grinds it with the aid of gastroliths where breakdown of cellulose-rich material occurs and enhances enzymatic access to nutrients. The size of the gizzard varies across species, reflecting dietary requirements of different species of Anatidae. For example, in case of mallards, the stomach mean weight is about 75 g, which is approximately 2.3% of the total body weight. The proventriculus is tubular, with average length of 60 mm and 15 mm in diameter, while the ventriculus is oval, measuring length and width of 67 mm and 46 mm, respectively (Hassan et al., 2012). Species-specific variations further illustrate this diversity. In the red-crested pochard, the tongue is 41 mm long with five large conical papillae on each side; the glottis measures 8 mm, and the esophagus extends 16–16.5 cm in the cervical region. The proventriculus lies dorsal to the left hepatic lobe, while the ventriculus measures 78 × 45 mm^2 and accounts for nearly 9.84% of body mass (914 g). The duodenum forms a 15-cm loop (Boarcas et al., 2024). The small intestine consists of spiral jejunal loops, with the ileum (120 mm × 13 mm) connected to two ceca by ileo-cecal ligaments. Each cecum, approximately 20 cm long, narrows at the ileo–ceco–colic junction. In tufted ducks (*Aythya fuligula*),

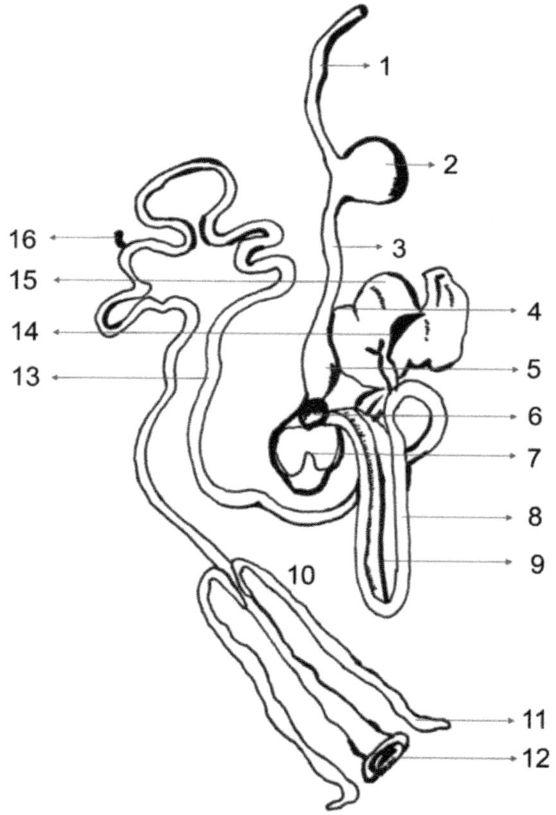

Figure 3. The avian digestive tract (1. esophagus, 2. crop, 3. post-crop esophagus, 4. proventriculus, 5. isthmus, 6. and 7. muscular stomach, 8. distal duodenum, 9. pancreas, 10. colon, 11. ceca, 12. vent, 13. jejunum, 14. gallbladder, 15. liver, 16. Meckel's diverticulum). Reproduced from "Gastrointestinal anatomy and physiology (Sturkie's Avian Physiology, 7th eds.)," Proszkowiec-Weglarz, M., London, 2021.

the ceca are particularly well-developed and open at the transition between the small and large intestine (McLelland, 1989; Działa-Szczepańczyk & Charuta, 2009). The colon, about 100 mm long and 15 mm wide, dilates slightly near the coprodeum (Boarcas et al., 2024).

The liver, another vital organ, weighs 58–113 g in ducks and 85–171 g in geese, representing 3.1–4.1% and 1.5–3.6% of body weight, respectively. Typically, the right lobe is larger, while the left is bean-shaped, with a gallbladder situated beneath both lobes. Beyond bile secretion, the liver plays central roles in carbohydrate metabolism (glycogenesis, glycogenolysis, gluconeogenesis), lipid metabolism (lipogenesis, cholesterol synthesis), and storage of vitamins, iron, triglycerides, and glycogen, along with hemoglobin turnover (Nickel et al., 1977).

From the gizzard, semi-digested food enters the duodenum, where pancreatic enzymes, bile, and bicarbonate complete digestion. Pepsin, trypsin, chymotrypsin, and carboxypeptidase break down proteins, while bile emulsifies lipids and bicarbonate neutralizes acidity to maintain optimal pH. Nutrient absorption occurs mainly in the jejunum and ileum, marked by a Meckel's diverticulum at their junction. The paired ceca enable microbial fermentation of undigested matter and water reabsorption, while the colon continues water absorption before the cloaca expels mixed fecal and urinary wastes (Lucy & Karthiayini, 2022).

5. Urinary System

The urinary system in Anatidae comprises a pair of kidneys and a ureter, which opens into the urodeum of the cloaca. In spur-winged goose, urinary system anatomically showed the kidney as a brownish, flattened organ situated intra-abdominally, occupying the synsacral fossa. The kidney is characterized by the presence of cranial, middle, and caudal lobes, respectively. The kidneys are made up of millions of renal corpuscles. The ureter is a tubular organ emerging from the kidneys and enters into the cloaca at the dorsomedial aspect, whereas cloaca forms a connective link between the urinary, digestive, and reproductive systems (Majama et al., 2024). The cranial lobe of the kidney in the mallard ducks is small and round to oval in shape. Their kidneys are made up of two types of nephrons; one is the reptilian type and the other is the mammalian type. Reptilian-type nephrons are small, lack loops of Henle, and are constricted to the renal cortex region, whereas mammalian-type nephrons are larger with long or intermediate loops of Henle which are clumped into medullary cones. The middle lobe is elongated, and the caudal lobe is the most prominent and elongated (Majama et al., 2024). Due to the absence of loop of Henle in reptilian types of nephrons, the urine is highly diluted and uric acid is excreted out as the end product of the nitrogen metabolism in the ducks.

6. Reproductive System

Ducks have a complicated reproductive system, especially when it comes to their breeding pattern and genital structure. Waterfowls are characterized by a complex reproductive system due to the dynamicity of sexual selection, mating patterns, and distinctive ecological pressure. The reproductive system of ducks serves multiple functions: species-specific recognition, mate selection, reproductive success, and preventing extra-pair copulation beyond normal gamete production and fertilization. These specialized versions of duck reproductive anatomy and physiology reflect a decade of efficient evolutionary adaptation (Brennan et al., 2007).

6.1. Male Reproductive System

The Anatidae family exhibits very unique anatomical characteristics in their reproductive organ. Duck testes are elongated, bean-shaped, and pink in color, attached with seminiferous tubules in a compact organization. They are located asymmetrically on either side of the body's midline and cranioventral to kidney (Kareem et al., 2020; Brennan et al., 2007). The seasonal changes in morphological structure of male reproductive organs follow a synchronized annual cycle, which depends on photoperiod and environmental cues. Testicular recrudescence usually starts in late December to early January, and it develops over an estimated period of a month. The changes such as organ enlargement, spermatogenesis reactivation, and restoration of hormone production occur during this process. Testes grow to the largest size and heavy weight, and seminiferous tubules project peak diameter with an optimal-level spermatogenic activity. Seven distinct stages are shown in testicular development by histological analysis in the period of complete regression to full breeding condition. Stage 6 represents the pinnacle of reproductive readiness, characterized by maximum tubule diameter, abundant mature spermatozoa, and large central lumina filled with motile sperm. The regression phase begins early in the postnuptial molt period, with rapid initial changes occurring within 1.5 to 2.5 weeks. This process involves the mass expulsion of spermatozoa from seminiferous tubules, followed by the gradual reduction in organ size over the subsequent 1–2 months. The regressed testes return to their minimum size and weight, maintaining only basic cellular architecture until the next reproductive cycle (Johnson, 1961).

Spermatogenesis in ducks follows the typical avian pattern but exhibits unique temporal and quantitative characteristics. The process begins with spermatogonial stem cells located in the basal compartment of seminiferous tubules, which undergo mitotic divisions to produce primary spermatocytes. These primary spermatocytes cells then progress through meiotic divisions to produce secondary spermatocytes and eventually mature spermatozoa. The efficiency of sperm production varies significantly with season, age, and individual condition. During peak breeding season, daily sperm production can reach impressive levels, with testicular parenchyma showing optimal cellular organization and maximum functional capacity. The spermatogenic cycle in ducks is approximately 14–16 days, comparable to other avian species but with notable variations in cellular kinetics and production rates (Johnson, 1961). Sperm morphology in ducks exhibits species-specific characteristics, with variations in head shape, flagellar length, and mitochondrial arrangements. These differences reflect adaptations to specific fertilization environments and may play roles in sperm competition and fertilization success. The mature spermatozoa are initially stored in epididymis and vas deferens where they are fully developed and gain motility.

Anatidae are among the few species that possess an intromittent organ called a phallus. The male birds' intromittent organ is formed by an "out-pocketing" of the ventral wall of cloaca as shown in Figure 4. In erect condition, phallus is everted through the cloaca and becomes visible externally, whereas in non-erect phase, it remains within the cloacal or abdominal cavity inside a thin peritoneal sac. During ejaculation, semen passes along a recessed sulcus on the outer surface of the organ (Coker et al., 2002). The phallus is firmly divided into three parts: base, shaft, and glans. When fully erected, it can extent to 13–14 cm, but usually its average length is 6–8 cm. A fibrous cartilage body supports the base of the phallus and contains a narrow ejaculatory sulcus which guides the flow of semen (Gendy et al., 2016).

6.2. Female Reproductive System

The female reproductive system in the duck is mainly composed of the ovary and the oviduct. The ovary is positioned against the dorsal part of body wall, caudal to adrenal glands and toward the left of the spinal column (Brennan et al., 2007). Functionally, the ovary performs two major roles—first, the production of ova, and second, the synthesis of reproductive hormones such as progesterone, estradiol-17β, and small amount of testosterone (Linde et al., 2025). Morphologically, the ovary consists of numerous follicles of various developmental stages, with the largest follicles containing yolks destined for egg production. These developing follicles are enclosed by specialized tissue layers that serve as mechanical support and actively participate in endocrine regulation. The ovarian blood supply is carried out by the ovarian artery which aids in circulating essential nutrients, oxygen, and hormones that support follicular growth and maturation (Johnson, 1961).

During the breeding season, the ovary undergoes a striking transformation, characterized by nearly a hundredfold increase in size and weight compared with its regressed state during the non-breeding phase. This dramatic enlargement results from the rapid proliferation and maturation of multiple follicles, extensive vascularization, and heightened steroidogenic activity. The largest yolk-filled follicles that emerge during this period form the essential nutrient reservoirs required for subsequent embryonic development (Johnson, 1961).

The oviduct, which lies caudally and dorsally within the abdominal cavity, displays remarkable variation in size based on the bird's reproductive status. During the laying period, the oviduct becomes highly voluminous with recorded lengths ranging from 33 to 59 cm. Anatomically, the oviduct maintains specific relationships with surrounding organs: it is rostrally connected to ovary, dorsally to left kidney, ventrally to gizzard and small intestine, and caudally to the cloaca. The oviduct is stably situated by dorsal and ventral oviductal ligaments which act as supportive suspensory structures.

Anatomically, the oviduct can be differentiated into five functionally specified regions such as infundibulum, magnum, isthmus, uterus, and vagina (Mahmoud et al., 2018). The initial segment of the oviduct is called the infundibulum which measures approximately 4–8 cm in length and features a funnel-like structure that captures the ovulated ovum and a tubular section that initiates the egg's passage through the female reproductive tract. Following infundibulum, magnum starts which is characterized as the longest, thickest, and most convoluted region, has pale white or creamy appearance, and is responsible for the accumulation of large quantity of egg-white proteins. The isthmus constitutes the third segment and is of 7–12 cm in length, comparably less convoluted, and extends distally from the magnum through a series of fine longitudinal folds connecting magnum to the uterus. The fourth region, the uterus is also referred as the shell gland which is a highly dilated region of the oviduct and positioned between the isthmus and vagina as shown in Figure 5. It is placed dorsolaterally to the ureters, synsacrum, and the caudal portion of left kidney, enhancing its compactness, whereas on its right side, it is anatomically associated with the terminal part of the colon. The vagina forms the terminal section of the oviduct exhibiting the spiral and highly convoluted structure facilitating both the sperm storage and egg passage.

Interestingly, the duck vagina demonstrates a unique morphological elaboration that appears to function as a structural mechanism to counter unwanted male intromission. These elaborations mainly include two key anatomical structures such as dead-end sacs (blind pouches) and clockwise spirals. The blind pouches vary in number from 0 to 3 depending on the species to species, are located near the cloacal orifice, and represent cul-de-sacs that branch away from the primary vaginal lumen. It is thought that their primary role is to either divert or misdirect the male intromission attempts, thereby providing females with a degree of control over insemination outcomes (Brennan et al., 2007).

The reproductive cycle of ducks is characterized by a precisely timed coordination between males and females, encompassing the phases of gonadal recrudescence, active breeding, and reproductive regression. Photoperiod serves as the important environmental cue regulating these cycles; however, other factors such as ambient temperature, nutrition, and social interactions also act as supplementary cues. In males, gonadal recrudescence often begins in late December to early January and expands up to 3 months to reach full reproductive maturity by mid to late March. In contrast, female reproductive development generally lags behind that of males by 2 to 3 weeks where the initial ovarian recrudescence starts in mid to late January, and gains full functional maturity over the subsequent 2 months.

The breeding season generally extends from March to July, with high reproductive activity during the spring months. Throughout this period, females exhibit enlarged reproductive organs and maintain optimal physiological conditions to support gamete production, fertilization, and successful reproduction. After the breeding season, the reproductive system regresses, reflecting the seasonal cyclicity typical of ducks. Although the exact timing and duration of the breeding cycle vary with species-specific traits, latitude, and local environmental conditions, the overall patterns remain relatively predictable within defined geographic regions (Okpe & Udoumoh, 2016).

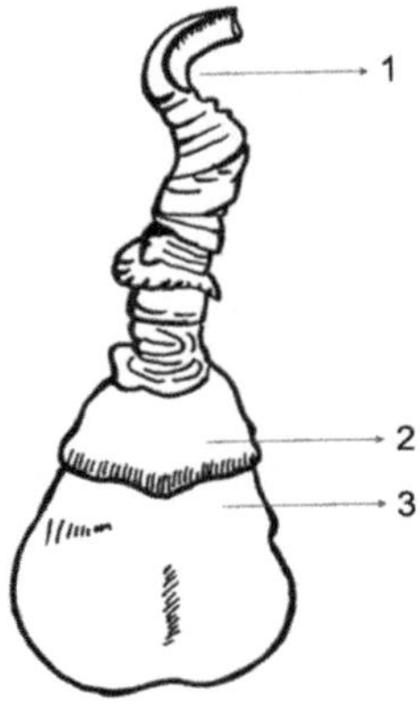

Figure 4. Male genitalia of mallard duck (1. phallus, 2. cloaca entrance, 3. cloacal muscles). Reproduced from "Bird genitalia (Current Biology)," Brennan, P. L., United States, 2022.

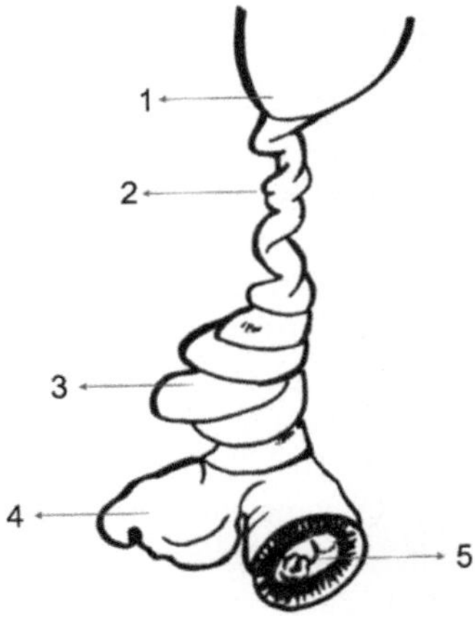

Figure 5. Female genitalia of mallard duck (1. shell gland, 2. clockwise spirals, 3. blind pouches, 4. intestine, 5. cloaca). Reproduced from "Bird genitalia (Current Biology)," Brennan, P. L., United States, 2022.

Hormonal Regulation of the Female Reproductive System: Estradiol-17β serves as the primary female reproductive hormone in ducks, exhibiting dramatic fluctuations throughout the reproductive cycle. Serum concentrations of estradiol-17β show progressive increases as females transit from juvenile to reproductive maturity, with peak levels occurring at the active laying period.

In juvenile Muscovy ducks, estradiol-17β levels average 76.7 ± 23.7 pg/ml, while infantile ducks show lower concentrations of 25.1 ± 6.9 pg/ml. During the laying period, concentrations increase up to 138.0 pg/ml, representing a 96.8 pg/ml of hormone increase over juvenile levels. The hormone shows strong positive correlations with oviduct length, average oviduct diameter, and dominant yolk follicle diameter. The brooding period is characterized by a significant decrease in estradiol-17β concentrations, dropping to 58.4 pg/ml, representing 79.6 pg/ml reduction from peak laying levels. This hormonal shift coincides with the rapid regression of reproductive organs and the redirection of physiological resources toward incubation and maternal care behaviors (Linde et al., 2025).

Progesterone plays a crucial role in reproductive regulation, showing considerable increases throughout the reproductive cycle. The infantile ducks show mean progesterone levels of 0.1 ± 0.01 ng/ml, whereas in juvenile, it rises up to 0.2 ± 0.05 ng/ml. The hormone reaches peak concentrations across the laying period, with first-egg layers showing 0.9 ± 0.1 ng/ml and peak production occurring after the sixth egg with levels of 1.3 ± 0.3 ng/ml. Progesterone concentrations show strong positive correlations with oviduct diameter, length, and dominant yolk follicle size, facilitating its significant role in maintaining the structural and functional integrity of the female reproductive system, during the process of oogenesis. Analogous to estradiol-17β, progesterone concentration decreases about 0.2 ± 0.1 ng/ml during the brooding phase. A study asserts that this decline specifies the cessation of active oviposition and the onset of incubation behavior, reflecting a physiological transition from reproductive phase to maternal care (Linde et al., 2025).

Hormonal Regulation of the Male Reproductive System: Testosterone is a crucial regulating hormone of reproductive physiology in the male ducks, for maintaining the testicular development and breeding behavior. Plasma testosterone concentration varies throughout the annual cycle, with the lowest concentrations in the non-breeding seasons, and reaches to peak levels in the spring breeding period (Stunden et al., 1998; Assenmacher et al., 1975). The relationship between testosterone levels, testicular size, and spermatozoa count is highly complex as both parameters show seasonal fluctuations. During the early phases of testicular recrudescence, testosterone levels increase and reach to maximum level in the peak breeding season. The positive correlation between testosterone and luteinizing hormone (LH) levels highlights their combined role in regulating fertilization. This hormonal alteration can influence the development of secondary sexual characteristics, aggressive behavior, and courtship displays (Penfold et al., 2000; Davis, 2002). Testicular regression and the onset of postnuptial molt synchronize with a subsequent decrease in testosterone concentration, enabling the organism to redirect its focus from reproduction to other associated activities such as feather replacement, seasonal adaptation, and other demands of the organism (Stunden et al., 1998).

Assisted Reproductive Technologies (ART) in Anatidae: The reproductive anatomy and physiology of Anatidae species exhibit several unique characteristics that directly influence the application of assisted reproductive technologies. Males possess an intromitted organ phallus which facilitates forced copulation and ensures successful sperm transfer during natural mating even in aquatic environment, and also presents challenges for artificial semen collection procedures. Female ducks' reproductive tract is equally complex, with multiple sperm storage tubules located in the utero-vaginal junction that can maintain viable sperm for extended periods (Bluhm et al., 1983). These sperm storage structures in oviduct allow females to maintain fertility for 3–4 weeks following a single insemination. This characteristic of female duck highly promotes the implications of artificial insemination (AI) protocols and breeding management strategies. The reproductive endocrinology of ducks follows typical avian patterns but with species-specific variations that impact the timing and success of assisted reproductive technologies. LH plays an important role in ovulation and testicular function, with plasma concentrations showing distinct seasonal patterns that correlate with breeding activity (Penfold et al., 2000). In northern pintails (*Anas acuta*), LH concentration increases during the pre-breeding period and remains elevated throughout the reproductive season, providing a hormonal framework for optimizing insemination timing. Prolactin level in duck is strongly correlated with broodiness and the cessation of egg-laying at the time of incubation, demonstrating it as a vital regulator. Understanding prolactin dynamics has enabled the development of management strategies to extend the productive laying period and improve the efficiency of artificial insemination programs (Oyebanjo et al., 2023). Follicle-stimulating hormone (FSH) concentrations also fluctuate seasonally, influencing spermatogenesis in males and follicular development in females.

6.3. Artificial Insemination Technologies in Anatidae

Semen Collection Techniques: Semen collection in ducks requires specialized methods adapted to their unique anatomy and behavior. The massage method which was initially developed for domestic chickens has been modified and found more effective for use in ducks with considerable high success (Watanabe, 1961). This massage technique involves gentle abdominal massage combined with manipulation and acclimatization of the cloacal area to stimulate semen ejaculation. However, the presence of the phallus in ducks needs careful handling to prevent injury and ensure effective semen collection. Training of male ducks is highly essential for effective semen collection; which typically requires 2–3 weeks of consistent daily handling before reliable semen collection by the trained professionals (Mahapatra et al., 2024). Collection frequency also impacts semen quality: every 2–3 days yields optimal sperm concentration and motility (Oguntunji et al., 2019).

Semen Evaluation and Processing: Evaluation of semen is critical for AI success. Standard parameters include volume, sperm concentration, motility, and morphology, with reference values varying among breeds (Giri et al., 2016). Muscovy duck semen shows higher sperm concentrations (3–4 billion sperm/ml) than common duck breeds (2–3 billion sperm/ml). Progressive motility of 85–95% is optimal, while <70% reduces fertility (Łukaszewicz, 2020). Computer-assisted sperm analysis (CASA) systems provide objective assessments of motility. Preservation techniques using poultry semen extenders (with antioxidants, antibiotics, and energy substrates) maintain viability for 24–48 hours at 4 °C (Penfold et al., 2001).

Insemination Procedures: Standard insemination involves restraining the female in a dorsal recumbent position and inserting a pipette 2–3 cm into the cloaca (Ash, 1962). Insemination volumes range from 0.02 to 0.05 ml, with 100–200 million motile sperm per dose. Timing is critical—insemination must occur 2–6 hours before ovulation for maximum success (Brun et al., 2008). In commercial mule duck production, insemination occurs twice weekly, while conservation programs may use single inseminations due to extended sperm storage (Ghanem et al., 2017).

Semen Cryopreservation: It is critical for genetic conservation. However, duck sperm are highly sensitive to freezing due to their membrane lipid composition (Stunden, 1996). Cryoprotectants such as dimethyl sulfoxide (DMSO) or glycerol (6–10%) and membrane stabilizers such as cholesterol-loaded cyclodextrins improve post-thaw viability (Blesbois, 2006). Slow-freezing methods (1–2 °C/min to –80 °C, followed by liquid nitrogen storage) produce the best outcomes (Antenucci, 2024). Species variation exists, with Muscovy sperm showing higher cryotolerance (>60% post-thaw motility) compared to other breeds. Fertility testing post-cryopreservation often requires heterospermic insemination trials due to challenges in establishing avian in vitro fertilization (IVF).

7. Circulatory System

Ducks have a complex cardiovascular system developed to handle the metabolic demands of diving, swimming, and flying. The circulatory system of ducks exemplifies the advanced physiological adaptations found in avian species, incorporating features that enable efficient oxygen transport, thermoregulation, and adaptation to aquatic environments. Understanding the cardiovascular anatomy and physiology of ducks is highly crucial as they provide insights into avian evolution, inform veterinary medicine practices, and help to explain the physiological basis for the remarkable diving abilities observed in many duck species.

Cardiac Anatomy and Structure: Like all birds, ducks possess a four-chambered heart consisting of two atria and two ventricles, providing complete separation of oxygenated and deoxygenated blood (Butler, 1985). This anatomical arrangement is essential for maintaining the high metabolic rates required for flight and thermoregulation. The right side of the heart receives deoxygenated blood from the systemic circulation and then pumps it to lungs through the pulmonary circulation. In contrast, the left side receives oxygenated blood from lungs and distributes it throughout the body via the systemic circulation. The duck heart exhibits several distinctive features compared to mammalian hearts. The avian heart is generally cone-shaped with a thin right ventricle and a thick left ventricle, with the wall thickness ratio typically being 1:4 (left to right). This asymmetry reflects the different pressure requirements of pulmonary versus systemic circulation. This unique feature continues from the newly hatched to the adult duck. The left ventricle generates a significantly higher pressure to pump blood throughout the entire body, necessitating increased muscular development (Rigdon & Frölich, 2010). In morphometric analysis of duck hearts, comparing wild mallards with domestic Pekin ducks revealed significant differences in heart size relative to body mass between wild and domestic forms. Wild mallard ducks demonstrated a relative heart weight of approximately 0.91% of body weight, while domestic Pekin ducks showed a relative heart weight of 0.76% of body weight. The absolute heart weight differences were even more pronounced. Wild mallard hearts averaged 10.0 ± 1.0 g in males and 8.6 ± 1.1 g in females, while domestic Pekin duck hearts averaged 19.5 ± 1.4 g in males and 18.9 ± 1.1 g in females. However, when adjusted for body mass, wild ducks maintained proportionally larger hearts, reflecting their higher metabolic demands and more active lifestyle (Bartyzel et al., 2005).

Valvular System and Conduction: The atrioventricular valves of duck hearts show unique structural features, such as the left atrioventricular valve is tricuspid rather than bicuspid as observed in mammals and the right atrioventricular valve possesses a single spiral flap of myocardium. These valves are connected with the Purkinje conduction system, indicating that they can close simultaneously with the ventricular contraction. The pulmonary artery and semilunar valves present at the base of the aorta also contain myocardial tissue extensions which are connected to the electrical conduction system. This arrangement suggests that these valves function as sphincters capable of regulating the blood flow dynamics during different phases of the cardiac cycle (Rigdon & Frölich, 2010).

Cardiac Output and Performance: Duck hearts show remarkable performance capabilities, with cardiac output values efficiently higher than those of mammals. A study on tufted ducks (*Aythya fuligula*) by Bevan and Butler (1992) established a strong positive correlation between heart rate and cardiac output, so rendering heart rate as a dependable indicator of cardiac performance. Normally, in rest, the heart rate ranges from 115 to 130 beats per minute, whereas during flight, it may increase up to 670 beats per minute. Ducks possess an exclusive physiological adaptation in which their heart rate and stroke volume do not act in the same manner as those of mammals, maintaining with their active lifestyle. While mammals primarily increase cardiac output through increased stroke volume, ducks achieve enhanced cardiac output predominantly through increased heart rate. This adaptation allows for rapid cardiovascular responses to changing metabolic demands, particularly important during transitions between rest and flight.

Circulatory Regulation and Blood Pressure: Duck circulatory systems maintain higher blood pressures than comparable mammals, with mean arterial pressures typically 20%–30% higher than mammalian values. This elevated pressure is necessary to ensure adequate perfusion of all body tissues, particularly during high-altitude flight when oxygen availability is reduced. The distribution of blood flow can be dynamically adjusted through vasoconstriction and vasodilation of peripheral vessels. During flight, blood flow is preferentially directed to the flight muscles (pectoralis and supracoracoideus), while flow to digestive organs and other non-essential tissues is reduced. This redistribution ensures that metabolically active tissues receive adequate oxygen and nutrient supply during periods of high demand (Bevan & Butler, 1992).

Blood Characteristics: Duck blood cells exhibit several distinctive characteristics compared to mammalian blood. Red blood cells are elliptical in shape and nucleated, measuring approximately 6 × 12 micrometers. The nucleated nature of avian red blood cells allows for ongoing metabolic activity and longer functional lifespan compared to mammalian erythrocytes (Gaehtgens et al., 1981). The concentration of red blood cells in duck blood typically ranges from 2.5 to 4 million cells per cubic millimeter, with hemoglobin concentrations optimized for the high metabolic demands of flight. The lifespan of duck red blood cells is approximately 28–45 days and plasma constitutes approximately 85% water with protein concentrations of 9%–11%, higher than typical mammalian values. Blood glucose levels are also elevated compared to mammals, ranging from 200–400 mg/dl, reflecting the high energy demands of avian metabolism. The plasma also contains specialized proteins and enzymes that support the unique physiological demands of aquatic birds, including enhanced oxygen-carrying capacity and buffering systems that help to maintain acid–base balance during diving activities (Ritchison, 2023).

Diving Physiology: Many duck species exhibit remarkable diving abilities, with some species capable of remaining submerged for extended periods while foraging underwater. These abilities require specialized cardiovascular adaptations to manage oxygen consumption and distribution during apneic periods. Ducks exhibit a characteristic "diving response" that includes bradycardia (reduced heart rate), selective perfusion of vital organs, and peripheral vasoconstriction. Bevan and Butler (1992) demonstrated that in tufted ducks, cardiac output decreases during diving compared to surface swimming. However, blood flow to essential organs such as the brain and heart is maintained through redistribution away from less critical tissues. The diving response is mediated by chemoreceptors that detect increased carbon dioxide levels during breath-holding. These receptors trigger neural responses that reduce heart rate and cause selective vasoconstriction of vessels supplying nonessential organs while maintaining or increasing blood flow to the cardiac muscle and central nervous system.

Thermoregulation in Aquatic Environments: Ducks must maintain core body temperature while in contact with water that is often significantly cooler than ambient air temperature. The cardiovascular system plays a crucial role in thermoregulation through counter-current heat exchange mechanisms and selective peripheral vasoconstriction. Blood vessels in the extremities, particularly the legs and feet, are arranged in counter-current patterns where warm arterial blood flowing to the extremities passes near cooler venous blood returning to the body core. This arrangement allows for heat recovery and minimizes heat loss to the environment (Midtgård, 1989).

8. Immune System

Waterbirds have highly developed immune systems, which allow them to survive in a diverse aquatic environment while maintaining resistance to numerous pathogens. The duck immune system has garnered significant scientific attention due to its ability to serve as a natural reservoir for avian influenza viruses without exhibiting severe clinical disease, contrasting sharply with the susceptibility observed in chickens and other domestic poultry (Webster et al., 1992; Olsen et al., 2006). The co-evolutionary relationship between ducks and influenza viruses has transformed a host–pathogen interaction which allows these birds to function as asymptomatic carriers of low-pathogenic avian influenza (LPAI) viruses while mostly surviving with high-pathogenic avian influenza (HPAI) virus infections (Evseev & Magor, 2019). The factors such as tissue tropism differences, specialized adaptive immune mechanisms, and robust innate immune responses enhance their resistance to disease and survival mechanism, ultimately making ducks as a unique clade in waterbirds. The duck immune system is categorized into primary (central) and secondary (peripheral) lymphoid organs where these primary lymphoid organs are mainly responsible for the generation and maturation of immune cells, whereas secondary organs facilitate immune responses activation and memory formation (Glick, 1978).

Primary Lymphoid Organs: The thymus is a paired, lobulated shape organ positioned in the subdermal connective tissue around the neck region. Each thymic lobe consists of five sub-lobes of varying shapes and sizes, appearing pale white to yellowish white in coloration. The organ exhibits an elongated and flattened morphology characteristic of waterfowl. Histologically, the thymus is enclosed by a thin connective tissue capsule from which a large number of delicate septa originate, dividing the organ into incompletely separated lobules. Each lobule displays the characteristic cortical-medullary organization, with a peripheral cortex containing compactly packed small lymphocytes and a central medulla with fewer cells of various sizes. The cortex has higher concentration of lymphocyte cells and is more basophilic than the medulla (Sultana et al., 2012). Within the medulla, pale-staining Hassall's corpuscles are arranged in concentric formations, representing specialized epithelial structures involved in T-cell selection and tolerance induction. The thymic architecture supports the development of T-lymphocytes through positive and negative selection processes, ensuring the production of immunocompetent T-cells while eliminating potentially autoreactive clones.

The bursa of Fabricius represents a unique avian structure essential for B-lymphocyte maturation and development, and humoral immunity. In ducks, it appears as a cylindrical, blind, and cecum-like elongation forming a dorsal diverticulum of proctodeal wall of the cloaca. The bursa exhibits a yellowish-white coloration and plays a pivotal role in B-cell maturation and immunoglobulin production. Histologically, the bursa of Fabricius consists of mucosal folds (plicae) projecting into lumen, where each plica is in streamlined shape that is wider in the mid portion and narrower in the apical and base portions.

Each fold of lamina propria is filled by a large number of follicles, each follicle possessing clearly defined margins and separation from adjacent lymphoid tissue by connective tissue elements. Each bursal follicle exhibits a characteristic structure with a peripheral cortex made up of densely packed small lymphocytes (appearing darkly stained) and central medulla containing fewer cells of variable sizes (appearing paler). A distinctive feature is the layer of undifferentiated epithelial cells occupying the periphery of the medulla region, and this fine capillary layer separates the cortex. This unique architecture facilitates the close interaction between developing B-cells and the specialized bursal microenvironment (Sultana et al., 2012).

Secondary Lymphoid Organs: Spleen in ducks is a triangular-shaped organ with dorsal and curved ventral surfaces, appearing as a rounded, reddish-brown structure positioned in the middle of gizzard and proventriculus. As a mixed lymphoid organ, the spleen contains both B-cell and T-cell zones, facilitating complex immune interactions. The spleen is enclosed by a thick capsule-like structure, with limited number of trabeculae extending into the parenchyma. It encompasses two core components: the red pulp and the white pulp, where segregation ridge of this pulp is very scare as compared to mammals. The white pulp is made up of network of reticular cells and fibers that facilitate plasma cells along with lymphoid nodules and sheathed arteries and lymphocytes of different sizes (Sultana et al., 2012). The red pulp contains venous sinuses and reticular cords highlighted by numerous macrophages, lymphocytes, and circulating blood cells. This structural organization facilitates the spleen's dual functions of immune surveillance and hematopoiesis. This reticular organization serves as a region for immune cell interaction, inducing antigen presentation and immune cell activation.

Ducks' cervical lymph nodes serve a pivotal role in immune development and responses. These lymph nodes serve as sites for antigen presentation, lymphocyte activation, and immune memory formation. The early expression of immunoglobulin genes in these organs highlights their necessity in B-cell maturation and antibody production (Bando & Higgins, 1996). Moreover, lymphoid tissues also involve gut-associated lymphoid tissues (GALT) in waterfowl which crucially aid in substantial food ingestion and protection against the consistent exposure of environmental pathogens. The cecal tonsils and other mucosal-associated lymphoid tissues (MALT) provide localized immune surveillance and response capabilities. The innate immune system in ducks acts as a first line of defense against microbial pathogens and plays an evident role in their remarkable resistance to viral infections, particularly avian influenza. This system comprises physical barriers, cellular components, and molecular recognition mechanisms that provide immediate protection against invading pathogens (Cornelissen et al., 2012).

Cellular Immunity: The cellular immune system in ducks exhibits numerous adaptations that facilitate to their pathogen resistance and survival advantage.

T-Lymphocyte Responses: Duck's T-lymphocytes highlight their distinctive characteristics in response to the viral infections. CD8+ T cells in ducks are retained for prolong time at infection sites in contrast to chickens, suggesting their pivotal role in successful viral clearance and these enhanced defense activities also contribute to increase the survival rate of ducks (Evseev & Magor, 2019; Dai et al., 2022). The T-cell repertoire in ducks demonstrates a high diversity and numerous variations in T-cell receptor genes, allowing them to recognize a broad range of antigenic epitopes. This diversity is maintained through both genetic and somatic mechanisms, ensuring comprehensive surveillance of potential pathogens (Evseev & Magor, 2019).

Major Histocompatibility Complex: Wild ducks possess large number of major histocompatibility complex (MHC) class I alleles, in contrast to domestic and poultry birds that show limited MHC class I diversity. This genetic diversity provides enhanced capability for presenting viral peptides to T-cells and contributing more effectively in cellular immune responses. The MHC class I molecules in ducks possess unique amino acid modification, notably cysteines at positions 95 and 112 within the peptide-binding groove, which potentially induce viral peptide binding and presentation, and contribute to the superior T-cell responses in ducks. The population-level peptide repertoire of ducks imposes important evolutionary pressure on avian influenza viruses and may influence viral evolution between the wild populations (Evseev & Magor, 2019).

Immune Memory and Secondary Responses: Ducks have high immunological memory that empowers them to combat viruses while also serves reservoir function. Memory B-cells and T-cells act as secondary responses to earlier encountered viruses or pathogens, while also enabling high survival against any type of infections. This memory response in ducks shows strain-specific characteristics, with strong protection against homologous re-infection but variable cross-protection against heterologous strains. This form of selective susceptibility may assist in sustaining the diversity of viruses within wild duck populations.

Molecular Mechanisms of Immune Regulation: The duck's immune regulation is carried out through the numerous molecular mechanisms which perform together to diversify the pathogens while maintaining homeostasis which involves complex signalling pathways, gene regulation networks, and post-translational modifications. The duck immune system is modulated by multiple signalling pathways that integrate pathogen recognition with appropriate effector responses. The retinoic acid-inducible gene I (RIG-I) signalling pathway represents a fundamental mechanism in ducks demonstrating more effective signalling in contrast to other avian species. Upon detecting viral RNA, RIG-I goes through a conformational change that uncovers its caspase activation and recruitment domains (CARDs). These domains eventually interact with the adaptor protein mitochondrial antiviral signalling (MAVS), which activates the transcription factors such as IRF3, IRF7, and NF-κB, creating a rapid and robust antiviral response in ducks (Evseev & Magor, 2019). The stimulator of interferon genes (STING) pathway provides another crucial signalling mechanism that associates with identifying viral nucleic acid upon rise in levels of interferons. Duck STING demonstrates enhanced activity compared to chicken STING, inducing to more effective antiviral responses system.

9. Conclusion

The anatomy and physiology of Anatidae demonstrate a complex integration of structural and functional adaptations that support their demanding lifestyles of flight, swimming, and often diving. The cardiovascular system, with its four-chambered heart, high arterial pressures, and unique valve structure, ensures efficient oxygen transport and quick adjustment to varying metabolic needs. The respiratory system, provided with rigid lungs and extensive air sac networks, facilitates continuous unidirectional air flow and superior gas exchange, enabling high-altitude flight and prolonged diving. Renal adaptations, including the coexistence of mammalian and reptilian nephron types, reflect their requirement to balance osmoregulation in aquatic environments. Hematological features, such as nucleated red blood cells with elevated hemoglobin levels and high glucose concentrations, further highlight the high metabolic activity of these waterbirds. Reproductive anatomy and physiology are equally specialized, with males showing seasonal changes in gonadal development and females possessing sperm storage tubules that enhance fertilization success. Endocrine regulation via the hypothalamic–pituitary–gonadal axis coordinates reproductive cycles with seasonal and environmental cues, ensuring breeding synchrony. Beyond reproduction, thermoregulatory strategies, such as counter-current heat exchange in extremities and dynamic cardiovascular adjustments, enable ducks to maintain homeostasis across diverse habitats and climates. Overall, the physiological complexity of Anatidae reflects the evolutionary refinement of multiple organ systems to meet the unique ecological and behavioral demands of this family. These insights deepen our understanding of avian biology. It also has practical implications for conservation physiology, veterinary science, and the development of assisted reproductive and genomic tools for both wild and domestic species.

References

Antenucci, E. (2024). Optimizing genetic resource conservation and semen cryobanking in aquatic and avian species with a focus on investigating the impact of ovarian fluid on fertilization success. Doctoral Dissertation, University of Molise. Retrieved from https://iris.unimol.it/handle/11695/145349

Ash, W. J. (1962). Studies of reproduction in ducks. *Poultry Science*, 41, 2002–2003.

Assenmacher, I., Astier, H., Daniel, J. Y., & Jallageas, M. (1975). Experimental studies on the annual cycles of thyroid and adrenocortical functions in relation to the reproductive cycle of drakes. *Journal of Physiology*, 70(5), 507–520.

Avilova, K. V. (2018). Spatial organization of the epithelial structures in the bill TIP organ of waterfowl (Anseriformes, aves). *Biology Bulletin Reviews*, 8(3), 234–244. https://doi.org/10.1134/s2079086418030027

Backues, K. A. (2015). Anseriformes. *In:* Miller, R. E., Fowler, M. E. (eds.), *Fowler's Zoo and Wild Animal Medicine*, Elsevier, 8, 116–126. https://doi.org/10.1016/b978-1-4557-7397-8.00016-5

Bando, Y., & Higgins, D. A. (1996). Duck lymphoid organs: Their contribution to the ontogeny of IgM and IgY. *Immunology*, 89(1), 8–12. https://doi.org/10.1046/j.1365-2567.1996.d01-703.x

Bartyzel, B., Karbowicz, M., & Bartyzel, I. (2005). A comparison of body and heart size between the mallard and Pekin duck. *Veterinarija ir Zootechnika*, 29(51), 22–25.

Bevan, R. M., & Butler, P. J. (1992). Cardiac output and blood flow distribution during swimming and voluntary diving of the tufted duck (*Aythya fuligula*). *Journal of Experimental Biology*, 168(1), 199–217. https://doi.org/10.1242/jeb.168.1.199

Blesbois, E. (2006). Advances in avian semen cryopreservation. *World Poultry Science Association Proceedings*, Italy.

Bluhm, C. K., Phillips, R. E., & Burke, W. H. (1983). Serum levels of luteinizing hormone, prolactin, estradiol and progesterone in laying and nonlaying mallards (*Anas platyrhynchos*) 1. *Biology of Reproduction*, 28(2), 295–305. https://doi.org/10.1095/biolreprod28.2.295

Boarcas, C. G., Mihai, S., Mustatea, A., Serbanescu, D., Cristian, M., Stoiculeasa, P., & Predoi, G. (2024). Considerations regarding the digestive system in the red-crested pochard (*Netta rufina*). *Revista Romana de Medicina Veterinara*, 4, 35–39.

Brennan, P. L. (2022). Bird genitalia. *Current Biology*, 32(20), R1061–R1062. https://doi.org/10.1016/j.cub.2022.09.015

Brennan, P. L., Prum, R. O., McCracken, K. G., Sorenson, M. D., Wilson, R. E., & Birkhead, T. R. (2007). Coevolution of male and female genital morphology in waterfowl. *PLoS One*, 2(5), e418. https://doi.org/10.1371/journal.pone.0000418

Bretz, W. L., & Schmidt-Nielsen, K. (1971). Bird respiration: Flow patterns in the duck lung. *Journal of Experimental Biology*, 54(1), 103–118. https://doi.org/10.1242/jeb.54.1.103

Brun, J., Mialon-Richard, M., Sellier, N., Batellier, F., & Brillard, J. (2008). Duration of fertility and hatchability of the common duck (*Anas platyrhynchos*) in pure- or crossbreeding with Muscovy drakes (*Cairina moschata*). *Theriogenology*, 69(8), 983–989. https://doi.org/10.1016/j.theriogenology.2008.01.015

Butler, P. J. (1985). Exercise in non-mammalian vertebrates: A review. *Journal of the Royal Society of Medicine*, 78(9), 739–747. https://doi.org/10.1177/014107688507800909

Coker, C. R., McKinney, F., Hays, H., Briggs, S. V., & Cheng, K. M. (2002). Intromittent organ morphology and testis size in relation to mating system in waterfowl. *The Auk*, 119(2), 403–413. https://doi.org/10.1093/auk/119.2.403

Cornelissen, J. B. W. J., Post, J., Peeters, B., Vervelde, L., & Rebel, J. M. J. (2012). Differential innate responses of chickens and ducks to low-pathogenic avian influenza. *Avian Pathology*, 41(6), 519–529. https://doi.org/10.1080/03079457.2012.732691

Dai, M., Sun, H., Zhao, L., Wu, Q., You, B., Xu, F. et al. (2022). Duck CD8+ T cell response to H5N1 highly pathogenic avian influenza virus infection in vivo and in vitro. *The Journal of Immunology*, 209(5), 979–990. https://doi.org/10.4049/jimmunol.2101147

Davis, E. (2002). Male reproductive tactics in the mallard, *Anas platyrhynchos*: Social and hormonal mechanisms. *Behavioral Ecology and Sociobiology*, 52(3), 224–231. https://doi.org/10.1007/s00265-002-0513-z

Doğan, G. K., & Takcı, İ. (2020). A macroanatomical, morphometric and comparative investigation on the skeletal system of the geese (Anser anser) raised in the Kars region I: Skeleton axiale. *Research in: Agricultural & Veterinary Sciences*, 4(3), 81–95.

Działa-Szczepańczyk, E., & Charuta, A. (2009). Morphology and morphometry of caeca in the tufted duck *Aythya Fuligula*. *Zoologica Poloniae*, 54–55. http://dx.doi.org/10.2478/v10049-010-0003-1

Evseev, D., & Magor, K. E. (2019). Innate immune responses to avian influenza viruses in ducks and chickens. *Veterinary Sciences*, 6(1), 5. https://doi.org/10.3390/vetsci6010005

Eyton, T. C. (1838). *A Monograph on the Anatidae or Duck Tribe*. https://doi.org/10.5962/bhl.title.49546

Feduccia, A. (1975). Aves osteology. *Sisson and Grossman's The Anatomy of the Domestic Animals*, 2, 1790–1801.

Gaehtgens, P., Schmidt, F., & Will, G. (1981). Comparative rheology of nucleated and non-nucleated red blood cells. *Pflügers Archiv – European Journal of Physiology*, 390(3), 278–282. https://doi.org/10.1007/bf00658276

Gendy, E. E., El-Bably, S., & Shaker, N. (2016). Morphological studies on the male copulatory organ of balady duck (*Anasboscas domestius*) with special reference to its blood supply. *Veterinary Medical Journal Giza*, 62(4), 71–76. https://doi.org/10.21608/vmjg.2016.365348

Ghanem, H. M., Roushdy, K., Mahrous, K. F. & Hassan. (2017). Artificial insemination vs natural mating and genetic PRL/Pstl locus polymorphism and their effect on different productive and reproductive aspects in duck. *Advances in Animal and Veterinary Sciences*, 5(4), 179–184.

Giri, S. C., Sastry, K. V. H., Sahoo, S. K., & Kumar, P. (2016). Semen collection, evaluation and fertile period in Khaki Campbell and white Pekin breeds of ducks. *The Indian*

Journal of Animal Sciences, 86(10), 1121–1124. https://doi.org/10.56093/ijans.v86i10.62379

Glick, B. (1978). The immune response in the chicken: Lymphoid development of the bursa of fabricius and thymus and an immune response role for the gland of harder. *Poultry Science*, 57(5), 1441–1444. https://doi.org/10.3382/ps.0571441

Grimmett, R., Inskipp, C., and Inskipp, T. (2016). *Birds of the Indian Subcontinent*. India: Oxford University Press.

Hassan, S. A., & Moussa, E. A. (2012). Gross and microscopic studies on the stomach of domestic duck (*Anas platyrhynchos*) and domestic pigeon (*Columba livia domestica*). *Journal of Veterinary Anatomy*, 5(2), 105–127. https://doi.org/10.21608/jva.2012.44877

Johnson, O. W. (1961). Reproductive cycle of the mallard duck. *Ornithological Applications*, 63(5), 351–364. https://doi.org/10.2307/1365295

Kareem, D. A., Eman Sami Jassem, E. M., Daaj, S. A., & Al-Khalad, W. J. (2020). Morphological and histological study of the testes in adult duck. *Plant Archieves*, 20, 751–755.

Landolt, R., Vanden Berge, J. C., & Zweers, G. A. (1989). The cervical column of the mallard (*Anas platyrhynchos* L.) and the chicken (*Gallus gallus* L.). *Fortschritte der Zoologie*, 35, 74–78.

Linde, M., Wehrend, A., & Farshad, A. (2025). Estradiol-17ß, progesterone, and oviductal changes in muscovy ducks (*Cairina moschata forma domestica*) during reproductive phases. *BMC Veterinary Research*, 21(1), 1–13. https://doi.org/10.1186/s12917-025-04737-5

Lovette, I. J., & Fitzpatrick, J. W. (2016). *Handbook of Bird Biology*. John Wiley & Sons.

Lucy, K. M., & Karthiayini, K. (2022). Anatomy and Physiology of Ducks. *In:* Jalaludeen, A., Churchil, R. R., Baéza, E. (eds.), *Duck Production and Management Strategies*. Springer, Singapore. https://doi.org/10.1007/978-981-16-6100-6_4

Łukaszewicz, E. (2020). Reproductive season and male effect on quantitative and qualitative characteristics of Muscovy duck (*Cairina moschata*) semen. *Poultry Science*, 99(12), 6390–6396.

Mahapatra, S., Nahak, A. K., Prasad, U. V. S. N., Mohanty, T. K., & Sahoo, G. (2024). Artificial insemination in duck. *Journal of Animal Science Advances*, 29(2), 54–61.

Mahmoud, H., Gad, M. R., Awaad, A. S., & Tawfiek, M. G. (2018). Morphological characteristics of the oviduct in Egyptian balady duck (*Anas boschas domesticus*) during laying cycle. *Journal of Veterinary Medical Research*, 25, 1–10. http://dx.doi.org/10.21608/jvmr.2018.43298

Majama, Y. B., Wulgo, A. M., Malah, M. K., & Kwabe, I. D. (2024). Gross anatomy and histology of the urinary system of the spur-winged goose (*Plectropterus gambensis*). *Veterinaria*, 73(1), 53–61.

McLelland, J. (1989). Anatomy of the avian cecum. *Journal of Experimental Zoology*, 252, 2–9. https://doi.org/10.1002/jez.1402520503

Melanie, E. (2025). The duck respiratory system: Airways, lungs, and air sacs in detail. *Ducks of Providence* (pet duck education blog). https://ducksofprovidence.com/duck-respiratory-system/

Mennega, A. & Calhoun, M. L. (1968). Morphology of the lower respiratory structures of the white Pekin duck. *Poultry Science*, 47, 266–280. https://doi.org/10.3382/ps.0470266

Midtgård, U. (1989). Circulatory adaptations too cold in birds. *In: Springer eBooks* (pp. 211–222). https://doi.org/10.1007/978-1-4757-0031-2_23

Nickel, R., Schummer, A., & Seiferle, E. (1977). *Anatomy of the Domestic Birds.* Verlag Paul Parey, Berlin, 202.

Oguntunji, A. O., Oladejo, A. O., Ayoola, M. O., Oriye, L. O., & Aina, A. B. J. (2019). Semen quality parameters of three duck genotypes in the humid tropics. *International Journal of Livestock Production*, 10(12), 379–385.

Okpe, G. C., & Udoumoh, A. F. (2016). Effect of season on structural composition of drake testes and daily spermatozoa production as determined by quantitative testicular histology. *Medycyna Weterynaryjna*, 72(10), 627–631. https://doi.org/10.21521/mw.5572

Olsen, B., Munster, V. J., Wallensten, A., Waldenström, J., Osterhaus, A. D. M. E., & Fouchier, R. A. M. (2006). Global patterns of Influenza A virus in wild birds. *Science*, 312(5772), 384–388. https://doi.org/10.1126/science.1122438

Onuk, B., Haziroğlu, R. M., & Kabak, M. (2010). The gross anatomy of larynx, trachea and syrinx in goose (*Anser anser domesticus*). *Kafkas Universitesi Veteriner Fakultesi Dergisi*, 16, 443–450.

Osiak-Wicha, C., Tomaszewska, E., Muszyński, S., Flis, M., Świetlicki, M., & Arciszewski, M. B. (2024). Comparative analysis of morphometric, densitometric, and mechanical properties of skeletal locomotor elements in three duck species (Anatidae: anatinae). *Animals*, 14(15), 2191. https://doi.org/10.3390/ani14152191

Oyebanjo, M., Osaiyuwu, O., & Salako, A. (2023). Prolactin, genetics, and bioinformatics: The trinity for improving duck egg production in Nigeria – A review. *Journal of Animal Science and Veterinary Medicine*, 8(4), 174–190.

Penfold, L. M., Harnal, V., Lynch, W., Bird, D., Bastien, H., Mirande, C. et al. (2001). Characterization of northern pintail (*Anas acuta*) ejaculate and the effect of sperm preservation on fertility. *Reproduction, Fertility and Development*, 13(4), 291–299.

Penfold, L. M., Wildt, D. E., Herzog, T. L., Lynch, W., Ware, L., Derrickson, S. E. et al. (2000). Seasonal patterns of LH, testosterone and semen quality in the northern pintail duck (*Anas acuta*). *Reproduction Fertility and Development*, 12(4), 229. https://doi.org/10.1071/rd00093

Proszkowiec-Weglarz, M. (2021). Gastrointestinal anatomy and physiology. *In:* Scanes, C. G. & Dridi, S. (eds.), *Sturkie's Avian Physiology* (7th eds.), (pp. 485–527). Elsevier, https://doi.org/10.1016/b978-0-12-819770-7.00010-4

Reda, M. (2017). Gross anatomical syringeal structures of goose (*Anser anser domesticus*). *Journal of Advanced Veterinary and Animal Research*, 4(4), 343–347. http://dx.doi.org/10.5455/javar.2017.d231

Rigdon, R. H., & Frölich, J. (2010). The heart of the duck. *Zentralblatt Für Veterinärmedizin Reihe A*, 17(1), 85–94. https://doi.org/10.1111/j.1439-0442.1970.tb00777.x

Ritchison, G. (2023). Cardiovascular and immune systems. *In: Fascinating Life Sciences* (pp. 885–1006). https://doi.org/10.1007/978-3-031-14852-1_6

Sarott, M. L. V., De Lima, T. G., & Lopes, E. Q. (2024). Anatomical description of the skeleton of the domestic duck (Tadorna tadorna – Linnaeus, 1758) found dead at the IBIMM Research Center in São Paulo – SP - Brazil. *South Florida Journal of Environmental and Animal Science*, 4(2), e4145. https://doi.org/10.53499/sfjeasv4n2-004

Stunden, C. E. (1996). Studies on reproductive characteristics and artificial insemination in captive mallard Ducks (*Anas platyrhynchos*). Master's Thesis, University of British Columbia. Retrieved from: https://open.library.ubc.ca/soa/cIRcle/collections/ubctheses/831/items/1.0087068

Stunden, C., Bluhm, C., Cheng, K., & Rajamahendran, R. (1998). Plasma testosterone profiles, semen characteristics, and artificial insemination in yearling and adult captive mallard ducks (*Anas platyrhynchos*). *Poultry Science*, 77(6), 882–887. https://doi.org/10.1093/ps/77.6.882

Sultana, N., Khan, M., Wares, M., & Masum, M. (2012). Histomorphological study of the major lymphoid tissues in indigenous ducklings of Bangladesh. *Bangladesh Journal of Veterinary Medicine*, 9(1), 53–58. https://doi.org/10.3329/bjvm.v9i1.11212

Wang, J., Zhang, Z., Zhou, K., Jiang, X., Huang, D., Qian, Z. et al. (2025). Analysis of morphological parameters of vertebrae in domestic geese and ducks. *Frontiers in Veterinary Science*, 12. https://doi.org/10.3389/fvets.2025.1531363

Watanabe, M. (1961). Experimental studies on the artificial insemination of domestic ducks and geese. *Bulletin of the National Institute of Agricultural Sciences, Series G*, 19, 1–74.

Webster, R. G., Bean, W. J., Gorman, O. T., Chambers, T. M., & Kawaoka, Y. (1992). Evolution and ecology of influenza A viruses. *Microbiological Reviews*, 56(1), 152–179. https://doi.org/10.1128/mr.56.1.152-179.1992

Yilmaz, B., Yilmaz R., Arican, I., & Yıldız, H. (2012). Anatomical structure of the syrinx in the mallard (*Anas platyrhynchos*). *Harran Üniversitesi Veteriner Fakültesi*, 1(2), 111–116.

Zarnik B. (1926). On the ethology of plesiosaurs with contributions to the mechanism of the cervical vertebrae of recent sauropsids. *Societas Scientiarum Naturalium Croatica*, 424.

CHAPTER

3

Ecological Niches and Habitats of Anatidae Family

Uzma Ansari[1], Farah Bano[2], Prateek[3], Ashish Kumar[3], Anshu Yadav[1] and Himanshu Mishra[1*]

[1] Department of Zoology, Buddha P.G. College, Deen Dayal Upadhyaya Gorakhpur University, Kushinagar 274403, Uttar Pradesh, India
[2] ICAR-National Bureau of Fish Genetic Resources (ICAR-NBFGR), Telibagh, Lucknow 226022, Uttar Pradesh, India
[3] Department of Zoology, University of Lucknow, Lucknow 226007, Uttar Pradesh, India

1. Introduction

The family Anatidae includes waterbirds such as ducks, geese, and swans. It has cosmopolitan distribution except Antarctica. It occupies a broad spectrum of habitats ranging from Arctic tundra and temperate wetlands to tropical lakes, estuaries, and man-made water bodies. Numerous anatid species are migratory, traversing extensive distances from frigid areas in search of sustenance and refuge (Ali, 2002). The International Ornithologists' Union (IOU) acknowledges 174 species of Anatidae over 53 genera, 32 of which contain a single species (Praveen et al., 2020). The traditional classification of the family has three subfamilies: Anseranatinae, Anserinae, and Anatinae (Popovkina & Poyarkov, 2002). The Anseranatinae is dedicated to the distinctive and primitive magpie goose, emphasizing its unique anatomical and behavioral traits in contrast to those of all other wildfowl (Davies & Frith, 1964). The subsequent two subfamilies broadly align with the conventional classification of wildfowl into swans and geese (*Anserinae*) and ducks (*Anatinae*) (Popovkina & Poyarkov, 2002).

*Corresponding author: mishra.himanshu2235@gmail.com

DOI: 10.1201/9781003669784-3

The extinct flightless *Garganornis ballmanni*, weighing 22 kg, is the largest anatid discovered to date. Numerous studies have predicted the distribution of Anatidae along with changes in their habitat location and use (Green, 1996; Shi et al., 2024), and a variety of ecological parameters have been found to be highly associated with Anatidae distribution (Green, 1996). For example, land use type, vicinity to water sources, and elevation altogether impact the distribution of wintering Anatidae (Dong et al., 2013). The majority of wildfowl dedicate significant time to resting on water surfaces, possessing anatomical and physiological adaptations such as buoyant body shape, dense plumage, and webbed feet, which are optimized for an aquatic lifestyle (McNeil et al., 1992).

Nevertheless, each species occupies a distinct ecological niche, characterized by specific habitat preferences, foraging strategies, and reproductive systems which are typically manifested through a suite of specialized adaptations (Futuyma & Moreno, 1988). As a result, while many morphological traits are identical between species, others differ significantly. Anatidae is obviously associated with aquatic settings and each species prefers a definite type of habitat, for example, lakes, marshes, streams, ponds, or reservoirs (Johnsgard, 2010). The suitability of a given aquatic habitat to sustain populations of a specific wildfowl species is contingent upon a complex interplay of biotic and abiotic determinants, including habitat heterogeneity, trophic resource availability, hydrological stability, and anthropogenic influence. These factors encompass water depth and quality, the type and extent of emergent or marginal vegetation, the capacity to sustain a minimum water level, the availability of preferred food resources, and composition of the substrate (Kauppinen & Vaananen, 1999).

Some species, particularly geese, frequently feed in locations far from their nesting or roosting grounds (Johnson et al., 2014). Few species, markedly geese, are known to undertake extensive foraging excursions, frequently feeding at sites located far from their nesting or roosting grounds (Johnson et al., 2014). Swans usually inhabit marshes, small lentic water bodies, and sluggish river systems, while geese demonstrate a preference for open grasslands, boggy tundra, moist meadows, expansive shallow lakes, and seasonally inundated floodplains (Grimmett et al., 2016). Shelducks predominantly dwell in saline or brackish environments, including estuarine zones and tidal mudflats, whereas sheldgeese display a preference for open grassland ecosystems, located at considerable distances from aquatic habitats (Johnsgard, 2010).

The term ecological niche refers to the multidimensional space encompassing a species functional role within an ecosystem, including its habitat use, diet, temporal activity, and interactions with other organisms (Hutchinson, 1957). In Anatidae, niche dimensions are influenced by a complex interplay of morphological traits, such as body size, bill shape, behavioral strategies, and physiological needs (Johnsgard, 1962). Anatidae often live in a variety of water bodies, including rivers, lakes, and marshes, and feed on algae, aquatic plants, insects, crustaceans, fish, as well as other aquatic creatures (Habib, 2018). Shoals are vital habitats for benthic creatures, mollusks, and plankton, providing a varied range of food resources (Sun et al., 2023). The plenty and availability of food resources have a noteworthy impact on the distribution of Anatidae.

In recent years, conservation efforts have gradually more targeted Anatidae species to diminish the adverse effects of recreational hunting and to cope escalating agricultural damage caused by their populations (Van Roomen & Madsen, 1992). Species within the family Anatidae have emerged as important flagship taxa for promoting wetland conservation initiatives (Kear, 1990). Concerns regarding the decline of waterfowl populations were instrumental in the formulation of the first international treaty for nature conservation, the Ramsar Convention on Wetlands of International Importance, especially as waterfowl habitat in 1971 (Matthews, 1993).

The conservation of Anatidae requires an integrated understanding of their ecological niches, particularly how species utilize and share space and resources across dynamic and human-influenced landscapes. This chapter aims to provide an outline of the ecological niches and habitat relationship of Anatidae species with a focus on niche differentiation, foraging strategies, habitat types, and conservation implication in the context of wetland loss, climate change, and human–wildlife interactions.

2. Anatidae: Taxonomy and Classification

The taxonomic classification of the family *Anatidae* remains intricate and has been a focus of ongoing revision and debate. Phylogenetic relationships within and among its constituent groups are inferred from a comprehensive suite of morphological, anatomical, and behavioral characteristics, including body proportions, plumage differentiation in adults and juveniles, tarsal scutellation patterns, vocalizations, syrinx and tracheal morphology, molting strategies, biogeographic distribution, and species-specific behavioral traits (Johnsgard, 1961).

Wildfowl exhibit morphological and behavioral adaptations appropriate for an aquatic lifestyle, often spending extended periods resting on the water surface (McNeil et al., 1992). Each species occupies a distinct ecological niche pigeonholed by specific habitat preferences, foraging strategies, reproductive systems, and other life-history traits, which are frequently reflected in a suite of specialized morphological, physiological, and behavioral adaptations (Polechova & Storch, 2008). The family Anatidae is taxonomically structured into three subfamilies, encompassing 13 tribes, 43 genera, and a total of 148 extant species (Popovkina & Poyarkov, 2002).

2.1. Subfamily Anseranatinae

The subfamily is represented by a single, monotypic tribe, genus, and species, the magpie goose (*Anseranas semipalmata*), which exhibits evident morphological, behavioral, and phylogenetic divergence from other members of the family.

Owing to these distinctions, several authorities have advocated for its placement in a separate family, *Anseranatinae* (Johnsgard, 2010). The magpie goose reveals weakly webbed feet, and its remarkably enlarged hind toes are linked with semi-terrestrial adaptations and frequent perching behavior, reflecting its inimitable locomotors and habitat-use patterns (Johnsgard, 1961).

Like other members of the family Anatidae, the magpie goose undergoes a single annual molt of its primary and secondary flight feathers (Pyle, 2005). A prominent morphological feature distinguishing this species is the markedly hypertrophied trachea present in both sexes, which can exceed 150 cm in length in adult males (Johnsgard, 1961). The third distinguishing trait of magpie geese is their breeding habits (Davies, 1963). The birds frequently form a trio of a male and two females who lay their eggs in the same nest, with all three birds sharing incubation responsibilities (Skutch, 1957).

2.2. Subfamily Anserinae

The subfamily encompassing geese, swans, and whistling ducks, akin to the previously described group, exhibits negligible sexual dimorphism in plumage, vocalizations, and behavioral traits. Moreover, pair bonds within these taxa have a tendency to be robust and persist over multiple breeding seasons or for life (Johnsgard, 1961).

In both tribes (Dendrocygnini and Anserini), the male helps build the nest and rear the young, but only in some species the male participates in incubation (Johnsgard, 1961). The Dendrocygnini is more isolated than the Anserini, as showed by the paucity of intertribal hybrids involving whistling ducks (Johnsgard, 1960). Like the magpie goose, Anserinae species have only one molt per year that occurs after the breeding season and replaces all feathers (Skutch, 1957). However, the flying feathers are shed immediately, making the birds not capable to fly for a period of roughly three to six weeks (Terrill & Shultz, 2023). The sexes in all species have essentially identical plumage, but differ slightly in body weight and, on occasion, voice.

2.3. Subfamily Anatinae

The largest subfamily is Anatinae, which comprises sheldgeese, shelducks, and all other ducks (Johnsgard, 1961). Most Anatinae species differ from the former subfamilies in that pair connections are weaker and must be re-established each year, often with different mates. Different authorities classify it into a variety of tribes, nine of which are recognized (Johnsgard, 1961). In some species, the male's breeding plumage closely resembles that of the female, but the sexes usually have very different breeding plumages (Johnsgard, 1962). Similarly, the adult voices of both sexes tend to be considerably different in these species due to sexual changes in the structure of the syrinx and, in certain cases, the trachea. In contrast to former subfamilies, the front surface of lower tarsus bears a scale pattern (scutellated) ordered linearly (Johnsgard, 2010).

3. Global Distribution and Habitats

Anatidae species inhabit a wide spectrum of wetland ecosystems, encompassing high Arctic tundra, temperate bogs, rivers, estuaries, freshwater and saline lakes, ponds, swamps, coastal lagoons, and intertidal habitats such as mudflats and bays, extending even into open marine environments. They also exploit anthropogenic wetlands, including rice paddies, agricultural landscapes, wastewater treatment facilities, and aquaculture ponds. Wetlands, which serve as critical habitats for these birds, are commonly characterized by high levels of biological productivity and ecological richness.

Subantarctic islands, such as the Auckland Islands and South Georgia, are home to a number of duck species (Shirihai, 2008). In remote oceanic islands such as Micronesia, the Galapagos Islands, and the Hawaiian Islands, ducks have made their way. There, they are commonly vagrants rather than inhabitants (Pratt et al., 1987). Some species of ducks, mostly those that breed in the Northern Hemisphere's temperate and arctic regions, migrate, but do not live in the tropics. Certain ducks are nomadic, searching for polls and makeshift lakes that arise after localized severe rain, especially in Australia where rainfall is unpredictable (Pacific Black Duck, 2008).

Swans, on the other hand, are rarely seen in tropical regions and are primarily found in temperate ones. One extinct species was discovered in New Zealand and the Chatham Islands, one species was discovered in Southern South America, and four or five species are found in the Northern Hemisphere and Australia, respectively. Tropical Asia, Northern South America, Central America, and all of Africa are devoid of them. The mute swan is a single species that has been brought to North America, Australia, and New Zealand. According to Kear (2005), the mute swan is a partial migrant, living in parts of Western Europe but migrating exclusively in Asia and Eastern Europe.

3.1. Habitats

Habitat is a fundamental term in animal ecology and conservation, as wildlife cannot exist without habitat and its accompanying resources (Block & Brennan, 1993). In ecological terms, habitat denotes the environment inhabited by numbers of a species at any stage of their life cycle, providing the essential resources for survival and reproduction (Block & Brennan, 1993; Jones, 2001). The most fundamental and widely used measure of habitat utilization for individual species is their occurrence in specific habitats. Quantitative evidence of habitat relationships arises when species presence or abundance varies in proportion to measurable habitat characteristics (Brown, 1984). Nevertheless, habitat connectivity must not be conflated with habitat or resource selection, because selection is influenced by availability, which combines both the accessibility and acquisition of habitats and their associated resources (Block & Brennan, 1993; Jones, 2001).

Habitat selection has been defined as the inconsistent use of habitats relative to their availability and resource quality, independent of extrinsic constraints such as predation, competition, territoriality, density dependence, or anthropogenic influences (Fretwell & Lucas, 1970). Habitat selection has likely evolved because individuals exploiting specific habitats and resources gain immediate or short-term advantages, which translate into long term benefits for survival, reproduction, and overall fitness. This process is primarily driven by natural selection, although other evolutionary mechanisms such as gene flow may also contribute (Jones, 2001).

Northern Hemisphere waterfowl species including ducks, geese, and swans (Anatidae) utilize a three-dimensional habitat encompassing air, land, and water throughout their yearly cycle to meet physiological needs and improve fitness (Mayr, 1970). Wetlands are vital habitats for birds, providing places for nesting, breeding, roosting, raising young, feeding, resting, shelter, and social interactions (Weller, 1999). Anatidae species inhabit a wide spectrum of wetland habitats, encompassing high Arctic tundra, temperate bogs, rivers, estuaries, freshwater and saline lakes, ponds, swamps, coastal lagoons, intertidal mudflats, bays, and, in some cases, open marine environments. Additionally, these birds can also be found in artificial wetlands, such as rice fields, agricultural landscapes, sewage treatment sites, and aquaculture ponds. The wetlands they use are often very productive ecosystems, which mean that even small areas can support significant densities of waterbirds (Miyabayashi & Mundkur, 1999).

Anatidae exhibit a strong ecological alliance with aquatic environments, with individual species representing specific habitat preferences such as lakes, marshes, rivers, ponds, or reservoirs (Holopainen et al., 2015). Water depth and quality, the kind and abundance of emergent or fringe vegetation, stable water levels year-round, bottom substrate, and the availability of certain foods all influence whether a certain water body can support a specific wildfowl species (Holopainen et al., 2015).

In general, dabbling ducks primarily from the Anatini tribe tend to inhabit shallower waters compared to diving ducks, such as those from the Aythyini, Mergini, and Oxyurini tribes (Johnsgard, 1961). Swans predominantly inhabit marshes, small lacustrine environments, and sluggish riverine systems, while geese show a preference for open grasslands, boggy tundra, moist meadows, expansive shallow lakes, and seasonally immersed floodplains (Sigman, 1991). Shelducks usually live in saltwater or brackish environments such as estuaries and mudflats, while sheldgeese are more commonly coupled with open grasslands, often located at considerable distances from water sources (Johnsgard, 2010).

Furthermore, eiders and steamer ducks are some of the most marine-adapted members of the Anatidae family, usually residing in coastal regions for most of the year. On the contrary, certain highly specialized species such as the torrent duck (*Merganetta armata*) and blue duck (*Hymenolaimus malacorhynchos*) are found in swift, turbulent rivers characterized by rapids and waterfalls (Johnsgard, 2010). Many species are reliant on forested environments, with certain taxa, such as perching ducks from the tribe *Cairinini* and some arboreal-nesting members of the *Mergini* tribe, exhibiting specific reliance on tree cavities for nesting (Johnsgard, 2010).

In addition, individuals, habitat, and resources may not completely control the spatial distribution of waterfowl and other sociable animals. Although social factors may influence habitat and resource utilization, migrating waterfowl tend to prioritize wide geographic regions, particularly those important during non-breeding and breeding seasons (Baldassarre et al., 2006). In first-order inhabited regions, ducks use second-order habitat selection, focussing on specific wetland systems (Cowardin et al., 1979) and potentially nearby landscapes such as arable forage fields or nesting grounds. Individuals then make third-order selections, by selecting specific wetlands or other local sites, followed by fourth-order selections of microhabitats where they can nest, forage, roost, or perform other activities required for acquiring food, mates, or other resources (Wiens, 1973; Baldassarre et al., 2006).

Non-migratory species, which do not undergo seasonal movements between geographic regions (e.g., subtropical species such as the mottled duck [*Anas fulvigula]*), exhibit first-order habitat selection encompassing their local to regional annual home range and associated habitats. Later habitat selection occurs at finer scales second to fourth order within their home range, analogous to the hierarchical selection patterns observed in migratory species (Jones, 2001).

Some Anatidae have more specific spring staging habitats and studies have been conducted to discover which are more appealing to different species (Heitmeyer & Vohs, 1984). Weather conditions in winter and spring can change marsh availability, affecting nesting phenology and success by limiting food sources for pre-nesting and nesting birds such as mottled duck (*Anas fulvigula*) (Arzel et al., 2006). Therefore, some species, such as mallard (*Anas platyrhynchos*), rely on spring water conditions for breeding success (Krapu et al., 1983).

In general, the Anatini ducks require much shallower water than the diving ducks of the Aythyini, Mergini, and Oxyurini. Furthermore, they prefer gently sloping land with some vegetation surrounding it where they can build a nest or seek refuge in the event of danger, as well as smaller bodies of water than diving ducks (Carboneras, 1992). Habitat preferences in members of group/species of family Anatidae are specified in Table 1.

3.2. Classification of Habitats

Anatidae family is distinctly associated with mostly aquatic ecosystems, with each species exhibiting a predilection for certain habitats such as lakes, marshes, streams, ponds, or reservoirs (Carboneras, 1992). They occupy a wide variety of aquatic and wetland habitats, ranging from the frigid Arctic tundra to tropical lakes, estuarine environments, rivers, marshes, and even anthropogenic wetlands such as rice fields and sewage treatment ponds (Ali, 2002).

Table 1. Habitat preference in members of family anatidae (Carboneras, 1992)

S. No.	Group/Species	Habitat Preference
1	Swans	Marshes, shallow lakes, or slow-flowing rivers
2	Geese	Grasslands, boggy tundra, damp meadows, large shallow lakes, and floodplains
3	Shelducks	Salt or brackish water, such as estuaries and mudflats
4	Sheldgeese	Grasslands, sometimes well away from water
5	Eiders and steamer ducks	Most marine of the Anatidae and they occur along the coast during most of the year
6	Perching duck (West Indian whistling duck [*Dendrocygna arborea*])	Perching in trees. Holes in trees for nesting
7	Torrent duck (*Merganetta armata*) and the blue duck (*Hymenolaimus malacorhynchos*)	Highly specialized birds that live along fast-flowing rivers with rapids and waterfalls
8	Sea duck, harlequin duck (*Histrionicus histrionicus*)	Found mainly in coastal waters during the winter

3.2.1. Freshwater Wetlands

Freshwater habitats such as lakes, ponds, rivers, floodplains, marshes, and swamps are the most common environments for Anatidae. Many dabbling ducks (e.g., mallards, teals) and geese rely on shallow freshwater wetlands rich in aquatic plants and invertebrates during both breeding and non-breeding seasons (Owen & Black, 1990). Aquatic plants play an imperative role by contributing food, shelter, and breeding grounds for various animal species (Dvork & Best, 1982). They also maintain the livelihoods of adjoining human populations (Banerjee et al., 2012; Ghosh, 2005) and function as significant bio-indicators of water body health (Linton & Goulder, 2000). Diverse submerged vegetation creates numerous microhabitats that host varied invertebrates, offering protection from predators and attachment sites (Dvork & Best, 1982; Diehl, 1992). A richer aquatic plant cover is positively linked to greater bird species diversity and abundance (Andrikovics et al., 2006).

3.2.2. Coastal, Estuarine, and Marine Habitats

Anatidae occupy a variety of coastal habitats including estuaries, mudflats, bays, lagoons, and open sea, and have evolved adaptations such as salt-excreting glands, webbed feet for swimming and diving, as well as specialized bills suited to their foraging methods (dabbling, diving, or filter feeding). Examples include northern pintails (*Anas acuta*) in coastal wetlands, Canada geese (*Branta canadensis*), mute swans (*Cygnus olor*), and long-tailed ducks (*Clangula hyemalis*).

3.2.3. Urban and Peri-Urban Wetlands

Wetlands in urban and peri-urban areas often feature diverse habitat types due to natural and human-induced changes causing habitat heterogeneity. These wetlands serve as feeding, stopover, and breeding sites for waterfowl. Among these, ducks (order Anseriformes) are particularly responsive to environmental patchiness and use different parts of wetland mosaics differently. Duck species select microhabitats and resources in varied ways across wetlands, based on their behavior and spatial preferences during feeding and breeding.

3.2.4. Grasslands and Agricultural Lands

Geese often forage on grasslands, farmlands, and meadows, including crop fields.

4. Ecological Niches

The term "ecological niche" refers to a species' place within an ecosystem, reflecting both the range of circumstances required for the species to survive and its ecological relevance in the environment. It was first defined by Joseph Grinnell in 1917, focusing on a species habitat and its behaviors (Grinnell, 1917). Later, in 1957, G. Evelyn Hutchinson expanded the idea to describe the full range of environmental factors influencing a species or population (Hutchinson, 1957). The ecological niche comprises the entire range of interactions between a species and its biotic and abiotic environment, making it a fundamental term in ecology (Polechova & Storch, 2008).

Wetlands contain a variety of habitat types because of their variability caused by anthropogenic and/or natural disturbances (Tews et al., 2004). Wetlands have an important role in waterfowl nesting, feeding, and resting (Connor & Gabor, 2006). Anatidae is a group of species that are very sensitive to environmental patchiness and use multiple habitat types in a wetland mosaic (Austin, 2002). Duck species frequently select available resources and microhabitats in diverse ways all over wetlands, depending on their behavior or geographic position for breeding and feeding (Paracuellos, 2006).

Species differ in the quantity and prevalence of habitat types employed relative to their availability. Mallard is more prevalent in reed beds and exhibits a greater niche breadth index score in wetland ecosystem. This duck is recognized as a generalist and edge species, feeding on emergent vegetation in reed beds and flooded meadows (Krapu et al., 2006). Garganey is more common in flooded meadows and prefers open shallow water with abundant submerged plants (Poysa, 1983).

Teal, shoveler, and wigeon look more specialized with the latter two species being exclusively associated with flooded pastures. These species prefer open, shallow-water habitats rich in submerged vegetation, where they forage on sub-emergent and emergent plants (Poysa, 1983; Holm & Clausen, 2006). In this regard, flooded pastures seem to be a more favorable habitat for these species than flooded meadows, where vegetation (Juncus sp. and Carex sp.) covers a large proportion of the surface and moistened areas are limited.

Anatid species have a wide range of dietary preferences, from specialization in terrestrial or aquatic vegetation, tubers, crustaceans, and fish to extremely generalized feeding methods (Johnsgard, 2010). This variation in diet corresponds with differences in feeding environments and is closely tied to distinct feeding mechanisms, which are reflected in their morphology.

Furthermore, species within the Anatidae family show significant variance in the functional structure of their feeding apparatus, which is tailored to congregate diverse nutritional and environmental demands (Goodman, 1954). While aquatic and terrestrial eating need different bill forms, resource usage is influenced by more than just bill shape. For example, jaw muscle size corresponds with dietary specialization, indicating the biomechanical needs for processing diverse food kinds (Kooloos & Zweers, 1989). Consequently, anatids display substantial diversity in bill form and function, illustrating an adaptive radiation that enables them to exploit a wide range of ecological niches. There are several roles of niches.

4.1. Niche as the Description of a Species' Habitat Requirements

The specific environmental conditions needed for a species' survival.

4.2. Functional Role

The position or role a species has within its ecosystem is vital. Multiple species can fulfil similar roles, such as predators of small mammals. This reflects a species' part in food webs and ecological processes.

4.3. Community Interactions

The niche includes abiotic factors such as temperature and humidity, as well as biotic factors such as competition, which influence where a species can live. Hutchinson's view highlights that niches are dynamic and influenced by interactions among species, shaping how communities are structured.

4.4. Ecological Roles

Birds play a crucial role in ecosystems and are among the most valuable forms of life on Earth, adapting well to both watery and land habitats. There are approximately 9,000 bird species worldwide, with over 1,250 species found in India (Karr, 1976). They are key contributors to maintaining the balance of ecosystems (Mariappan, 2013). Waterbirds play an imperative role in delivering various ecosystem services, particularly supporting services such as seed dispersal and nutrient cycling, which are decisive for sustaining ecological communities, agriculture, and fisheries (Green & Elmberg, 2014). They also supply regulating services, such as pest control and provisioning services, which embrace the supply of resources directly utilized by humans, such as food and raw materials. Furthermore, waterbirds provide cultural services, demonstrating their importance in recreation, spiritual rituals, and human well-being (Green & Elmberg, 2014).

4.5. Nutrient Cycling

Waterbirds such as ducks, geese, and swans help ecosystems function by moving nutrients, dispersing seeds, and regulating insect populations (Mariyappan et al., 2023). They act as plant-eaters, browsers, and predators, sometimes controlling pests (Green & Elmberg, 2014). Because of their important ecological roles, waterbirds also serve as useful indicators of environmental change (Amat & Green, 2010). Additionally, these birds have provided humans with food resources such as meat, eggs, and feathers for a long time and contribute cultural and economic benefits through activities such as bird watching (Cooper & Loomis, 1991).

4.6. Seed Dispersal

Dabbling ducks (subfamily Anatinae) are among the most extensively studied seed-dispersing waterbirds (Soons et al., 2016). These omnivores eat greedily a wide range of plants and seeds (Soons et al., 2016). They have an extensive inclusive distribution, with a population around 200 million (Wetlands International, 2012), and are very mobile, often migrating over long distances. This mobility makes them important agents for spreading plant seeds across large areas, which is a crucial ecosystem service (Green & Elmberg, 2014).

Dabbling ducks facilitate seed dispersal through two primary mechanisms: externally, via epizoochory, when seeds adhere to their bills, feet, or feathers and are subsequently dislodged; and internally, via endozoochory, when ingested seeds survive passage through the digestive tract and are later excreted intact (Wenny et al., 2016). Due to the water-repellent nature of their feathers and their frequent preening behavior, external seed attachment is relatively rare (Brochet et al., 2010), rendering internal dispersal the predominant mechanism.

4.7. Biodiversity of Other Taxa

Waterbirds significantly impact aquatic biodiversity in various beneficial ways. Geese, for example, help to preserve species-rich grasslands by encouraging cohabitation and managing competition among plant species (Jasmin et al., 2008). Their grazing also helps boost primary productivity (Bazely & Jefferies, 1985; Cargill & Jefferies, 1984). Waterfowl grazing has been exposed to enhance both the diversity and productivity of submerged macrophytes (Hidding et al., 2010).

In seagrass habitats, the feeding activities of geese and wigeon (*Anas penelope*) help reduce mud buildup, yielding long-term benefits for seagrass beds (Nacken & Reise, 2000). Likewise, swan grazing on water crowfoot (*Ranunculus penicillatus pseudofluitans*) in chalk streams can improve the plant's overwinter survival by minimizing its hydraulic resistance, thereby reducing the likelihood of dislodgement during high-flow events (Wood et al., 2012).

Conversely, intensive grazing by geese can have ruinous ecological impacts, including accelerated soil erosion, elevated soil salinity, reduced nitrogen retention, and various other forms of habitat degradation (Gauthier et al., 2006; Buckeridge & Jefferies, 2007).

4.8. Indicator Species

Members of family Anatidae are often served as flagship species for conservation efforts (Boere et al., 2006). Their broad appeal to the public has helped raise awareness about conservation challenges and stimulated initiatives to safeguard these birds, their populations, and their habitats. As a result of this attention and protection, many other wetland species which typically receive less public recognition and support have also benefited (Green & Elmberg, 2014).

Certain waterbird species function as reliable bio-indicators of aquatic biodiversity (Amat & Green, 2010). For instance, population dynamics of ducks and coots are closely linked to variations in submerged macrophyte abundance (Wicker & Endres, 1995), which significantly contributes to the structural and functional integrity of aquatic ecosystems (Carpenter & Lodge, 1986). The crested coot (*Fulica cristata*) has been recognized as a valuable indicator species for high aquatic plant species richness (Green & Elmberg, 2014).

4.9. Economic and Cultural Significance

Although cultural ecosystem services often provide significant economic advantages, their value is seldom quantified. Using contingent valuation methods, researchers analyzed the value individuals place on goose populations in Scotland, indicating that the economic rewards from protecting geese exceed the costs caused by agricultural harm such as crop loss by up to 700 times (MacMillan et al., 2004). Furthermore, geese tourism has exposed to be a major contributor to the economy of Islay, a Scottish island (Edgell & Williams, 1992).

5. Niche Dimension in Anatidae

The Anatidae family encompasses wide variety of aquatic birds, including ducks, geese, and swans. Their ecological niches are shaped by several key dimensions, allowing diverse species to coexist.

5.1. Feeding Ecology and Diet

Wildfowl inhabiting estuarine or marine environments generally consume a diet dominated by animal matter particularly mollusks and crustaceans when present in such habitats, whereas species occupying more terrestrial habitats and consequently remaining distant from water for extended periods tend to adopt a predominantly or entirely vegetarian diet. The diurnal foraging patterns of wildfowl are influenced by multiple factors, including species identity, seasonal variation, geographic region, specific habitat, type of food, and feeding strategy (Carboneras, 1992).

Adult true geese maintain an almost entirely vegetarian diet year-round, occasionally supplementing it with small invertebrates. In contrast, such animal matter constitutes a much larger proportion of the diet of juveniles, which may at times rely almost exclusively on small aquatic insects, mollusks, or crustaceans (Carboneras, 1992). Swans are primarily, though not exclusively, herbivorous, consuming a diet dominated by aquatic vegetation and algae, which they supplement with a variety of animal matter, including aquatic insects, mollusks, amphibians, fish eggs, and other similar resources.

While certain species exhibit a generalized herbivorous diet, consuming a wide array of plant materials including leaves, buds, stems, seeds, roots, and tubers such as potatoes (Carboneras, 1992), others demonstrate more specialized feeding strategies. Steamer ducks, for instance, predominantly exploit animal prey, primarily consuming mollusks, especially mussels (*Mytilus*) and crustaceans such as amphipods, with occasional ingestion of small fish. In contrast, sheldgeese and shelducks are primarily herbivorous, feeding on grasses, seeds, grains, bulbs, tubers, new shoots, and algae. Sheldgeese are totally vegetarian, and their chicks eat virtually exclusively plant stuff.

Herbivorous birds, such as geese, have limited dietary habits due to their digestive physiology (Sedinger, 1997). Geese's short meal retention period (Mattocks, 1971) leads to low destruction of plant cell walls (Buchsbaum et al., 1986). To meet their nutritional needs, geese must consume high-quality vegetation for extended periods of time, ranging from 40% to 80% of the day depending on the season (Gawlik & Slack, 1996).

Adults and young have varied nutrient needs: goslings need more nitrogen to flourish, whereas adult females require energy-rich food (soluble carbohydrates or fat) to restore their physiological reserves. These discrepancies appear to remain in the autumn, when immature geese, still in the process of growing, have been observed foraging in nitrogen-rich settings rather than those preferred by adults (Raveling et al., 1991). During the summer, goslings and adult females require large amounts of high-quality food to meet their nutritional requirements (Thomas & Prevett, 1982).

Geese have strong selective foraging behavior, typically seeking plant species with higher nutritional concentrations, notably nitrogen, to suit their dietary needs (Therkildsen & Madsen, 1999). Intensive grazing can alter plant community composition by changing the availability and nutritional quality of fodder resources (Kerbes et al., 1990). These grazing-induced alterations are frequently dynamic, reflecting seasonal fluctuations in plant phenology and nutrient profiles with possible cascading impacts on wetland productivity and trophic interactions. The quality and availability of food plants have a substantial impact on gosling development and survival throughout their first year (Black & Owen, 1989).

Birds feeding in aquatic or terrestrial habitats require different bill structures, and jaw muscle size is also linked to dietary habits (Kooloos & Zweers, 1989). The variation in bill morphology among anatids is frequently regarded as a classic example of adaptive radiation (Black & Owen, 1989). Their bills are typically moderately long, flattened, and blunt at the tip, with a small nail at the upper tip and fine comb-like structures called lamellae along the edges. Some species have highly specialized bills such as mergansers and shovelers, while others, such as scoters, have less specialized bills (Johnsgard, 2010).

These bill variations help species exploit different food sources within the same habitat, reducing competition. Birds are grouped into various sub-guilds based on their feeding behaviors. Holmes and Recher (1986) found that these guilds differ depending on habitat features and food availability. Even within a guild, members often have somewhat different feeding requirements, which helps minimize competition during times of scarce resources. How and where birds forage plays a key role in how they partition resources (Wiens, 1989). Family Anatidae has been classified primarily into three feeding guilds. Feeding ecology (guild structure) of members of family Anatidae is described in Table 2.

Table 2. Feeding ecology and three main foraging techniques in family Anatidae

S. No.	Habitat	Feeding Ecology	Foraging Techniques
1	Freshwater (lakes, ponds, marshes)	Herbivore (consumes aquatic vegetation, seeds, roots, stems, leaves, flowers)	Dabbling (mallard, northern pintails, gadwall, green winged teal)
2	Saltwater (coastal lagoons, estuaries, intertidal zones)	Carnivores (aquatic invertebrate, mollusks, crustaceans, insects, small fish)	Diving (common pochard, tufted duck, redhead)
3	Man-made (rice fields, aquaculture ponds)	Omnivores (insects, mollusks, small animals, fruits, and seeds)	Grazing and filter feeding (bar-headed goose, greylag goose, northern shoveler)

5.1.1. Omnivores

Some Anatidae species exhibit an omnivorous diet, consuming a variety of food items such as insects, small vertebrates, fruits, and seeds; examples include the common pochard, tufted duck, and Egyptian goose. They forage across a variety of habitats, including terrestrial forests and open landscapes, functioning both as predators and as agents of seed dispersal, thereby influencing insect population dynamics and facilitating plant regeneration. For instance, the common pochard (*Aythya ferina*) predominantly consumes aquatic vegetation, seeds, and small invertebrates. Tufted ducks (*Aythya fuligula*) dine on mollusks, water insects, and plants, while Egyptian geese (*Alopochen aegyptiaca*) eat grasses, seeds, leaves, insects, and small aquatic animals (Carboneras, 1992).

5.1.2. Herbivores

Birds primarily feed on aquatic and terrestrial vegetation, including leaves, roots, tubers, and algae. Their foraging strategies involve grazing on pasture grasses, extracting aquatic plants from shallow waters, and excavating tubers from muddy substrates. The greylag goose (*Anser anser*) primarily grazes on grasses and aquatic plants in wetlands and agricultural fields, whereas the mute swan (*Cygnus olor*) feeds predominantly on submerged aquatic vegetation by upending in shallow water. The magpie goose, in contrast, is a semi-terrestrial species that grazes on grasses and consumes seeds (Carboneras, 1992).

5.1.3. Carnivores

Species that feed on fish, frogs, worms, and insects exhibit diverse foraging strategies. Piscivorous species, such as fish-catching birds, often dive or swim to secure their prey, while insectivorous species typically glean insects from the water's surface. Others may probe mud or aquatic vegetation in search of invertebrates. The common merganser (*Mergus merganser*), for example, is a specialized piscivorous diving duck equipped with serrated bills that facilitate the capture and retention of slippery prey (Carboneras, 1992).

5.2. Foraging Behaviour

Foraging refers to the activity of searching for and collecting food, either to consume right away or to store for later (Stephens et al., 2008). Different bird species gather their food in various ways, influenced by their dietary needs and the shape of their beaks. Anatids such as ducks, geese, and swans display a wide range of feeding habits, from specializing in terrestrial or aquatic plants, tubers, invertebrates, and fish to being general feeders (Johnsgard, 1978, 2010). Research in ecomorphology shows how differences in their physical traits support their feeding adaptations (Nudds & Bowlby, 1984).

Wildfowl employ diverse feeding strategies, mainly including surface feeding (known as dabbling), diving underwater, and grazing (Chatterjee et al., 2020). The detailed information of foraging strategies has been provided in Table 2. They may also forage by wading, probing with their bills, filtering through mud, tipping upside down (upending), sifting through debris on the bottom, chasing fish, reaching for seeds or grasses above the water, or even breaking apart termite nests (Table 3). Traditionally, dabbling and diving ducks have been classified based on taxonomy rather than ecological distinctions. Members of the tribe Anatini typically feed by dabbling at the water's surface, whereas those of the tribe Aythyini are commonly recognized as diving foragers; thus, these tribes have been referred to as "dabbling" and "diving" ducks, respectively (Carboneras, 1992). Ducks, in particular, illustrate a wide range of feeding behaviors and adaptability in how they gather food (Zydelis & Richman, 2015).

Table 3. Description of foraging techniques observed in waterbirds (Kumar, 2024)

Foraging Category	Technique	Description
Surface feeding	Surface feeding	Stretching the neck and keeping the bill parallel to the water while skimming the water's surface
	Skimming	Flying low over water surface in order to pick food items
	Filtering	Bill retained on water surface to filter food particles by submerging mandible only
	Upending	Tail and legs above water while it is partially submerged vertically
Aerial feeding	Catching/hawking	Catching of flying insects in air
Diving	Plunge diving	Diving head first into water from flight/ perch to catch fish
	Foot plunging	Vertical dive with wings half folded
Wading-visual feeding	Standing still	Visual inspection while standing in mud
	Walking slow	Slow cautious walking to attack prey
	Picking	Collecting visible food from surface
	Striking	Swift hit after stalking motion
	Stabbing	Rapid stabbing using mandibles
Wading-tactile feeding	Probing	Inserting beak into mud and browsing
	Pecking	Shallow penetration into mud repeatedly
	Sweeping bill	Lateral sweep of bill
	Scything	Side-to-side cutting motion in shallow water
	Tactile technique	Walking with partially open bill underwater
	Foot trembling	Rapid shaking of one leg to disturb prey

Magpie geese eat mostly seeds, which they collect on land by using their feet to bend tall grasses and retrieving seedheads with their bills. Additional feeding strategies recorded in this species include grazing and mud filtering (Carboneras, 1992). Geese obtain a large percentage of their food from grazing, primarily ingesting grasses, sedges, herbs, and mosses. They live in arid agricultural terrain, especially in the winter and eat largely grain, bulbs, and tubers, particularly potatoes, which they dig up with their feet and beak (Carboneras, 1992).

5.2.1. Dabbling

Dabbling ducks such as mallards and pintails typically feed in shallow waters by tipping forward to reach plants and small animals (Nelms et al., 2012). This group encompasses many species, each with unique physical features, ecological needs, diets, and feeding techniques. Their webbed feet are crucial in supporting these activities. Dabblers primarily consume food from the water surface or shallow areas, using their webbed toes to aid balance and movement.

5.2.2. Diving

Diving ducks such as canvasbacks and scaups inhabit deeper waters, diving below the surface to feed on aquatic plants and animals (Tome & Wrubleski, 1988). Compared to dabblers, geese, and swans, divers have denser, more compact bodies that reduce buoyancy and allow longer underwater stays (Lovvorn & Jones, 1991). To decrease buoyancy further, they press their feathers close to their bodies before diving, minimizing trapped air. Their wings are shorter and more compact, enabling tight folding to reduce water resistance during dives. Once underwater, divers use both their wings and feet to propel themselves as they search for food (Townsend, 1909). These adaptations enable them to access food in deep water, where dabblers cannot reach.

5.2.3. Grazing

Geese often obtain food by grazing on grasses, sedges, herbs, and mosses (Giroux et al., 1998). In colder seasons, they frequent dry farmlands, consuming grains, bulbs, and tubers such as potatoes, which they dig up with their feet and bills. Magpie geese mostly consume seeds collected on land, using their feet to bend long grasses and their bills to harvest seedheads (Frith & Davies, 1961). They also engage in grazing and mud-filtering, with diets mostly consisting of plant material including grass blades and plant bulbs. The Cape Barren goose grazes on short pasture grasses, tussocks, sedges, as well as broad-leaved plant leaves and seeds.

5.2.4. Filter Feeding

Some dabbling ducks, such as mallards and northern shovelers, have broad, flat bills with fine, comb-like features known as lamellae (Gurd, 2005). These enable them to filter out minute plants, seeds, and insects from the water, making them effective feeders in shallow waters and wetlands rich in such nutrients. Shovelers have specially developed bills for filter feeding, and they frequently feed collaboratively by swimming in small circles to stir up food while filtering dirt churned up by one another. The pink-eared duck similarly has a specialized bill for filter eating and follows comparable feeding habits.

6. Morphological Adaptations

Most species are similar in various morphological traits, however differ significantly in others. Most Anatidae possess a broad, extended body; however, some specialized divers have a more rounded shape (De Mendoza et al., 2024). This distinctive body form enhances buoyancy, but makes walking on land difficult; nevertheless, they are far more proficient on the ground than many seabirds.

6.1. Bill

Members of the Anatidae family show considerable disparity in the morphological specialization of their feeding apparatus, reflecting evolutionary adaptations to diverse foraging strategies and habitat-specific dietary requirements (Nudds & Bowbly, 1984; Wainwright, 1996). Distinct bill morphologies are required for aquatic versus terrestrial feeding, although these variations may also imitate adaptations to resource utilization. In turn, patterns of resource use may be affected by the size and configuration of jaw musculature (Kooloos & Zweers, 1989). Anatids exhibit substantial miscellany in bill morphology and hence often regarded as a classic example of adaptive radiation (Owen & Black, 1990).

Bill morphology among wildfowl exhibits varying degrees of specialization, ranging from highly specialized forms in species such as mergansers and shovelers to more generalized structures in scoters (Johnsgard, 2010). Overall, wildfowl species demonstrate considerable variation in both body size and bill configuration (DuBowy, 2000). The bill exhibits a robust, conical morphology and contains serrated lamellae along its interior surfaces. These structures are especially evident in plankton-feeding taxa, where they facilitate filtration and are adapted into sharper cutting edges in *ichthyophagous* species (Johnsgard, 1962). Sawbills (*Mergus*) are among the utmost specialized, with serrated teeth for catching fish. The distal end of the upper mandible bears a shield-like keratinized structure known as the nail, which exhibits greater rigidity than the surrounding rhamphotheca and is morphologically adapted for specialized functions such as grazing or molluscivory.

Variations in bill morphology provide vital insights into avian ecological adaptations, evolutionary trajectories, functional anatomy, and craniomandibular biomechanics. For example, the peculiar morphology of the freckled duck's bill is similar to that of stiff-tailed ducks. Its feeding habits are very peculiar, closely resembling those of stiff tails, as it uses its bill to filter material from the substrate (Carboneras, 1992). Furthermore, shovelers have bills designed for filter feeding, allowing them to capture small planktonic creatures largely from the water's surface. They frequently feed cooperatively, swimming in small circles to concentrate food particles and filter suspended materials from disturbed substrates in their wake (Carboneras, 1992).

6.2. Neck

All species exhibit a relatively elongated neck most prominently in swans and geese, though also apparent in certain duck species and possess proportionally small heads (Baldassarre, 2014).

6.3. Legs

Ducks, geese, and swans acquire robust legs, the skin of which is entirely scaled (Johnsgard, 2010). They are situated at the distal extremities of the body, particularly among those tribes considered the most highly adapted to aquatic environments within the family, whose members exhibit the most distinct limitations in terrestrial mobility (Johnsgard, 2010). The rear toe is diminished and raised, while the three anterior toes are predominantly interconnected by webbing, although in Anseranas and certain Anas ducks, the webbing is just partial (Johnsgard, 2010).

6.4. Wings

The wings are comparatively short, pointed, and strong (Johnsgard, 1961). To gain or sustain velocity in the air, heavier birds with significant wing loading must beat their wings continuously and quickly (Brown, 1963). Consequently, the wing muscles are inherently substantial and robust; they are firmly anchored to the prominently keeled sternum, contributing to the broad-breasted appearance of wildfowl (Brown, 1963). In some species, however, the structure reveals a divergent morphological specialization; such as, it is slender and acuminate in males of the northern pintail (*Anas acuta*) and long-tailed duck (*Clangula hyemalis*), while in members of the tribe Oxyurini (stiff-tails), it is elongated and narrow, often held erect and functioning as a rudder during underwater diving (Johnsgard, 1961).

6.5. Feathers

Anatidae members acquire a dense layer of feathers for insulation, indispensable for aquatic birds (Mulcahy & Heard, 2025). To maintain their waterproofing capabilities, feathers must remain in optimal condition, prompting birds to assign many hours daily to feather maintenance through preening (Terrill & Shultz, 2023). Members of this family hold well-developed uropygial (oil) glands with dense feathering, which assist the secretion and distribution of oil across the plumage (Elder, 1954). Salt glands, specialized excretory organs situated above each eye, are present in all species and function to filter the blood, excreting excess salt as a highly concentrated solution that drips from the tip of the bill (Ritchison, 2023). This is only true for species or individuals that spend a significant amount of time in coastal or brackish water.

7. Social Behavior

Ducks, geese, and swans are well documented for their gregarious behavior (Baldassarre, 2014). They characteristically aggregate in flocks, with group sizes ranging from a few individuals to several hundred thousand, contingent upon species assemblage, seasonal variation, and geographic locality (Emlen, 1952). The set of behaviors consistently exhibited in social contexts encompasses foraging, communal roosting, and resting, occurring both in aquatic and terrestrial environments, as well as coordinated movements associated with migratory activity (Newton, 2024).

7.1. Flocking

Anatidae members often come together in flocks that can range from just a few individuals to vast gatherings of hundreds of thousands, depending on the species, the time of year, and their location. Flocking is most typical outside of the breeding season. For geese and swans (and some duck species), these flocks are usually made up of family groups or a few families combined. In contrast, large groups of some duck species may mainly be composed of one sex, as males and females often occupy different areas when they are not breeding.

Forming flocks offers significant advantages, such as improved safety from predators and easier food discovery (Beauchamp, 2013). By observing and joining flock members who have found food, birds can feed more efficiently than if they searched alone. If a predator appears, one of the vigilant birds will sound an alarm, causing the whole flock to take off quickly and making it difficult for the predator to catch a single individual (Vine, 1971). Predators are more successful only if they manage to separate one bird from the group, which wildfowl are generally good at avoiding (Lima, 2009).

In geese and swans and to a lesser extent in some duck species, the typical flock usually consists of a single family group or sometimes a few families. In contrast, large flocks of many other duck species are often dominated by individuals of a single sex, as males and females frequently occupy different areas during much of the non-breeding season (Emlen, 1952). Flocking behavior offers considerable advantages by reducing individual predation risk through safety in numbers and by improving foraging efficiency using collective information sharing (Vine, 1971).

7.2. Vocalization

Wildfowl also have well-developed vocal abilities, which are vital for communication in many situations. The magpie goose, for instance, is very noisy and calls out with loud honks whether flying, swimming, or on land, often triggering responses from other geese nearby (Cummings, 2016). Whistling ducks are named for their sharp, whistling calls that are commonly heard when they feed, rest, or fly (Johnsgard, 2010). True geese are well known for their powerful, carrying calls especially notable when they gather in large flocks (Boyd, 1953). These honking or barking vocalizations are particularly prominent during migration, when flocks high in the sky can be heard from a long distance away.

8. Breeding

Most wildfowl species attempt to breed just once yearly, usually choosing the most favorable time according to environmental conditions (Johnson & Grier, 1988). This is especially common in areas with clear seasonal changes, where both food and habitat are predictably available at certain times. In temperate regions, many wildfowl migrate because breeding sites rich in food and water during spring and summer often freeze over in winter and can't support all the birds after breeding (Davis et al., 2014). In tropical regions, where temperature and daylight don't vary much, the year is divided more by wet and dry seasons. Usually, wildfowl begin breeding when rains start so their chicks will hatch when water is plentiful (Marshall, 1959). In areas where water is always available, like along tropical rivers, breeding doesn't necessarily follow the seasons, and eggs may be found at any time of year, even within one species.

9. Migration

Migration is the regular, repeated, seasonal movement of groups of animals between different geographic areas, naturally returning to their original locations. This phenomenon is especially common in birds, many of which follow annual migration cycles. While some migrations involve long-distance travel, others are less conspicuous. Birds migrate for numerous reasons such as accessing regions with abundant food, enjoying milder climates, and facing less competition for safe nesting sites.

Numerous Anatidae populations migrate regularly between northern breeding wetlands and southern non-breeding habitats, often traversing international boundaries. Other species move locally, either within or across national lines, typically in reaction to shifts in water supply. To complete their annual cycle, they rely on a huge network of wetlands throughout their territory. Although migration tactics vary greatly between species and populations, Anatidae generally have similar underlying migratory orientations. Anatidae can be found in two major flyways across Asia. They are known as the Central-South Asian flyway and the East Asian flyway (Miyabayashi & Mundkur, 1999).

Birds breeding in the Northern Hemisphere are mostly migratory, often undertaking long-distance latitudinal migrations between northern breeding grounds and southern wintering areas (Kirby et al., 2008). Examples include species such as northern pintail (*Anas acuta*), Eurasian teal (*Anas crecca*), and Eurasian wigeon (*Mareca penelope*) that move south in winter. Birds traveling shorter, less directed distance for food or safety engage in vagrant or irregular migration. The bar-headed goose (*Anser indicus*), one of the world's highest-flying birds, winters in different Indian wetlands (Hawkes et al., 2011). Some species also migrate laterally, moving east to west or vice versa.

Waterbirds are species dependent on wetlands during some part of their annual cycle and are found within the conservation priority areas. Besides waterbirds, other wetland-dependent birds such as kingfishers, raptors, and passerines also benefit from conservation efforts. These birds use diverse habitats across their yearly cycle. The continental flyway traverses broad semi-arid zones characterized by a scarcity of wetlands, particularly at critical staging areas, where habitat overlap among various migratory waterbird assemblages is frequently observed (Kirby et al., 2008). Certain species such as northern pintail (*Anas acuta*), Eurasian teal (*Anas crecca*), and Eurasian wigeon (*Mareca penelope*) undertake long-range migrations (Rasmussen & Anderton, 2005), whereas some migrate shorter distances, such as between the Himalayas and peninsular India.

India supports a rich avifaunal diversity, comprising over 1,360 bird species, of which approximately 257 are classified as waterbirds (Ali, 2002). Among these, nearly 81 species are identified as seasonal migrants; ducks, geese, waders, shorebirds, and cranes are among India's most prevalent and regular winter migrants (Jha & Mckinley, 2015). India, located centrally along the Central Asian flyway, hosts significant wintering populations of numerous waterbird species (Szabo & Mundkur, 2017). It is also an imperative breeding ground for pygmy cormorants (*Microcarbo pygmaeus*) and ruddy shelduck (*Tadorna ferruginea*), as well as several globally threatened species such as the Dalmatian pelican (*Pelecanus crispus*), Siberian crane (*Leucogeranus leucogeranus*), Oriental white stork (*Ciconia boyciana*), greater adjutant stork (*Leptoptilos dubius*), and white-winged wood duck (*Asarcornis scutulata*) (Kirby et al., 2008).

10. Conservation Status

The International Union for Conservation of Nature (IUCN) statues of species belonging to family Anatidae of class Aves illustrated in Figure 1. Most species are classified as "*Least Concern*," while only a small proportion fall under the "*Endangered*" category. Successful bird conservation relies on comprehensive knowledge of species distribution, relative population sizes, and trends across various geographic scales. IUCN Red List criteria provide the primary tool for assessing species' risk of extinction on regional and global levels, with commonly recognized categories such as "Critically Endangered," "Endangered," and "Vulnerable." Species that do not fit these categories are further evaluated under five conservation concern classifications developed by BirdLife International:

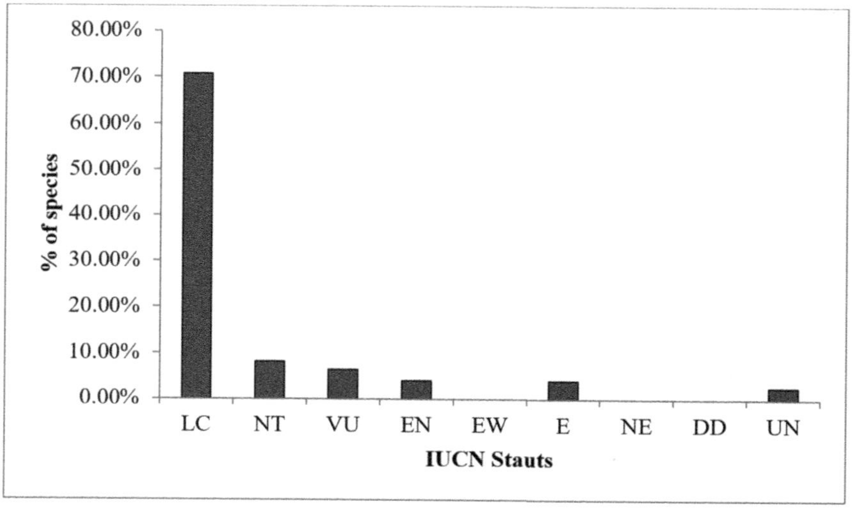

Figure 1. IUCN status of species belonging to family Anatidae of class Aves (IUCN, 2024).

"Declining," "Rare," "Depleted," "Localized," and "Secure." Apart from a few species lacking sufficient data, nearly all bird species receive at least a provisional status within the IUCN or BirdLife frameworks.

Migratory birds look numerous challenges, primarily stemming from anthropogenic activities, including habitat loss and degradation, climate change, pollution, illegal and unsustainable hunting, infrastructure expansion, and the spread of diseases (Kabir et al., 2024). These factors have caused widespread declines in both the diversity and populations of migratory species worldwide. Effective conservation of migratory birds therefore requires coordinated multinational efforts and cooperation through multilateral agreements. The Convention on Migratory Species (CMS), the Ramsar Convention on Wetlands, several migratory bird protection treaties, and laws such as the Neotropical Migratory Bird Conservation Act are among the most significant international accords.

Several regulations in India protect waterbirds, both resident and migratory. The National Wildlife Action Plan and the Indian Wildlife (Protection) Act of 1972 provide the legislative basis for safeguarding migrating birds and their wetland habitats. The Environmental Protection Act, Indian Forest Act, Pollution Control Acts, and Coastal Zone Regulation Act are important pieces of legislation that help manage and conserve wetland habitats and avian life. India is also a signatory to key international agreements, including the Ramsar Convention and bilateral treaties such as the India–Russia Migratory Bird Agreement, enhancing its commitment to safeguarding migratory waterbirds.

In essence, effective bird conservation depends on thorough assessment of species' conservation status, robust legal protections, and international collaboration. Addressing threats such as habitat destruction, pollution, hunting, and climate change through comprehensive policy frameworks and global cooperation is vital for preserving migratory bird populations and their habitats. India's integrated approach combining domestic legislation and international partnerships exemplifies efforts to conserve its diverse waterbird species and wetlands.

11. Conclusions and Future Prospects

The Anatidae family including ducks, geese, and swans represents outstanding ecological adaptability, occupying a vast array of wetlands and aquatic and terrestrial habitats worldwide, with the exception of Antarctica. Their evolutionary success is attributed to a mix of morphological and behavioral adaptations that allow different species to specialize in specific ecological functions and use a diverse variety of habitats and food sources. The ecological niches of Anatidae are well-defined by differences in habitat selection, dietary preferences, seasonal migrations, and their relationships with other species. Their diverse foraging techniques such as dabbling, diving, and grazing let them minimize competition and coexist within complex wetland systems, including those influenced by human activity. Migration is a fundamental aspect of Anatidae ecology, connecting populations and habitats across enormous regions. Wetlands play a critical role as breeding, staging, and wintering areas for these species. However, Anatidae populations face increasing challenges from habitat loss, wetland degradation, climate change, pollution, and human disturbances, particularly during their migratory journeys. Ensuring the long-term survival of these bird's calls for comprehensive, collaborative conservation strategies: the protection and restoration of wetlands, enforcement of environmental regulations, and sustainable management practices are all essential. Their popularity as flagship species for conservation has helped advance wetland protection efforts, benefiting many other species dependent on these vital ecosystems. Ongoing scientific research and coordinated management remain crucial for preserving both Anatidae populations and the health of wetlands worldwide.

Acknowledgments

The authors are thankful to the Principal, Buddha P.G. College, Kushinagar for providing all necessary administrative facilities. We are grateful to the Council of Science and Technology Uttar Pradesh (UPCST) for providing finical assistance through a major research project (Project ID-3768).

References

Ali, S. (2002). *The Book of Indian Birds*. 13th Edition. Bombay Natural History Society.

Amat, J. A., & Green, A. J. (2010). Waterbirds as Bioindicators of Environmental Conditions. *In:* C. Hurford, M. Schneider, & I. Cowx (Eds.), *Conservation Monitoring in Freshwater Habitats: A Practical Guide and Case Studies* (pp. 45–52). Springer, Dordrecht.

Andrikovics, S., Forro, L., Gere, G., Lakatos, G., & Sasvari, L. (2006). Water bird guilds and their feeding connections in the Bodrogkoz (Northeast Hungary). *Hydrobiologia*, 567, 31–42.

Arzel, C., Elmberg, J., & Guillemain, M. (2006). Ecology of spring-migrating Anatidae: A review. *Journal of Ornithology*, 147(2), 167–184.

Austin, J. E. (2002). Responses of dabbling ducks to wetland conditions in the Prairie Pothole Region. *Waterbirds*, 25(4), 465–473.

Baldassarre, G. (2014). *Ducks, Geese, and Swans of North America*. JHU Press.

Baldassarre, G. A., Bolen, E. G., & Saunders, D. A. (2006). *Waterfowl Ecology and Management*. Malabar, Florida, USA: Krieger.

Banerjee, S., Kar, D., Banerjee, A., & Palit, D. (2012). Utilization of some aquatic macrophytes in Borobandh: A lentic water body in Durgapur, West Bengal, India: Implications for socio-economic upliftment of local stakeholder. *Indian Journal of Applied & Pure Biology*, 27(1), 83–92.

Bazely, D. R., & Jefferies, R. L. (1985). Goose feces: A source of nitrogen for plant-growth in a grazed salt-marsh. *Journal of Applied Ecology*, 22, 693–703.

Beauchamp, G. (2013). *Social Predation: How Group Living Benefits Predators and Prey*. Elsevier.

Black, J. M., & Owen, M. (1989). Parent–offspring relationships in wintering barnacle geese. *Animal Behaviour*, 37, 187–198.

Block, W. M., & Brennan, L. A. (1993). The habitat concept in ornithology: Theory and applications. *Current Ornithology*, 11, 35–91.

Boere, G. C., Galbraith, C. A., & Stroud, D. A. (Eds.). (2006). *Waterbirds around the World: A Global Overview of the Conservation, Management and Research of the World's Waterbird Flyways*. The Stationary Office, 2006. ISBN:0114973334, 9780114973339

Boyd, H. (1953). On encounters between wild white-fronted geese in winter flocks. *Behaviour*, 85–129.

Brochet, A. L., Guillemain, M., Fritz, H., Gauthier-Clerc, M., & Green, A. J. (2010). Plant dispersal by teal (Anas crecca) in the Camargue: Duck guts are more important than their feet. *Freshwater Biology*, 55(6), 1262–1273.

Brown, J. H. (1984). On the relationship between abundance and distribution of species. *The American Naturalist*, 124(2), 255–279.

Brown, R. H. J. (1963). The flight of birds. *Biological Reviews*, 38(4), 460–489.

Buchsbaum, R., Wilson, J., & Valiela, I. (1986). Digestibility of plant constituents by Canada geese and Atlantic brant. *Ecology*, 67(2), 386–393.

Buckeridge, K. M., & Jefferies, R. L. (2007). Vegetation loss alters soil nitrogen dynamics in an Arctic salt marsh. *Journal of Ecology*, 95(2), 283–293.

Carboneras, C. (1992). Family Anatidae (ducks, geese and swans). *Handbook of the Birds of the World*, 1, 536–628.

Cargill, S. M., & Jefferies, R. L. (1984). The effects of grazing by lesser snow geese on the vegetation of a subarctic salt-marsh. *Journal of Applied Ecology*, 21, 669–686.

Carpenter, S. R., & Lodge, D. M. (1986). Effects of submersed macrophytes on ecosystem processes. *Aquatic Botany*, 26, 341–370.
Chatterjee, A., Adhikari, S., Pal, S., & Mukhopadhyay, S. K. (2020). Foraging guild structure and niche characteristics of waterbirds wintering in selected sub-Himalayan wetlands of India. *Ecological Indicators*, 108, 105693.
Connor, K. J., & Gabor, S. (2006). Breeding waterbird wetland habitat availability andresponse to water-level management in Saint John River floodplain wetlands, New Brunswick. *Hydrobiologia*, 567(1), 169–181.
Cooper, J., & Loomis, J. (1991). Economic value of wildlife resources in the San Joaquin Valley: Hunting and viewing values. *In: The Economics and Management of Water and Drainage in Agriculture* (pp. 447–462). Boston, MA: Springer US.
Cowardin, L. M. (1979). *Classification of Wetlands and Deepwater Habitats of the United States*. Fish and Wildlife Service, US Department of the Interior.
Cummings, J. L. (2016). *Geese, Ducks and Coots*. Wildlife Damage Management Technical Series, U.S. Department of Agriculture, Animal & Plant Health Inspection Service, Wildlife Services, August 2016.
Davies, S. J. J. F. (1963). The third distinguishing trait of magpie geese is their breeding habits. *Ibis*, 105(1), 76–98.
Davies, S. J. J. F., & Frith, H. (1964). Some comments on the taxonomic position of the magpie goose, *Anseranas semipalmata* (Latham). *Emu–Austral Ornithology*, 63(4), 265–272.
Davis, J. B., Guillemain, M., Kaminski, R. M., Arzel, C., Eadie, J. M., & Rees, E. C. (2014). Habitat and resource use by waterfowl in the Northern Hemisphere in autumn and winter. *Wildfowl*, 17–69.
De Mendoza, R. S., Carril, J., Degrange, F. J., & Tambussi, C. P. (2024). Specialized diving traits in the generalist morphology of Fulica (Aves, Rallidae). *Scientific Reports*, 14(1), 13966.
Diehl, S. (1992). Fish predation and benthos community structure—The role of omnivory and habitat complexity. *Ecology*, 73, 1646–1661.
Dong, Z., Wang, Z., Liu, D., Li, L., Ren, C., Tang, X. et al. (2013). Assessment of habitat suitability for waterbirds in the west Songnen plain, China, using remote sensing and GIS. *Ecological Engineering*, 55, 94–100.
DuBowy, P. J. (2000). Morphological correlates of community structure in North American waterfowl. *Community Ecology*, 1(2), 147–156.
Dvork, J., & Best, E. P. H. (1982). Macro-invertebrate communities associated with the macrophytes of Lake Vechten: Structural and functional relationships. *Hydrobiologia*, 95, 115–126.
Edgell, J. & Williams, G. (1992). The financial value and economic valuation of goose grazing in the European Community. *In:* van Roomen, M., & Madsen, J. (Eds.), *Waterfowl and Agriculture: Review and Future Perspective*. IWRB Special Publication 21. IWRB, Slimbridge.
Elder, W. H. (1954). The oil gland of birds. *The Wilson Bulletin*, 66(1), 6–31.
Emlen, J. T. (1952). Flocking behaviour in birds. *The Auk*, 69(2), 160–170.
Fretwell, S. D., & H. L. Lucas. (1970). On territorial behavior and other factors influencing habitat distributions of birds. I. *Acta Biotheoretica*, 19, 16–36.
Frith, H. J., & Davies, S. J. J. F. (1961). Ecology of the magpie goose, Anseranas semipalmata Latham (Anatidae). *CSIRO Wildlife Research*, 6, 91–141. doi:10.1071/CWR9610091.

Futuyma, D. J., & Moreno, G. (1988). The evolution of ecological specialization. *Annual Review of Ecology and Systematics*, 207–233.

Gauthier, G., Giroux, J. F., & Lochefort, L. (2006). The impact of goose grazing on arctic and temperate wetlands. *Acta Zoologica Sinica*, 52, 108–111.

Gawlik, D. E., & Slack, R. D. (1996). Comparative foraging behavior of sympatric snow geese, greater white-fronted geese, and Canada geese during the non-breeding season. *The Wilson Bulletin*, 108(1), 154–159.

Ghosh, S. K. (2005). *Illustrated Aquatic and Wetland Plants in Harmony with Mankind*. Standard Literature.

Giroux, J. F., Gauthier, G. I. L. L. E. S., Costanzo, G., & Reed, A. (1998).Impact of geese on natural habitats. *The Greater Snow Goose: Report of the Arctic Goose Habitat Working Group*. Arctic Goose Joint Venture Special Publication. Washington: US Fish and Wildlife Service, 32–57.

Goodman, D. C. (1954). *Functional Anatomy of the Feeding Apparatus in Waterfowl (Aves: Anatidae)*. University of Illinois at Urbana-Champaign.

Green, A. J. (1996). Analyses of globally threatened Anatidae in relation to threats, distribution, migration patterns, and habitat use. *Conservation Biology*, 10(5), 1435–1445.

Green, A. J., & Elmberg, J. (2014). Ecosystem services provided by waterbirds. *Biological Reviews*, 89(1), 105–122.

Grimmett, R., Inskipp, C., & Inskipp, T. (2016). *Field Guide to Birds of the Indian Subcontinent: India, Pakistan, Sri Lanka, Nepal, Bhutan, Bangladesh and the Maldives*. Bloomsbury Publishing.

Grinnell, J. (1917). The niche relationships of the California thrasher. *The Auk*, 34(4), 427–433.

Gurd, D. B. (2005). The ecology of adaptive radiation of dabbling ducks (Anas spp.). Ph.D. Thesis, Simon Fraser.

Habib, M. (2018). Waterfowl anatidae of the wetlands of Kashmir population trends habitat use and nesting success. Doctoral dissertation, Department of Ecology and Environmental Sciences, Pondicherry University.

Hawkes, L. A., Balachandran, S., Batbayar, N., Butler, P. J., Frappell, P. B., Milsom, W. K., & Bishop, C. M. (2011). The trans-Himalayan flights of bar-headed geese (*Anser indicus*). *Proceedings of the National Academy of Sciences*, 108(23), 9516–9519.

Heitmeyer, M. E., & Vohs Jr., P. A. (1984). Distribution and habitat use of waterfowl wintering in Oklahoma. *The Journal of Wildlife Management*, 51–62.

Hidding, B., Nolet, B. A., DE Boer, T., DE Vries, P. P., & Klaassen, M. (2010). Above- and below-ground vertebrate herbivory may each favour a different subordinate species in an aquatic plant community. *Oecologia*, 162, 199–208.

Holm, T. E., & Clausen, P. (2006). Effects of water level management on autumn staging waterbird and macrophyte diversity in three Danish coastal lagoons. *Biodiversity & Conservation*, 15(14), 4399–4423.

Holmes, R. T., & Recher, H. F. (1986). Determinants of guild structure in forest bird communities: an intercontinental comparison. *The Condor*, 88(4), 427–439.

Holopainen, S., Arzel, C., Dessborn, L., Elmberg, J., Gunnarsson, G., Nummi, P., & Sjoberg, K. (2015). Habitat use in ducks breeding in boreal freshwater wetlands: A review. *European Journal of Wildlife Research*, 61(3), 339–363.

Hutchinson, G. E. (1957). Concluding remarks. *Cold Spring Harbor Symposia on Quantitative Biology*, 22, 415–427.

Jasmin, J. N., Rochefort, L. & Gauthier, G. (2008). Goose grazing influences the fine-scale structure of a bryophyte community in arctic wetlands. *Polar Biology*, 31, 1043–1049.

Jha, K. K., & Mckinley, C. R. (2015). Composition and dynamics of migratory and resident avian population in wintering wetlands from northern India. *Notulae Scientia Biologicae*, 7(1), 1–15.

Johnsgard, P. A. (1960). Hybridization in the Anatidae and its taxonomic implications. *Condo*, 62: 25–33.

Johnsgard, P. A. (1961). The taxonomy of the Anatidae: A behavioural analysis. Paul Johnsgard Collection. 29. https://digitalcommons.unl.edu/johnsgard/29

Johnsgard, P. A. (1962). Evolutionary trends in the behaviour and morphology of the Anatidae. *Wildfowl*, 13(13), 130–148.

Johnsgard, P. A. (1978). The ornithogeography of the Great Plains states. Paul Johnsgard Collection. 44. https://digitalcommons.unl.edu/johnsgard/44

Johnsgard, P. A. (2010). Ducks, Geese, and Swans of the World: Contents, Preface, & Introduction. 2. Ducks, Geese, and Swans of the World by Paul A. Johnsgard. 2. https://digitalcommons.unl.edu/biosciducksgeeseswans/2

Johnson, D. H., & Grier, J. W. (1988). Determinants of breeding distributions of ducks. *Wildlife Monographs*, 3–37.

Johnson, W. P., Schmidt, P. M., & Taylor, D. P. (2014). Foraging flight distances of wintering ducks and geese: A review. *Avian Conservation and Ecology*, 9(2), 2.

Jones, J. (2001). Habitat selection studies in avian ecology: A critical review. *The Auk*, 118, 557–562.

Kabir, M. H., Shawon, R. A. R., Haque, M. K., Shyama, I. J., Islam, M. A., Saila, M., & Rahman, M. M. (2024). Enhancing conservation strategies for wild and migratory bird habitats in South Asia: A comprehensive review. *Asian Journal of Environment and Ecololgy*, 23, 73–88.

Karr, J. R. (1976). Seasonality, resource availability, and community diversity in tropical bird communities. *American Naturalist*, 105, 423–435.

Kauppinen, J., & Vaananen, V. M. (1999). Factors affecting changes in waterfowl populations in eutrophic wetlands in the Finnish lake district. *Wildlife Biology*, 5(2), 73–81.

Kear, J. (1990). *Man and Wildfowl*. T & AD Poyser, London.

Kear, J. (2005). *Ducks, Geese and Swans*. Bird Families of the World. Oxford University Press. ISBN 978-0-19-861008-3.

Kerbes, R. H., Kotanen, P. M., & Jefferies, R. L. (1990). Destruction of wetland habitats by lesser snow geese: A keystone species on the west coast of Hudson Bay. *Journal of Applied Ecology*, 242–258.

Kirby, J. S., Stattersfield, A. J., Butchart, S. H., Evans, M. I., Grimmett, R. F., Jones, V. R., & Newton, I. (2008). Key conservation issues for migratory land- and waterbird species on the world's major flyways. *Bird Conservation International*, 18(S1), S49–S73.

Kooloos, J. G. M., & Zweers, G. A. (1989). Mechanics of drinking in the mallard (*Anas platyrhynchos*, Anatidae). *Journal of Morphology*, 199(3), 327–347.

Krapu, G. L., Klett, A. T., & Jorde, D. G. (1983). The effect of variable spring water conditions on mallard reproduction. *The Auk*, 100(3), 689–698.

Krapu, G. L., Pietz, P. J., & Brandt, D. A. (2006). Mallard brood movements, wetland use, and duckling survival during and following a prairie drought. *The Journal of Wildlife Management*, 70(5), 1436–1444.

Kumar, P. (2024). Foraging ecology of wetland birds in heterogeneous habitats: A case study of the Dighal wetlands, Haryana, India. *Journal of Applied and Natural Science*, 16(4), 1627.

Lima, S. L. (2009). Predators and the breeding bird: Behavioral and reproductive flexibility under the risk of predation. *Biological Reviews*, 84(3), 485–513.

Linton, S., & Goulder, R. (2000). Botanical conservation value related to origin and management of ponds. Aquatic Conservation: *Marine and Freshwater Ecosystems*, 10, 77–91.

Lovvorn, J. R., & Jones, D. R. (1991). Effects of body size, body fat, and change in pressure with depth on buoyancy and costs of diving in ducks (Aythya spp.). *Canadian Journal of Zoology*, 69(11), 2879–2887.

Macmillan, D., Hanley, N., & Daw, M. (2004). Costs and benefits of wild goose conservation in Scotland. *Biological Conservation*, 119, 475–485.

Mariappan, N., Kalfan, B. K. A., & Krishnakumar, S. (2013). Assessment of bird population in different habitats of agricultural ecosystem. *International Journal of Scientific Research in Environmental Sciences*, 11, 306–316.

Mariyappan, M., Rajendran, M., Velu, S., Johnson, A. D., Dinesh, G. K., Solaimuthu, K., & Sankar, M. (2023). Ecological role and ecosystem services of birds: A review. *International Journal of Environment and Climate Change*, 13(6), 76–87.

Marshall, A. J. (1959). Internal and environmental control of breeding. *Ibis*, 101(3–4), 456–478.

Matthews, G. V. T. (1993). The Ramsar Convention on Wetlands: its history and development. Gland: Ramsar Convention Bureau.

Mattocks, J. G. (1971). Goose feeding and cellulose digestion. *Wildfowl*, 22(22), 107–113.

Mayr, E. (1970). *Populations, Species and Evolution*. Harvard University Press.

McNeil, R., Drapeau, P., & Goss-custard, J. D. (1992). The occurrence and adaptive significance of nocturnal habits in waterfowl. *Biological Reviews*, 67(4), 381–419.

Miyabayashi, Y., & Mundkur, T. (1999). *Atlas of Key Sites for Anatidae in the East Asian Flyway*. Tokyo, Japan: Wetlands International.

Mulcahy, D., & Heard, D. (2025). Ducks, Geese, and Swans (Waterfowl). *Zoo Animal and Wildlife Immobilization and Anesthesia*, 583–605.

Nacken, N., & Reise, K. (2000). Effects of herbivorous birds on intertidal seagrass beds in the northern Wadden Sea. *Helgoland Marine Research*, 54, 87–94.

Nelms, K. D., Porter, M. D., & Gray, M. J. (2012). Managing small impoundments for wildlife. *Small Impound. Manag. N. Am. Am. Fish. Soc. Bethesda MD*, 391–420.

Newton, I. (2024). Learning and social influence on bird migration. *Ardea*, 112(2), 171–199.

Nudds, T. D., & Bowlby, J. N. (1984). Predator–prey size relationships in North American dabbling ducks. *Canadian Journal of Zoology*, 62(10), 2002–2008.

Owen, M., & Black, J. (1990). *Waterfowl Ecology*. Black and Son Ltd a division of Routledge, Chapman and Hall, Inc. ISBN: 0-216-92675-0

Pacific Black Duck. *www.wiresnr.org. Archived from the original on May 1, 2008.* Retrieved 2018-04-27.

Paracuellos, M. (2006). How can habitat selection affect the use of a wetland complex by waterbirds? *Biodiversity & Conservation*, 15(14), 4569–4582.

Polechova, J., & Storch, D. (2008). Ecological niche. *Encyclopedia of Ecology*, 2, 1088–1097.

Popovkina, A. B., & Poyarkov, N. D. (2002). History of phylogenetic and taxonomic studies of Anseriformes. *Casarca*, 8, 41.

Poysa, H. (1983). Resource utilization pattern and guild structure in a waterfowl community. *Oikos*, 295–307.

Pratt, H. Douglas, Bruner, Phillip L., Berrett & Delwyn G. (1987). *A Field Guide to the Birds of Hawaii and the Tropical Pacific*. Princeton, NJ: Princeton University Press. ISBN 0-691-02399-9.

Praveen, J., Jayapal, R., & Pittie, A. (2020). Taxonomic updates to the checklists of birds of India, and the South Asian region—2020. *Indian Birds*, 16(1), 12–19.

Pyle, P. (2005). Molts and plumages of ducks (Anatinae). *Waterbirds*, 28(2), 208–219.

Rasmussen, P. C., & Anderton, J. C. (2005). *Birds of South Asia: The Ripley Guide* (Vol. 2, pp. 1–378). ISBN 84-87334-67-9.

Raveling, D. G., Nichols, J. D., Hines, J. E., Zezulak, D. S., Silveira, J. G., Johnson, J. C., & Weldon, J. A. (1992). Survival of cackling Canada geese, 1982–1988. *The Journal of Wildlife Management*, 63–73.

Ritchison, G. (2023). Urinary system, salt glands, and osmoregulation. *In: In a Class of Their Own: A Detailed Examination of Avian Forms and Functions* (pp. 1185–1252). Cham: Springer International Publishing.

Sedinger, J. S. (1997). Adaptations to and consequences of an herbivorous diet in grouse and waterfowl. *The Condor*, 99(2), 314–326.

Shi, J., Meng, L., Xia, S., Liu, S., & Zhou, L. (2024). Habitat suitability and determinants for Anatidae in multi-watershed composite wetlands in Anhui, China. *Animals*, 14(7), 1010.

Shirihai, Hadoram (2008). *A Complete Guide to Antarctic Wildlife*. Princeton, NJ, US: Princeton University Press. ISBN 978-0-691-13666-0.

Sigman, M., Jordan, S., Quinlan, S., & Graves, M. (1991). *Wetlands & Wildlife: Alaska Wildlife Curriculum Teacher Information Manual*. Parts I–II. Alaska State Department of Fish and Game; U.S. Fish and Wildlife Service.

Skutch, A. F. (1957). The incubation patterns of birds. *Ibis*, 99(1), 69–93.

Soons, M. B., Brochet, A. L., Kleyheeg, E., & Green, A. J. (2016). Seed dispersal by dabbling ducks: An overlooked dispersal pathway for a broad spectrum of plant species. *Journal of Ecology*, 104(2), 443–455.

Stephens, D. W., Brown, J. S., & Ydenberg, R. C. (Eds.). (2008). *Foraging: Behaviour and Ecology*. University of Chicago Press.

Sun, X., Shen, J., Xiao, Y., Li, S., & Cao, M. (2023). Habitat suitability and potential biological corridors for waterbirds in Yancheng coastal wetland of China. *Ecological Indicators*, 148, 110090.

Szabo, J. K., & Mundkur, T. (2017). Conserving wetlands for migratory waterbirds in South Asia. *In: Wetland Science: Perspectives from South Asia* (pp. 105–127). New Delhi: Springer India.

Terrill, R. S., & Shultz, A. J. (2023). Feather function and the evolution of birds. *Biological Reviews*, 98(2), 540–566.

Tews, J., Brose, U., Grimm, V., Tielborger, K., Wichmann, M. C., Schwager, M., & Jeltsch, F. (2004). Animal species diversity driven by habitat heterogeneity/diversity: The importance of keystone structures. *Journal of Biogeography*, 31(1), 79–92.

Therkildsen, O. R., & Madsen, J. (1999). Goose grazing selectivity along a depletion gradient. *Ecography*, 22(5), 516–520.

Thomas, V. G., & Prevett, J. P. (1982). The role of horsetails (Equisetaceae) in the nutrition of northern-breeding geese. *Oecologia*, 53(3), 359–363.

Tome, M. W., & Wrubleski, D. A. (1988). Underwater foraging behavior of canvasbacks, lesser scaups, and ruddy ducks. *The Condor*, 90(1), 168–172.

Townsend, C. W. (1909). The use of the wings and feet by diving birds. *The Auk*, 26(3), 234–248.

Van Roomen, M., & Madsen, J. (1992). *Waterfowl and Agriculture: Review and Future Perspective of the Crop Damage Conflict in Europe*. Special publication 21, International Waterfowl and Wetlands Research Bureau, Slimbridge, England.

Vine, I. (1971). Risk of visual detection and pursuit by a predator and the selective advantage of flocking behaviour. *Journal of Theoretical Biology*, 30(2), 405–422.

Wainwright, P. C. (1996). Ecological explanation through functional morphology: The feeding biology of sunfishes. *Ecology*, 77(5), 1336–1343.

Weller, M. W. (1999). *Wetland Birds: Habitat Resources and Conservation Implications*. Cambridge University Press.

Wenny, D. G., Sekercioglu, C., Cordeiro, N. J., Rogers, H. S., & Kelly, D. (2016). Seed dispersal by fruit-eating birds. *In:* C. H. Sekercioglu, D. G. Wenny, & C. J. Whelan (Eds.), *Why Birds Matter: Avian Ecological Function and Ecosystem Services* (pp. 107–131). University of Chicago Press.

Wetlands International. (2012). Waterbird Population Estimates, 5th edition. Summary Report. Wetlands International, Wageningen, The Netherlands.

Wicker, A. M., & Endres, K. M. (1995). Relationship between waterfowl and American coot abundance with submersed macrophytic vegetation in Currituck Sound, North Carolina. *Estuaries*, 18(2), 428–431.

Wiens, J. A. (1973). Pattern and process in grassland bird communities. *Ecological Monographs*, 43(2), 237–270.

Wiens, J. A. (1989). *The Ecology of Bird Communities* (Vol. 1). Cambridge University Press.

Wood, K. A., Stillman, R. A., Clarke, R. T., Daunt, F., & Ohare, M. T. (2012). The impact of waterfowl herbivory on plant standing crop: A meta-analysis. *Hydrobiologia*, 686, 157–167.

Zydelis, R., & Richman, S. E. (2015). Foraging behavior, ecology, and energetics of sea ducks. *Ecology and Conservation of North American Sea Ducks* (*Studies in Avian Biology*), 46, 241–266.

CHAPTER
4

Anatidae Family: Behavioral and Social Structure

Ali Akbar, Vinita Sharma and Stanzin Lado*
Department of Zoology, Central University of Jammu, Samba, Jammu and Kashmir, India

1. Introduction

The Anatidae family, which includes ducks, geese, and swans, is the most socially and behaviorally advanced group in the avian class. Anatidae are found on every continent except Antarctica and have successfully adapted to a wide range of habitats, ranging from the Arctic tundra to tropical wetlands. This global distribution has driven the evolution of diverse behaviors that help Anatidae species to survive, reproduce, and interact with one another in challenging ecosystems (Johnsgard, 1962; Green, 1996). Nesting and foraging behaviors play a crucial role in their survival, which is influenced by prey and predator relationships, anthropogenic activities, and climatic factors (McKinney, 1965). These behaviors not only impact their survival but also determine fecundity, clutch size, and brood survival (Arzel et al., 2006).

Another distinguishing characteristic of the Anatidae is the level of sexual dimorphism, particularly in the area of plumage and vocalization. Males in most species have brighter feathers and are apt to give elaborate displays to attract mates, while the females are dull and often show more camouflage. These characteristics are more than appearances; however, they play crucial roles in species recognition and mate choice (Johnsgard, 1963). Both vocal and visual communications are important. From the alarm call to indicate danger to the display of the courtship dance to the synchronized takeoff, these are signals that help to coordinate behaviors within flocks, maintain bonds, and lessen conflict (McKinney, 1965; Johnsgard, 1962).

*Corresponding author: stanzin.zool@cujammu.ac.in

DOI: 10.1201/9781003669784-4

Numerous factors, such as the availability of food, competition, and the energy required for rearing juveniles, determine the phase of pairing. For example, long-term pair bonds are observed in some species; however, the pair bond lasts for one winter (Johnsgard, 1960). Pair bonding also determines dominance and access to resources. Migration is another major force in the lives of Anatidae. Most species undertake long journeys between breeding and wintering grounds, which requires not just endurance but precise coordination and energy management. The migration pattern determines the success of breeding and is directly related to environmental conditions along its routes (Owen, 1996; Arzel et al., 2006). Recent studies suggest alloparenting, homosexual pair bonding, and complex dominance hierarchies. These behaviors would indicate that the birds are not only capable of being adaptable but are socially flexible, which helps them cope with rapid environmental changes (Johnsgard, 1963; Green, 1996). In summary, this chapter primarily focuses on the vast array of social behavior and patterns within the family Anatidae, how they choose mates, construct nests, raise offspring, and forage for food. Long-distance migration, lifestyle, animal interaction, mating patterns, and flock size are also discussed. Collectively, these behaviors show a complete description of how ducks, geese, and swans survive, adapt, and thrive in an ecosystem that is constantly in flux.

2. Distribution of Anatidae Across the Globe

Apart from Antarctica, all the continents are home to the Anatidae family, which includes ducks, geese, and swans. Anatidae species vary extensively in the Northern Hemisphere, especially through regions such as North America, Europe, and North Asia. These are major breeding sites, especially for the temperate and subarctic regions such as Scandinavia, Canada, and Russia. At seasons, the majority of species migrate and use major flyways such as the Mississippi, East Asian–Australasian, and Pacific routes to migrate from their breeding and wintering sites (Green, 1996; Zeng et al., 2019). According to Tu et al. (2009) and Cao et al. (2010), major sites of overwintering waterfowls are the Poyang Lake and the Yangtze River floodplain, with numerous individuals, even endangered species such as the tundra swan (*Cygnus columbianus*) and the swan goose (*Anser cygnoides*), spending their lives here. However, pressure from the destruction of their breeding sites and wetlands, as well as changes in land use, takes its toll on these species.

Though the Southern Hemisphere usually has fewer species, its evolutionary records are more diverse. Lower species numbers are also observed in regions such as Australia, New Zealand, and Southern South America, but most of them are of ancient lineages; examples include the freckled duck in Australia and the Chiloé wigeon from South America (Zeng et al., 2019). Africa provides key wintering sites for migrant species from Eurasia, though with smaller numbers of local species. Locally, species from the African sub-Saharan region are the Egyptian goose and the spur-winged goose. Both breeding and wintering species such as the tufted duck and greylag goose are widespread throughout Europe, particularly from the east and north (Green, 1996).

3. Evolutionary History of the Anatidae Family

Fossil records reveal that the Anatidae appeared during the Late Eocene (approximately 38–34 million years ago), a time when sea levels decreased and the seas around the continents had receded. This evolutionary change most likely facilitated the establishment of shallow freshwater habitats in Asia, which serve as the primary niches of the earliest anatids (Zelenkov, 2020). The most critical transitional fossil of the time is the duck-like *Romainvillia,* with characteristics from the Cretaceous–Paleogene-aged fossil *Presbyornithidae* and the modern Anatidae. *Romainvillia* has also been found in Kazakhstan and France, with an earlier and worldwide distribution that was most likely initiated by ecological potential from newly formed wetland ecosystems (Zelenkov, 2018). These early anatids had body morphologies well suited for foraging and filter-feeding, from shallow pools of water characteristics shared by numerous current duck lineages. The family significantly diversified throughout the Miocene Period (approximately 23–25 million years ago), facilitating the emergence of taxa such as Tadorna (approximately 17–15 mya), and faunal compositions similar to that of the modern period developing approximately 9–6 Ma (Zelenkov, 2020).

Morphological and behavioral development of Anatidae has specific patterns of specialization toward generalization. Earlier species would have had unspecialized, simple bill shapes and basic plumage patterns. Over the span of time, selection pressure favored the emergence of increasingly complex traits, including sexually differentiated plumage, sophisticated vocalization, and special-purpose foraging devices in the form of spoon-bills and serration along the edge for straining. Characteristically, these directions of evolution also showed a pattern from long-term pair bonds toward sporadic mating, with an association toward greater sexual dimorphism. Additionally, molting behaviors evolved from a single annual molt to biannual cycles, a change likely linked to mating and camouflage strategies (Johnsgard, 1962). One of the most compelling pieces of evidence for shared ancestry in Anatidae is the high frequency of hybridization. Over 418 interspecific hybrids have been recorded across 126 of the family's 149 species, with approximately 20% being fertile. This genetic compatibility supports the idea that current species evolved from a common, unspecialized ancestral form, with features that manifested differently across descendant lineages. This ancestral duck is hypothesized to have generalist traits, with modern species expressing specialized versions through microevolutionary changes. No macromutations or large-scale chromosomal changes are required to explain this diversification; standard mechanisms of mutation, drift, and selection (Scherer & Hilsberg, 1982). Evolutionary relationships of 39 taxa of the Anatidae family using nucleotide sequences are given in Figure 1.

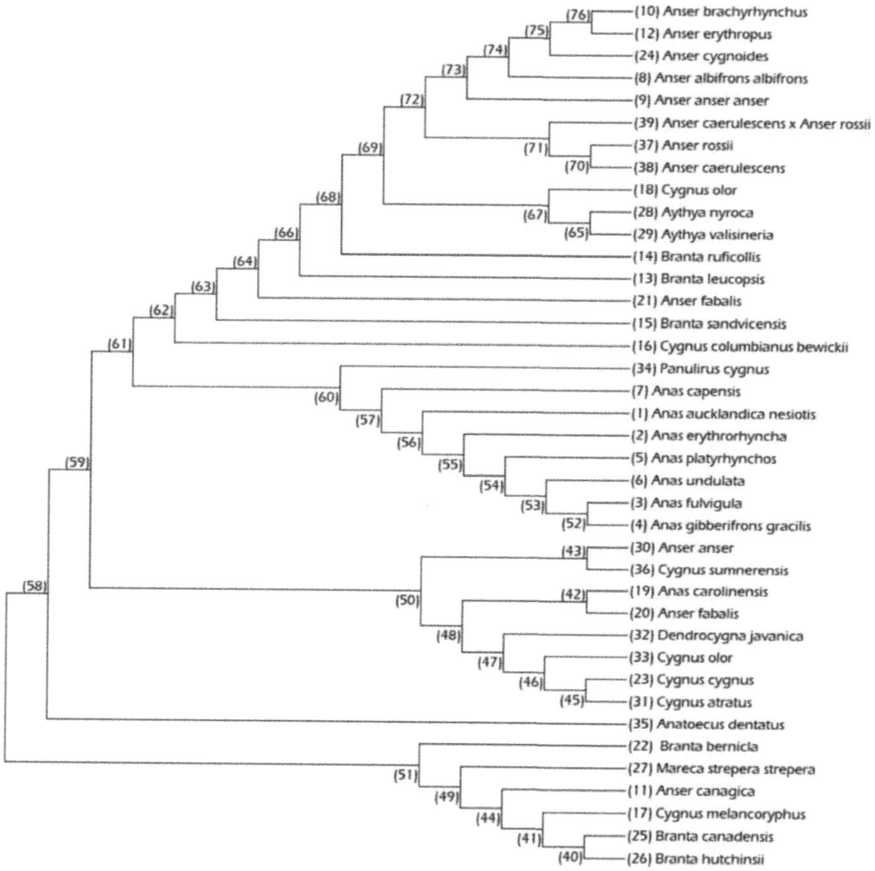

Figure 1. Evolutionary relationships of 39 taxa of the Anatidae family using nucleotide sequences.

The evolutionary history was inferred using the "neighbor-joining method" (Saitou & Nei, 1987). The optimal tree with the sum of branch length = 111.225 is shown. The percentage of replicate trees in which the associated taxa clustered together in the bootstrap test (1,000 replicates) are shown (Felsenstein, 1985). The evolutionary distances were computed using the "maximum composite likelihood method" (Tamura et al., 2004), and are in the units of the number of base substitutions per site. The analytical procedure encompassed 39 nucleotide sequences. The pairwise deletion option was applied to all ambiguous positions for each sequence pair, resulting in a final dataset comprising 16,815 positions. Evolutionary analyses were performed in MEGA12 (Kumar et al., 2024) utilizing up to 8 parallel computing threads.

4. Nesting and Foraging Behavior

The Anatidae family, consisting of swans, geese, and ducks, is the most ecologically diverse among birds. Evidence for this diversity can be seen in their distribution on different continents and freshwater as well as in marine environments, indicating their incredible resilience. Anatidae species, ranging from tundra in the Arctic to tropical wetlands, evolved a variety of behavioral, morphological, and physiological characteristics in order to tap a variety of environmental niches.

4.1. Ducks

4.1.1. Habitat Diversity and Nesting Behavior

Ducks show a rich variety of nesting behaviors, often responding to predation pressure, availability, and environmental factors. For example, while selecting breeding sites, there is a risk–benefit consideration in Finland's breeding ducks between nest predation probability and invertebrate food quantity. Due to greater invertebrate diversity and reduced predation probability on nests, seasonal ponds, beaver, and artificial ponds are favored over permanent lakes (Holopainen et al., 2024). Nesting ducks from higher altitudes, such as mallards, gadwalls, and blue-winged teal, often nest in areas with rich opportunities for grassland and wetland habitats, as shown in a study in Canada's Prairie Pothole Region by Devries et al. (2023).

Body size is also involved in nesting ecology, with smaller species of ducks nesting more often in areas with greater solar radiation and greater distances from sources of food in order to minimize energetic expenditure (Shutler et al., 1998). Nesting location also differs between species; dabbling ducks such as mallards and gadwalls, being extremely adaptable, often nest in a variety of environments, including marshes, prairie, and agricultural land. Even though nest success can be lower in areas with high predation pressure, mallards, in particular, are well documented for their ability to thrive in human-altered environments (Maxson & Riggs, 1996; Yetter et al., 2009). Dabbling ducks forage in freshwater or shallow waters, typically because they can easily get vegetation as well as invertebrates from there. Diving ducks such as ring-necked ducks and lesser scaup are more specialized in terms of environmental use and typically nest in deeper, more open wetlands. These ducks often take in aquatic invertebrates as a part of their diet, with resulting resource availability directly influencing nesting quality (DuBour, 2019; Hammell et al., 2022). As an indicator of prey availability, lesser scaup ducklings in tundra wetlands develop more slowly in lakes with low richness in amphipods (DuBour, 2019). Common pochard and lesser scaup are two of the diving ducks whose populations have declined in recent decades. Clutch size, female condition, and the availability of nearby nests are some of the factors that influence nesting success in these ducks. Common pochards in France, for example, exhibit higher nest survival in areas where they nest in open, as well as wet meadow, yet willows negatively influence survival (Folliot et al., 2017). Mallards (*Anas platyrhynchos*) and teal (*Anas crecca*) express a strong preference for Carex-dominated formations as newly hatched young. As they mature, they make use of a wide range of environments, whereas older juveniles have a greater reliance on open water and floating vegetation (Nummi & Pöysä, 1995). Both predator avoidance and foraging strategy are possibly linked with this age-related change in preference for different environments. Additionally, teal ducks extensively utilize flooded breached zones, representing a comparatively low proportion of the total coastal environment. This underlines the importance of microhabitats with low species diversity, yet rich in invertebrate prey. Where there is a changing regime in water levels, as well as vegetation diversity, this versatility proves crucial.

4.1.2. Food Habits and Foraging Behavior

Ducks are omnivorous, with differing diet compositions depending on age, species, and environmental factors. Dabbling ducks mainly eat plants or vegetation from aquatic environments, insects, as well as small invertebrates, whereas diving ducks mainly eat more animal material, including mollusks as well as crustaceans (Holopainen et al., 2015; Ando et al., 2022). Lesser scaup ducklings in northern wetlands mainly feed on amphipods, but they also include other mollusks as well as aquatic invertebrates in their diet (DuBour, 2019). Ducks in agricultural landscapes modify their feeding habits in accordance with accessible resources. Wintering ducks in Japan, including gadwalls, subsist mainly on agricultural plants such as rice and lotus, while other individuals might utilize plants from land as well as from water sources as a means to minimize competition (Ando et al., 2022). Wintering ducks in Algeria, such as gadwalls and pochards, consume seeds from plants such as *Myriophyllum* and *Potamogeton* with minimal macroinvertebrate consumption (Ayaichia et al., 2018).

Duck food sources significantly depend on the availability of resources. To grow and survive, northern pintail ducklings in boreal wetlands prefer invertebrate-dense environments such as shallow lakes and ponds (Holopainen et al., 2024). Habitat diversity is shown through greater survival in northern pintail ducklings in grass-dominated environments as they foraged on perennial cover and on seasonal wetlands (Johns et al., 2024). Resource partitioning is an important component that helps sympatric duck species coexist. Large-bodied ducks, for example, in agricultural areas, focus on crop plants, while smaller species utilize aquatic as well as ground plants as alternative plant resources for avoiding competition (Ando et al., 2022). When foraging pressure is greater, for example, in February, at a time of high abundance of ducks, this segregation in diet is most evident (Ando et al., 2022). Variation among food habits, foraging strategies, and bill structures of some members of the Anatidae family is described in Table 1.

Table 1. Food habits, foraging strategies, and bill structures of some members of the Anatidae family

Group	Common/Scientific Name	Feeding Habit	Bill Structure	Foraging Strategy	Source
Ducks	Northern shoveler (*Anas clypeata*)	Filter-feeding (plankton)	Spatulate, broad with fine lamellae	Surface dabbling	Li & Clarke, 2016
	Blue duck (*Hymenolaimus malacorhynchos*)	Benthic carnivore	Narrow, muscular, soft-tipped	Scraping and probing in rivers	Johnsgard, 1962
	Common merganser (*Mergus merganser*)	Piscivorous	Long, narrow, serrated	Active fish hunting by diving	Barnes & Thomas, 1987
	Black scoter (*Melanitta americana*)	Molluscivore	Heavy, rigid bill	Deep diving, crushing prey	Avilova, 2018
	Ruddy duck (*Oxyura jamaicensis*)	Omnivorous diver	Flattened, compact bill	Diving and stirring the benthos	Johnsgard, 1962
Geese	Canada goose (*Branta canadensis*)	Omnivorous grazer	Broad, triangular bill	Terrestrial grazing, dabbling	Li & Clarke, 2016
	Greylag goose (*Anser anser*)	Herbivorous grazer	Wide, sturdy bill	Grass cropping	Pecsics et al., 2017
	Swan goose (*Anser cygnoides*)	Tuber grazer	Robust, digging-adapted bill	Substrate excavation	Li & Clarke, 2016
	Upland goose (*Chloephaga picta*)	Grazing herbivore	Medium triangular bill	Terrestrial grazing	Barnes & Thomas, 1987
	Cackling goose (*Branta hutchinsii*)	Herbivore	Compact bill	Grass and aquatic grazing	Cantlay et al., 2023

(Contd.)

(*Contd.*)

Group	Common/Scientific Name	Feeding Habit	Bill Structure	Foraging Strategy	Source
Swans	Whooper swan (*Cygnus cygnus*)	Aquatic herbivore	Long, broad with fine lamellae	Upending for submerged plants	Kurk, 2008
	Bewick's swan (*Cygnus columbianus*)	Herbivore	Slender, fine lamellae	Feeding on tubers, pondweeds	Li & Clarke, 2016
	Black-necked swan (*Cygnus melancoryphus*)	Herbivore	Broad lamellate bill	Upending in macrophyte beds	Barnes & Thomas, 1987
	Tundra swan (*Cygnus columbianus*)	Herbivore	Extended flat bill	Grazing and aquatic foraging	Avilova, 2018
	Black swan (*Cygnus atratus*)	Herbivore	Flat, red-tipped bill	Grazing aquatic vegetation	Johnsgard, 1962

4.2. Swans

4.2.1. Habitat Diversity and Nesting Behavior

Swans are present in all the continents, with the majority of the species found in temperate and arctic regions of North America and Eurasia (Johnsgard, 2016). They usually occupy the wetlands, rivers, lakes, and coasts, feeding on aquatic flora, cereals, and grasses. Swans like shallow water bodies such as marshes, ponds, and estuaries where they forage for aquatic grasses and small animals (McKelvey, 1981; Kissane, 2019). Other types of swans, such as the black-necked swan (*Cygnus melancoryphus*), are found in coastal marshes and salt marshes where they forage for Zostera and Ulva, coastal marine grasses (Mathiasson, 1973; Norambuena et al., 2023). Other swans, such as the mute swan (*Cygnus olor*), are tolerant of human adaptations and found in urban parks and agricultural lands (Sears, 1989). In winter, most swan species migrate to grasslands and agricultural lands and engage in feeding on winter wheat, potatoes, and root vegetables (Dirksen et al., 1991).

Furthermore, most swans are herbivorous and feed mainly on aquatic grasses, crops, and aquatic plants. Yet, some swans occasionally ingest small animals, for example, insects, frogs, and fish, especially during breeding or migration times, where there is high energy demand (Brazil, 2002). Swans display seasonal diet changes, such as in summer, swans emphasize aquatic vegetation, for example, submerged macrophytes and algae, which are found profusely in wetlands (Kissane, 2019; Kouzov et al., 2024). In winter, swans indulge in terrestrial grasses, cereal crops, and root vegetables. For instance, Bewick's swans (*Cygnus columbianus bewickii*) from the Netherlands shift from pondweed tubers to grasses and winter wheat as the season advances (Dirksen et al., 1991). Swans feed on crops during their migration processes; for example, corn and rice, which yield high-energy food for their travel (Liu et al., 2022). Additionally, mute swans (*Cygnus olor*) are opportunistic feeders on algae, aquatic plant matter, and bread provided by humans in cities. The latter can create nutritional imbalances (Sears, 1989). However, whooper swans (*Cygnus cygnus*) consume aquatic flora such as Phragmites and *Sparganium* but also consume crops such as potatoes and winter cereals (Brazil, 1981; Rees et al., 1997). Black swans (*Cygnus atratus*) forage on aquatic plants submerged underwater and saltmarsh vegetation, with *Lamprothamnium macropogon* being prominent in their diet (Kissane, 2019).

4.2.2. Habitat Choice and Foraging Habits

The foraging habits of swans are sophisticated and usually follow feeding sites based on food abundance, depth of the water, and human habitation. For instance, the shallow lakes and paddy fields of China are the favorite of Bewick's swans, who forage there for tubers and cereals with efficiency (Yu et al., 2020). The Scottish and Icelandic whooper swans forage on farmland, mainly in the stubble fields and grasslands, for the abundance of waste grain and pasture (Brazil, 1981; Rees et al., 1997).

4.3. Geese

4.3.1. Habitat Diversity and Nesting Behavior

Geese are very versatile birds with a diverse range of breeding and feeding habits. Habitat types and dietary preferences are strongly associated with breeding and feeding strategies and differ between and among species and geographic locations. Most goose species, for example, favor breeding and wintering habitat types such as sedge marshes and paddy fields. These offer high food abundance in terms of grass and sedge, which can be used for feeding and aiding the bird in the harshest months (Fan et al., 2020). Breeding geese such as the pink-footed geese in the Arctic prefer wet tundra and bottom river valleys. These provide high-quality vegetation and open water to aid in predator escape. Such breeding grounds remain in high demand in the breeding season as goslings need high-quality food (Fox et al., 2009; Anderson et al., 2012). Certain geese, such as the lesser snow geese, remain in degraded habitat types due to philopatry even in low food abundance. Such persistence may result in changed feeding behavior and even decreases in breeding success (Pezzanite et al., 2005).

4.3.2. Foraging Behavior

Geese have dietary adaptability and can change their diet seasonally with food availability. Greylag and bar-headed geese in India change from seeds and grasses to crops such as wheat and peas as the season advances (Middleton & Valk, 1987). Where there is limited biomass present in the upper layer, like in early spring in the Arctic, geese such as the pink-footed geese forage below the surface for plant storage organs with high nutrient content (Fox et al., 2006; Anderson et al., 2012). High-quality foods, such as timothy grass in artificial grasslands, are the necessary nutrients that contribute to energy reserves in preparation for breeding (Fox, 1993). Food availability and protection from predators determine where geese choose to nest. Barnacle geese choose places in the uplands in the pre-breeding and nesting seasons, and pink-footed geese change to the valleys of the rivers post-hatching for greater food accessibility as well as safety of nesting (Fox et al., 2009). Habitat degradation has an impact on nesting success, as in lesser snow geese, where sparse vegetation contributes to younger gosling sizes and possibly decreased reproduction success (Pezzanite et al., 2005).

5. Sexual Dimorphism Among Species

Sexual dimorphism in plumage patterns in waterfowl arises from selection on existing traits, often involving the loss or elaboration of patterning in one sex. These traits may serve social or signalling functions and are not necessarily linked to reproductive roles alone (Gluckman, 2014). Recent research on this subject has revealed that many species that are thought to be monogamous are dimorphic, whereas numerous strongly polygynous species are monomorphic (Selander, 2017). In a majority of Anatidae species, males develop spectacular, iridescent, or structurally complex feathers, and these are often on or in close association with the regions near or around the head, neck, or wing. Such traits are commonly utilized in courtship display, which helps in prospecting females. Male wood ducks (*Aix sponsa*), for instance, possess showy plumage with elaborate markings, while male common goldeneyes (*Bucephala clangula*) possess bright orange heads with black-and-white patching on the wings. Such traits have probably resulted from sexual selection, where females have a preference for males with colorful or more flashy plumage (Johnsgard, 1962; Pfennig & Hurlbert, 2012).

Moreover, female Anatidae, on the other hand, have dull, camouflage plumage that helps in concealing themselves in their habitat as well as in repelling predators when they nest or raise young. This is characteristic of ground-nesting ones, such as in the northern pintail (*Anas acuta*) and the green-winged teal (*Anas carolinensis*). The dull appearance in females is a direct outcome of their incubation and care of young ones, where camouflage is essential for securing a successful result (Johnsgard, 1962; Bennett & Owens, 2002). Plumage patterns are also used in species detection and reproductive isolation. As an example, unique wing specula in most Anatidae species have been conjectured as apparent signals promoting recognition of conspecifics. Such traits are typically more pronounced in males and a result of both natural and sexual selection (Johnsgard, 1962; Pfennig & Hurlbert, 2012).

Furthermore, some Anatidae have seasonal plumage dimorphism. For example, males of the eider (*Somateria mollissima*) have a highly conspicuous black-and-white breeding plumage that is replaced with a dull gray-brown plumage when non-breeding. The seasonal adjustment is due to requirements for display during breeding and reduced requirements when not breeding (Johnsgard, 1962; Greenberg et al., 2013). Sexual size dimorphism (SSD) is common in Anatidae, with males generally being larger than females. That is particularly true for species where males are actively involved in intense competition for mates or territories. For example, male mute swans are longer-winged and longer-bodied than females, which aids them during territorial disputes (Friedman & Remeš, 2016; Włodarczyk et al., 2016). However, species such as the marbled teal (*Marmaronetta angustirostris*) are less extreme in their size dimorphism. This observation suggests SSD is the result of the cumulative action of several factors, such as mating systems, feeding ecology, and predator defense (Green, 2000; Friedman & Remeš, 2016). Examples of plumage and size dimorphism in the Anatidae family are described in Table 2.

The structure of the wing and tail also varies between the sexes of some Anatidae species. Males are longer winged and have long elaborate tails, which are displayed during courtship. For example, male harlequin ducks have colorful tail patterns not present in the females (Johnsgard, 1962; Pfennig & Hurlbert, 2012). Sexual selection is the primary force behind the sexual dimorphism of

Table 2. Examples of plumage and size dimorphism in the Anatidae family

Group	Name	Plumage Dimorphism	Body Size Dimorphism	Bill Shape/Size	Behavior/ Vocalization	Reference
Ducks	Muscovy duck (*Cairina moschata*)	Males more colorful, larger caruncles	Males nearly 2× heavier	Longer, broader bill in males	Males more aggressive, dominant	Oguntunji & Ayorinde,2014
	African Muscovy duck	Dimorphic wing and neck feathers	Males larger in wing, neck, and leg length	Longer beaks in males	Males defend territory	Yakubu et al., 2009
	Ivory Coast Muscovy duck	Subtle plumage differences	Males have larger frame	Broader beaks in males	Males show higher mating behavior	Loba-Messet et al., 2023
	Pekin duck	Low plumage dimorphism	Males slightly larger	Minimal differences	Vocalizations louder in females	Baeza et al., 2001
Geese	Greylag goose	Little to no plumage difference	Males approximately 10–15% heavier	Slightly deeper bills in males	Males show higher alertness and defense	Weidinger & van Franeker, 1998
	Canada goose (*Branta canadensis*)	No visible plumage dimorphism	Males larger in tarsus, wing, head	Minimal bill differences	Males display territorial calls	Pfennig & Hurlbert, 2012
	Cape Barren goose	Similar plumage	Males larger, longer necks	Slight head size difference	Males defend nesting area	Calabuig et al., 2007
	Snow goose	No color dimorphism	Males larger in body mass	Minimal differences	Males more vocal in flocks	Reeve & Pfennig, 2003
Swans	Mute swan (*Cygnus olor*)	Similar white plumage	Males larger neck and mass	Slight bill knob difference	Males perform threat displays	Calabuig et al., 2007

	Coscoroba swan	Subtle plumage differences by age	Males heavier, longer neck/tarsus	Males have deeper bills	Guarding behavior by males	Calabuig et al., 2011
	Whooper swan	Monochromatic plumage	Males larger in all morphometrics	No clear dimorphism	Males vocalize more in pairs	Reeve & Pfennig, 2003
	Trumpeter swan	Same plumage in both sexes	Males slightly larger	Slightly larger bills	Males defend territory	Price et al., 1998

Anatidae. Males in polygynous species, where males congregate over multiple mates, typically acquire exaggerated secondary sexual characteristics, such as large body size or bright plumage, to impress females. For instance, the males of the common eider (*Somateria mollissima*) display an elaborate courtship dance to impress the females, and their vibrant plumage has been speculated as the result of sexual selection (Dunn et al., 2001). Monogamous species, on the other hand, such as the mute swan (*Cygnus olor*), show weak sexual dimorphism since the pressure for strong male–male competition is lowered (Dunn et al., 2001; Włodarczyk et al., 2016).

6. Group Coordination and Vocal Communication in the Anatidae Family

The Anatidae flourishes through efficient group coordination and auditory signals. Auditory signals such as alarms call, directions call, or affection call maintain the Aves coordination and prevent conflicts. This type of vocalization ensures survival in unpredictable and changing environmental conditions.

6.1. Kin Relationship and Hierarchy in the Anatidae Family

The kin relationships within the Anatidae family are complex and characterized by unique social behaviors and reproductive strategies, which involve brood parasitism, cooperative breeding, and social structures. Kin-based behaviors in Anatidae are closely related to ecological factors such as resource availability and breeding opportunities within the habitat (Eadie & Lyon, 2011; Andersson et al., 2019). A study results indicate that heart rate (HR) responses are much higher during social encounters than during non-social events. For instance, the HR of geese increases more when they witness agonistic encounters than when they witness non-social ones. This observation suggests that the geese may view social situations as more stressful or relevant (Smith et al., 2000; Scheiber et al., 2008). The HR reactions of observers differ depending on whether the subjects are related or unrelated. Individuals with higher social status often elicit more intense HR reactions from observers, highlighting the importance of social hierarchy (Smith et al., 2000; de Vries et al., 2003; Klaassen et al., 2006; Scheiber et al., 2008).

6.1.1. Conspecific Brood Parasitism (CBP) and Host–Parasite Relationship

In the Anatidae family, CBP is a reproductive tactic in which a female lays eggs in the nests of other females of the same species (Andersson et al., 2019). CBP frequently involves closely related females, with host–parasite pairings being the most closely related. This type of relationship can contribute to overall fitness because hosts may benefit from producing related parasite progeny (Eadie & Lyon, 2011). For example, waterfowl, a group with female-biased natal philopatry and geographically related females, show CBP. If host costs are large, parasites instead may target unrelated females, but relatedness between the parasite and host can result in inclusive fitness for both (Andersson et al., 2019). Female-biased natal philopatry in waterfowl results in local kin selection, facilitating cooperative breeding and brood merging among relatives (Eadie & Lyon, 2011; Andersson et al., 2019). Hosts may have better success rates when cared by relatives, as demonstrated in wood ducks (*Aix sponsa*), where main parasite success is related to host kinship (Nielsen et al., 2006). Therefore, cost–benefit analysis plays an essential role in animal behavior.

6.1.2. Cooperative Breeding and Kinship

In Anatidae, cooperative breeding involves individuals raising offspring that are not their own, demonstrating kin relationships. Female common eiders frequently establish kin-based social groups during breeding, with more relatedness between females inside groups than the general population mean (Hario et al., 2012; Andersson et al., 2019). Kin selection predicts that individuals would exhibit action favoring the survival and fertility success of their relatives even when such action incurs a direct fitness cost. Females in the Anatidae family are eager to care for related children, whether through cooperative breeding or brood parasitism (Eadie & Lyon, 2011; Andersson et al., 2019). Their evolutionary process is pushing them to take action, since it is anticipated that the benefits of doing so will outweigh the costs. As a result, throughout the animal kingdom, cost–benefit analysis is essential.

6.1.3. Social Structure and Kinship

Kinship influences the social structures of the Anatidae, especially in species with female-biased natal philopatry. Female greylag geese (*Anser anser*) build strong family connections, with adult female siblings resting closer together than unrelated individuals (Frigerio et al., 2001). This social cohesiveness is believed to minimize social stress and agonistic interactions, emphasizing the adaptive advantages of kin-based sociality.

6.2. Vocal Communication in the Anatidae Family

In the Anatidae family, vocal communication is characterized by a range of sound generation techniques driven by physical adaptations, which include a range of specializations in the trachea and syrinx that are linked to sound production (Johnsgard, 1971). The tracheal tube enlarges irregularly in males of several duck species, especially pochards (*Aythyini*) and sea ducks (*Mergini*). In many species, males have specially designed tracheal and syringeal structures that act as a resonating tube. Male ducks have a bulb-like, air-filled protuberance from the left syringeal skeleton called a bulla on the syrinx (Rüppell, 1933; Johnsgard, 1961; Frank et al., 2007). It plays an important role in courtship vocalizations, which helps in enhancing their vocalizations, especially during mating displays, and acts as a Helmholtz resonator, amplifying specific frequencies that are crucial for attracting females (Mishkind et al., 2025).

Furthermore, the tracheal anatomy in the Anatidae family influences their sound, which has developed to distinguish between male and female species, improving communication and attracting mates. Due to these unique anatomical characteristics, ducks, geese, and swans can make loud, far-reaching sounds that are particularly helpful as contact calls (Johnsgard, 1971; Carboneras, 1992). Females and young ones typically make contact calls with harsher, lower-pitched notes that they use to constantly indicate their presence. This provides young birds more confidence to swim around in search of their food, knowing that the female is close, even if they can't see it. Certain sounds in the Anatidae family are associated with specific displays or sexual behavior. Moreover, aggressive behavior often involves loud and long-range calls. These sounds indicate guarded territory and are accompanied by a display of ownership. Various calls are made during courting and copulation. These low-pitched voices are unique to each species and prevent interbreeding between closely related species (Carboneras, 1992).

In geese, vocal communication is essential for social interactions and group dynamics. Geese use a wide range of vocalizations to coordinate activities, maintain social cohesion, and signal distinct behaviors. This communication is critical for their survival and social structure since it aids in conflict resolution and improves collaboration within flocks (Steen et al., 2012). The magpie goose is a vocal bird that honks loudly, similar to genuine geese, when in flight, from water, or on land. These honks sometimes elicit a response from mates or other conspecifics. Additionally, sheldgeese have significant sexual diversity in their calls, despite relative similarity among species. Males make a high-pitched whistle that is repeated quickly, while females make harsher calls, including cackles, growls, quacks, and grating sounds. The whistling ducks get their name from the thin, whistling noises they frequently produce while flying, foraging, or resting on land. When these cries are repeatedly made, a flock of birds will often call together in a shrill twittering, though sometimes they produce harsher, shorter calls (Carboneras, 1992).

7. Mating Behavior

Courtship and mating behaviors effectively ensure normal reproduction performance in animals (Zhang et al., 2019). Courtship and mating behavior success not only maintains populations and sustains animal production, but it also contributes to successful fertilization. Greater copulation frequencies are associated with greater probabilities of successful egg fertilization (Gumułka & Rozenboim, 2013). Animal courtship can occur through vocalizations, physical display, specific behavioral expressions, nesting, and some other means (Ota et al., 2015).

7.1. Mating Behaviors in Ducks

Duck courtship is a crucial step to mating and includes a range of actions, displays, and behaviors. Ducks frequently use complex courtship behaviors such as vocalizations, visual displays, and certain postures to attract females. Male northern pintails (*Anas acuta*) court females by displaying their colorful scapular feathers and white breasts, which are preferred by females (Sorenson & Derrickson, 1994). Vocalization also contributes substantially during duck courtship. Males have a special vocal organ called the syrinx, which is able to generate a variety of sounds and contributes to enhancing certain frequencies of courtship vocalization. Vocalization has also been thought to play an important role in mating attraction and pair-bonding development (Mishkind et al., 2025).

Ducks use multiple mating behaviors, such as forced copulations, polygyny, and monogamy. The most frequent mating behavior of ducks is monogamy, while paired males are frequently involved in extra-pair copulations (EPCs) as a secondary mating behavior to maximize their probabilities for successful breeding. Male mallards (*Anas platyrhynchos*) have forced copulations with non-mate females (McKinney et al., 1983; Gauthier, 1988). Male ducks have evolved enlarged phalluses with morphological specializations such as ridges and knobs, in order to perform forced copulations. As a defense, females have evolved sophisticated vaginal structures, such as clockwise coils and dead-end sacs, which can help prevent or resist forced intromission (Brennan et al., 2007; Brennan et al., 2010). Whereas, female resistance to forced copulation is a typical trait in ducks. This resistance can take the form of active avoidance, such as fleeing or diving, and passive resistance, such as remaining motionless during copulation attempts (McKinney et al., 1983; Adler, 2010).

Male duck mating behavior is significantly regulated by hormones, especially testosterone. Courtship conduct and the frequency of forced copulations are positively connected with testosterone levels. Paired males with greater testosterone levels in captive mallards were more likely to attempt forced copulations, even though they did not necessarily spend more time courting females (Davis, 2002). Chemical communication is another crucial component of duck mating behavior. Secretions that could function as sex pheromones are produced by the uropygial gland, which is situated above the tail. The volatile metabolites found in these secretions, such as indole acetaldehyde and picolinic acid, are more prevalent in males than in females. The precise significance of these chemical cues in duck reproduction is still being studied, although they may be involved in mate selection and sexual communication (Liu et al., 2022).

7.2. Mating Behavior in Geese

Geese are typically monogamous, creating long-term pair connections necessary for successful reproduction. Geese mating behavior consists of a set of coordinated activities between males and females that are necessary for copulation and subsequent reproductive success. Mutual activities, such as the triumph ceremony which reinforces mate connections and territorial displays, help to sustain these bonds (Akesson & Raveling, 1982; Lank & Rockwell, 1995). The timing and preferences of mating vary among geese breeds. For example, Hungarian geese usually mate in the morning, but domesticated geese such as Sichuan and Zhedong white geese prefer to mate in the late afternoon (Bao et al., 2022).

In general, during courtship displays, the gander attracts a goose by neck-over-neck, tidbitting, or strutting. During copulation attempts, the gander displays mounting in which it approaches a crouched female from the front of her body or forces her down and places one or both feet on the dorsal surface of her torso; with trampling it makes small feet movement on the top of the dorsal surface of the female's torso, and depresses tail and cloacae areas toward the female. Following mounting and treading, if the copulation is successful, the gander contacts the goose's cloacae. The tail of the male becomes twisted underneath that of the female and thrusts toward her. When the male loses his equilibrium and falls off or dismounts after lowering his tail, it is an indication of an unsuccessful copulation (Gumułka & Rozenboim, 2013).

7.3. Mating Behavior in Swans

Swans are well-known for their enduring monogamous relationships, maintaining prolonged pair and family bonds that significantly influence their reproductive and social behavior (Johnsgard, 2020). This monogamy is believed to be a response to their comparatively slow life histories marked by immense parental investment in offspring and low rates of reproduction (Soni, 2022; Bowers, 2023). Maintaining a close relationship with partners throughout the non-breeding season benefits Bewick's swans in a variety of ways, including increased success in aggressive encounters and reduced threats from other flock members (Scott, 1980). Swans do not engage in forced copulation because it conflicts with the roles of paired males in defending breeding territories and mates (McKinney et al., 1983). Both males and females of several swan species have ornamentation that influences their choice of mating. The black swan (*Cygnus atratus*), for instance, has curled feathers on its wings that are prominently shown during social interactions by both sexes (Kraaijeveld et al., 2004).

Moreover, swans frequently form pairs through elaborate displays and rituals. Bewick's swans (*Cygnus columbianus bewickii*) form pairings during winter, establishing their bond through mutual preening and hostility (Scott, 1980). These behaviors are used to establish dominance within the pair and to fortify the pair bond. Both males and females participate in breeding efforts once a couple is formed, though the distribution of labor may differ depending on the species. In mute swans, males are concerned with defending their territory and their families, whereas the incubation duties are largely the responsibility of the females (Włodarczyk & Minias, 2015). On the other hand, male black swans are longer at the nest compared to the females and are also actively participating in incubation (Brugger & Taborsky, 1994).

Seasonal variations also take place in the courtship display and the ornamentation of most species of swans throughout the year. For instance, the bill knob's size, a sexually selected trait, varies dynamically during the breeding season of mute swans. Males have the smallest bill knob at molt, when the birds are changing their feathers, and the maximum size of bill knob at cygnet hatching, when male parental care for the young is most important. These seasonal changes in ornamentation are predicted to be the result of differential selective pressure during the breeding season. For example, during breeding, males with the big bill knobs would be preferentially able to hold their territories and secure mates. Near the end of the breeding season, when the selective pressure for breeding is lower, the cost of maintaining large ornaments can be larger than their benefit, and the ornament will become reduced (Horrocks et al., 2009).

8. Social Interactions in Anatidae

Social interactions within the family Anatidae exhibit considerable complexity and are shaped by an array of ecological and social determinants. These interactions are integral to their foraging strategies, social hierarchies, and overall viability. The subsequent sections delineate fundamental elements of social interaction pertinent to this avian family. The Anatidae family has specific behavioral characteristics in social interactions. To increase their foraging efficiency, most Anatidae use social cues, often relying on the presence of conspecifics to locate food sources. Because significant aggregations can reduce competitive interference through behavioral adaptations, hence promoting coexistence, social dynamics may offer competitive benefits (Monier, 2024). The availability of food has a significant impact on social relations; limited resources can increase the number of encounters and alter group behavior (Guo et al., 2024).

Furthermore, mandarin ducks, which belong to the Anatidae family, exhibit group foraging and sociability. The significant level of friendliness displayed by the mandarin duck (*Aix galericulata*) has a significant impact on its behavioral patterns. These birds tend to forage in cohesive groups because of their social setup, which protects them from potential predators. For their general survival and well-being in their native environments, such behavioral patterns are crucial (Shurtleff & Savage, 1996). Comfort motions, such as bathing and preening, are also a part of social contact and serve both practical and cultural purposes. These behaviors could indicate readiness for flight or the formation of social bonds among conspecifics (McKinney, 1965). These are behaviors carried out within social groups that enhance group cohesiveness and group communication. Social facilitation is the underlying concept by which the species of the mandarin behaves. One individual's behavior can instigate conspecifics to follow or complement the same behavior, and that is how the phenomenon works. If one duck, for instance, initiates the activity of preening, then it can instigate other ducks present to start the same activity. In foraging, the same phenomenon of social facilitation manifests, and the larger the number of ducks present, the more efficient the foraging (Bruggers & Jackson, 1977; Clayton, 1978). Foraging and vigilance are innate behaviors of the mandarin ducks (Clayton, 1978).

Social interactions of organisms are largely determined by the surrounding environment. Ducks from areas with vegetation are less alert compared to those living in open waterfronts. Perceived predation pressure and accessible resources are some of the factors explaining these behavioral variations. Social interactions of the mandarin ducks are determined by their environment and their innate sociability. Their interactions are essential not only for the survival of individuals but also play a great role in enhancing the populations' viability at an ecosystems level, such as the situation with Richmond Park, where the social interactions of the mandarin ducks are complex and determined by their sociable nature, group dynamics, and the environmental factor, all of them playing an essential role in determining their behavior and survival strategies (Pagel & Dawkins, 1997). Pre-laying behavior of *Anas platyrhynchos* can be largely governed by social ranking (Barrett et al., 2019; Cooper & Appleby, 1996). For example, *Gallus gallus domesticus* alpha males have lesser aggressive encounters when fighting for the nest. Similarly, in ducks, social ranking can be predicted by how they behave near breeding area (Freire et al., 1997).

The habitat of the *Anser* species, such as the greylag geese and Canada geese, is strongly influenced by social interactions, which shape the survival mechanisms, family relationship, and physiological adaptations. These interactions are marked by sophisticated social structures and behaviors that enhance the well-being of the individual and the group. The social environmental conditions influence *Anser anser's* anatomical structure and physiological activity. Furthermore, greylag geese have their heart rate raised by social interactions similar to that raised by energetically demanding activities such as taking off while flight (Scheiber et al., 2005). Families in geese have unique social relationship, with offspring living with their parents for longer periods. This relationship encourages social coordination, which may reduce stress and enhance effectiveness in competitive environment. Larger family groups have lower stress responses, which increases their overall resilience (Blackwell et al., 2019). When faced with possible threats, such as moving cars, *Branta canadensis* uses social behavior to direct its escape reactions. In social groups, they assess dangers by combining individual and group cues, which can alter their alertness and the range at which they take off, thereby affecting their survival strategies (Briffa & Elwood, 2001).

Social interactions have a significant impact on both individuals directly involved and the spectators, or passive viewers, of the encounters. When bystanders observe social events, they may experience changes in their physiological response, such as changes in heart rate, and behavioral patterns (Peake, 2005; Scheiber et al., 2005; Zulandt et al., 2008). In a study on geese's social behavior, the researcher divides events into two main categories: "social" and "non-social." The movement of geese and aggressive encounters are the examples of social behavior, whereas traffic or loud noises are the examples of non-social events. This distinction is crucial since it affects the geese's physiological response (Wascher et al., 2008).

Swans' social interactions have a significant influence on their behavior, particularly when it comes to foraging and the formation of social hierarchies. According to empirical research, these interactions can dictate movement patterns, dominance hierarchies, and resource accessibility, which are critical for their survival and energy expenditure optimization. Because singleton swans often retreat from particular locations due to social constraints, they may counteract this tendency by moving to search for unexplored resources, highlighting the complex interplay between social dynamics and foraging effectiveness (Badzinski, 2003).

Tundra swans display a clear hierarchy of dominance, with family groups exercising authority over non-familial units. Family groups engage in conflicts less frequently but win most of the time, suggesting a systematic approach to social relationships that consumes less energy (Peiman & Robinson, 2010). Swans mostly engage in aggressive encounters with conspecifics, according to some research. Swans that share the same ecological niche and use the same food resources are more likely to clash with one another (Wood et al., 2015). For waterbird population, management and conservation initiatives are essential for social interactions. For instance, smaller waterbird species may experience displacement or negative consequences, and their survival rates and habitat use can be greatly influenced by the aggressive behaviors displayed by larger species (Wood et al., 2020).

8.1. Migratory Pattern and Wintering Strategy

Animals make numerous crucial decisions during migration, such as where to go, how long to remain, and when to depart. Each of these decision has an impact on the organism and enhances their fitness (McNamara et al., 1998; O'Neal et al., 2018). Due to breeding constraints, long-distance migratory birds travel faster in the spring than in the autumn. However, due to environmental factors such as wind and food availability, certain species move more slowly in the spring (Lee et al., 2023). Waterfowl are most common, widely distributed, and capable of both short- and long-distance travel, frequently over continents (Lurz et al., 2002).

Swans are often less migratory than geese, with some species such as the mute swan moving mainly to areas with harsh winters. Environmental factors and food availability influence swan migration patterns in the autumn and spring, notably among whooper swans. Conversely, geese are renowned for their extensive, well-planned migrations. The spring and autumn migrations of geese are distinguished by their varying lengths and distances. Variations in temperature, daylight hours, and food availability can affect geese migration timing. Anatidae species usually have clearly defined flyways predictable paths used across several generations. Most populations from Northern Hemisphere breed in Arctic or temperate areas during spring and summer and then migrate southward to warmer wintering grounds. For example, Eurasian populations follow the East Asian–Australasian flyway and North American species follow the Mississippi and Pacific flyways. Some geese and ducks experience "molting migrations" at mid-year and undertake solo journeys to isolated wetlands without predators, to replace and cast their feathers and thereby becoming flightless for a while (Owen, 1996). Migration also does not happen entirely on a direct route and always involves multiple stopover points during which birds replenish their energies. These stopover points are very important for survival and breeding success but are increasingly threatened by land-use changes. Sex and age at times vary from migratory behavior such that males and females conclude their winters in different geographic regions known as differential migration. Despite the appearance of migration-friendly weather, autumn-migrating ducks usually spend several days at mid-migration stopovers (O'Neal et al., 2018). Hunting pressure and habitat condition significantly affect the timing of duck migration (Legagneux et al., 2009).

Wintering behavior among Anatidae is regulated by thermoregulation, foraging access, and predator risk. Most species congregate around unfrozen lakes or ponds due to geothermal activity or human intervention. Geese migrate longer distances during winter foraging (mean approximately 7.8 km) than dabbling ducks (approximately 5.1 km), most likely due to their differing feeding preference and weight (Johnson et al., 2014). Energy minimization is maximized with numerous ducks employing nocturnal foraging and daytime resting at secure roosts. Some species have also learned how to survive on agricultural landscapes, feeding on fields by day and roosting on nearby wetlands. Such adaptability allows their use of anthropogenic habitats but also exposes them to hunting and human disturbances. Climatic anomalies such as extremely mild winters can still alter wintering ranges and postpone the onset of migration.

8.2. Reproductive and Foraging Behavioral Changes

Habitats can become degraded through catastrophic weather events or even through the action of species present in a habitat (Johnson & Collinge, 2004). When the habitat deteriorates, some species move to less affected locations, while others stay but behave differently, possibly making up for the habitat shift (Bruun & Smith, 2003; Rockwell et al., 2003). Habitat degradation has negatively impacted the foraging behavior and reproductive success of lesser snow geese, as studied by Pezzanite et al. (2005), as there is a decline in gosling size and mass, which correlates with reduced juvenile survival and overall reproductive success. Lesser snow geese in degraded habitats have higher motor activity and pecking rates, while adults spend less time on vigilance and comfort behaviors. Despite these behavioral changes, gosling size continues to drop, indicating a worse reproductive success. Adult snow geese have become more mobile, but they are also pecking more. Despite smaller bite sizes, the adaptability can be the compensatory reaction to the smaller size and quantity of the food, enabling them to consume more food in less time. Overall, the foraging pattern is adjusted by accommodating the new habitat. As an example, with the patchier distribution of the food, geese are spending most of their time searching for food rather than consuming. This behavior change signifies the challenge created by the impoverished environment. The survival in impoverished conditions is given credit by philopatry, even though it can be harmful to the respective environments (Pezzanite et al., 2005). Ground reclamation of the coastal Yancheng wetlands has reduced acceptable locations for wintering Anatidae, centralizing high-quality patches of their habitats and shifting the spatial distribution southward (Xia et al., 2024).

8.3. Dominance Hierarchies and Aggression Modulation

Dominance hierarchies reduce conflict about breeding and resources. Aggressive behaviors and social interactions shaped by certain traits, social processes, and environments build and uphold these hierarchies. The family Anatidae uses social signals, experience, and physical characteristics to determine dominance. They affect the chances that a particular individual will win a contest, and thus their social position within the hierarchy (Tibbetts et al., 2022). Dominance hierarchies are formed and maintained primarily through aggression. Some factors, such as social relation stability and advantage from various positions, could influence it (Holekamp & Strauss, 2016; Silk et al., 2019).

Duck dominance hierarchies are normally linear and accommodate social group organization and reduced conflict over resources. Greater levels of testosterone among dabbling ducks are found corresponding with larger ranks for dominance, thus confirming that hormone levels are strongly linked with aggressive behavior and social position (Poisbleau et al., 2005). According to the suppression hypothesis, dominant individuals preserve their position by consistently displaying aggressiveness toward subordinates, conditioning them to accept lower ranks (Forkman & Haskell, 2004). Domestic fowl establish aggressive dominance through initial encounters, and once hierarchies are formed, aggressiveness levels diminish, resulting in a stable social unit (Banks, 1984). Tundra swans form a dominance hierarchy dependent on family or group size. Families outperform nonfamily groupings in conflicts, demonstrating a dominance structure that promotes larger or more cohesive groups. When competing among other social units, swans display typical agonistic behavior at lesser intensities. This behavior helps secure dominance without increasing conflict situations that can prove energetically costly (Badzinski, 2003). In greylag geese, individual variations in aggression are a strong indicator of dominance rank. As a personality attribute, aggression is constant across environments and time, which helps to maintain social hierarchies (Kleindorfer et al., 2024).

8.4. Alloparenting and Adoption

Alloparenting or rearing young not their own is a fascinating behavior found among many birds, including the Anatidae family, such as ducks, geese, and swans. Adoption as a process of alloparenting involves accepting unrelated young and rearing and provisioning them. Alloparenting among duck species from the genera *Anas* and *Aythya* can take various forms, including brood amalgamation, when multiple females combine their broods and build a single group. This behavior is particularly prevalent in species that nest in high densities, such as the common pochard (*Aythya ferina*). Brood amalgamation can occur either before hatching, when females lay eggs in the nests of others, or after hatching, when females abandon their young to the care of other conspecifics (Beauchamp, 1997; Traylor et al., 2008).

According to a study on greylag geese by Kalmbach et al. (2005), goslings may actively seek out high-quality foster parents to enhance their chances of survival and fitness. The study also found that lone goslings could choose between dominant and subordinate foster families, with the majority of them selecting the dominant family when it came to adoption. Adoption occurs often in wild populations of black swans (*Cygnus atratus*), which display alloparenting. The mean relatedness between foster parents and alien cygnets was far higher than would be predicted by chance, according to a study that used microsatellite analysis. This implies that cygnets migrate from low-quality pairings to high-quality, related pairs, indicating that kin selection influences adoption. Because they could give better care and boost their chances of successful reproduction, high-quality partners were more likely to adopt cygnets (Kraaijeveld, 2005).

Adoption is also considered an adaptive behavior for goslings to maximize their chances of survival, but the cost and benefit of adoption are more multifaceted for parents (Kalmbach, 2006). Studies on the alloparenting behaviors of geese, especially greylag geese (*Anser anser*), have focused on adoption. According to Kalmbach et al. (2005), in greylag geese, lone goslings had the option of choosing between dominant and subordinate foster families, and the majority of them opted for the dominant family, which implies that to increase their chances of surviving and becoming fit, goslings might actively look for good foster parents.

8.5. Social Learning and Foraging Traditions

Foraging behavior in Anatidae is largely dictated by an integration between ecological and social learning processes and visual adaptations. Observation and imitation help individuals learn about sources of food and foraging behavior that can lead to developing cultural traditions. Foraging ecology among species of Anatidae is strongly correlated with their visual field shape. For both aquatic and terrestrial environments, foraging behavior impacts binocular field parameters such as maximal binocular field width and vertical binocular field extent that are paramount in discovering sources of food and predators (Cantlay et al., 2023). Young geese and ducks, for instance, learn feeding behavior through imitating experienced adults; such a process is driven through individual learning and cultural transmission (Giraldeau et al., 1994; Post et al., 2017). Therefore, these mechanisms can be fully comprehended by researching the neuroethology of birds.

Additionally, species such as mallard (*Anas platyrhynchos*) learn foraging behaviors in shallow water by social learning. This includes behaviors such as upending and purifying water for aquatic plants and animals (Li & Clarke, 2016; Mukherjee et al., 2021). Geese learn to graze on terrestrial vegetation by copying experienced adults. Such behaviors include crop and grass trimming. Some goose species, such as the greylag goose (*Anser anser*), learned how to exploit agricultural areas through a social learning process (Kotrschal et al., 1993; Mulder et al., 1995; Van Der Leeuw et al., 2003). Specialized foraging behaviors such as filter-feeding and trampling are prevalent among swans (Cygninae), and most are learned through social interactions. Bewick's swans (*Cygnus columbianus bewickii*) have a generation-old habit of trampling aquatic vegetation to find buried tubers (Klaassen et al., 2006; Nolet et al., 2014). Some swan species, such as the whistling swan (*Cygnus columbianus*), mastered filter-feeding on water life through imitation and observation (Bailey, 1974; Badzinski, 2003).

8.6. Same-Sex Pair Bonding

The family Anatidae has also been discovered for its social complexities, such as monogamous pair relationships and cooperative breeding behaviors, which theoretically could allow for same-sex pair bonds in specific situations. Male–male relationships among males of black swans (*Cygnus atratus*) engage in nest incubations and rearing juvenile offspring, defending territory boundaries and perform coordinated mating displays. In some cases, these pairs are able to outperform heterosexual pairs in terms of chick survival rates (Kraaijeveld et al., 2004). Greylag geese (*Anser anser*) also create long-term male–male friendships, with research finding such couples can endure for more than a decade, bringing mutual social and physiological benefits, including lower stress reactions and enhanced social standing within flocks (Wascher et al., 2008). Long-term same-sex pair relationships among duck species are rare because of their largely seasonal breeding behaviors, but female–female mounts and affiliative behaviors have actually been studied, especially in situations with skewed sex ratios (MacFarlane et al., 2010).

8.7. Response to Habitat Degradation

The Anatidae family is heavily impacted by habitat loss, particularly in inland lentic marshes and forests, and responds to habitat degradation in a variety of ways, including behavioral shifts, population dynamics, and conservation issues. Degradation of habitat, mostly due to loss and fragmentation, has an impact on Anatidae species range, reproductive success, and feeding habits. A number of variables, including regional distribution, habitat usage, and migration patterns, affect these changes. Different species react differently to habitat deterioration; some change their behaviors to adapt, while others become less numerous (Green, 1996).

8.8. Effect on Population Dynamics

In the Anatidae family, the high population density, widespread distribution, and migratory behavior are considered biological indicators of the wetlands. For example, the Anatidae family of the Bakhtegan wetlands has also significantly suffered from the destruction of their habitat, which includes decreased water levels and parching of the wet regions. Consequently, their populations and richness have decreased. Because of habitat loss and drying wetlands, the number of Anatidae in Iran's Bakhtegan wetlands has drastically decreased, from over 169,000 individuals in the late 20th century to just 6,623 in recent years (Hosseini Tayefeh et al., 2023). However, another study by Michot (1996) found that the Anatidae family, particularly ducks, showed no significant population changes in response to habitat degradation between 1969 and 1994, possibly due to increased habitat management efforts or a lack of a threshold where marsh loss impacts food availability.

9. Conclusion

The social and behavioral complexity of the Anatidae family is a remarkable evolutionary response to a variety of ecological challenges. These birds exhibit a high level of social intelligence and flexibility in their mating displays and nesting habits, including migration plans and novel social behaviors such as same-sex partnering and alloparenting. This understanding is crucial for us as we continue to study the Anatidae family. Their social learning and responsiveness to environmental cues present a fascinating opportunity to observe these interactions in a natural environment. This helps us understand the concept of gene–environment interaction. By gaining a deeper understanding of how genes and the environment influence these complex behaviors, we can develop more strategic and effective protocol to protect diverse species and their evolving environments. Moreover, understanding both proximate and ultimate mechanisms is crucial to gaining insight into animal behavior. This is particularly relevant given the ongoing challenges of habitat loss and climate change, which continue to alter natural habitats and ecosystems.

Declaration

The authors declare that there is no conflict of interest.

References

Adler, M. I. (2010). Sexual conflict in waterfowl: Why do females resist extrapair copulations? *Behavioral Ecology*, 21(1), 182–192.

Akesson, T. R., & Raveling, D. G. (1982). Behaviors associated with seasonal reproduction and long-term monogamy in Canada geese. *The Condor*, 84(2), 188–196.

Anderson, H. B., Godfrey, T., Woodin, S. J., & van der Wal, R. (2012). Finding food in a highly seasonal landscape: Where and how pink-footed geese *Anser brachyrhynchus* forage during the Arctic spring. *Journal of Avian Biology*, 43(5), 415–422. https://doi.org/10.1111/j.1600-048X.2012.05699.x

Andersson, M., Åhlund, M., & Waldeck, P. (2019). Brood parasitism, relatedness and sociality: A kinship role in female reproductive tactics. *Biological Reviews*, 94(1), 307–327. https://doi.org/10.1111/brv.12455

Ando, H., Ikeno, S., Narita, A., Komura, T., Takada, A., Isagi, Y. et al. (2022). Temporal and interspecific dietary variation in wintering ducks in agricultural landscapes. *Molecular Ecology*, 31(20), 5275–5289. https://doi.org/10.1111/mec.16584

Arzel, C., Elmberg, J., & Guillemain, M. (2006). Ecology of spring-migrating Anatidae: A review. *Journal of Ornithology*, 147(2), 167–184. https://doi.org/10.1007/s10336-006-0054-8

Avilova, K. V. (2018). Spatial organization of the epithelial structures in the bill tip organ of waterfowl (Anseriformes, Aves). *Biology Bulletin Reviews*, 8(3), 234–244. https://doi.org/10.1134/S2079086418030027

Ayaichia, F., Samraoui, F., Baaziz, N., Meziane, N., Samraoui, B., & Samraoui, B. (2018). Sitting ducks: Diet of wintering wildfowl in Lake Tonga, northeast Algeria. *Wetlands Ecology and Management*, 26(2), 231–243. https://doi.org/10.1007/s11273-017-9567-6

Badzinski, S. S. (2003). Dominance relations and agonistic behaviour of tundra Swans (*Cygnus columbianus columbianus*) during fall and spring migration. *Canadian Journal of Zoology*, 81(4), 727–733. https://doi.org/10.1139/z03-049

Baeza, E., Dessay, C., Wacrenier, N., Marche, G., & Le Bihan-Duval, E. (2001). Sexual dimorphism for growth in Muscovy ducks and changes in plasma hormones associated with genetic selection. *INRA Productions Animales*, 14(2), 119–127.

Baeza, E., Williams, J., Guémené, D., & Duclos, M. J. (2001). Sexual dimorphism for growth in Muscovy ducks and changes in insulin-like growth factor I (IGF-I), growth hormone (GH) and triiodothyronine (T3) plasma levels. *Reproduction Nutrition Development*, 41(2), 173–179.

Bailey, R. O. (1974). Hierarchy of waterfowl feeding with whistling swans. *The Auk*, 91(3), 488–493. https://doi.org/10.2307/4084468

Banks, E. M. (1984). Relations between aggressive social dominance and resource utilization in domestic fowl. *Political Psychology*, 5(4), 617–628. https://doi.org/10.2307/3791231

Bao, Q., Zhang, Y., Yao, Y., Luo, X., Zhao, W., Wang, J. et al. (2022). Characteristics of the mating behavior of domesticated geese from *Anser cygnoides* and *Anser anser*. *Animals*, 12(18), 2326. https://doi.org/10.3390/ani12182326

Barnes, G. G., & Thomas, V. G. (1987). Digestive organ morphology, diet, and guild structure of North American Anatidae. *Canadian Journal of Zoology*, 65(7), 1812–1817.

Barrett, L., Malecki, I., & Blache, D. (2019). Differences in pre-laying behavior between floor-laying and nest-laying Pekin ducks. *Animals*, 9(2), 40. https://doi.org/10.3390/ani9020040

Beauchamp, G. (1997). Determinants of intraspecific brood amalgamation in waterfowl. *The Auk*, 114(1), 11–21. https://doi.org/10.2307/4089061

Bennett, P. M., & Owens, I. P. F. (2002). Ecological basis of sexual dimorphism. *In: Sexual Selection and Conservation* (pp. 130–150). Oxford University Press. https://doi.org/10.1093/oso/9780198510888.003.0009

Blackwell, B. F., Seamans, T. W., DeVault, T. L., Lima, S. L., Pfeiffer, M. B., & Fernández-Juricic, E. (2019). Social information affects Canada goose alert and escape responses to vehicle approach: Implications for animal–vehicle collisions. *PeerJ*, 7, e8164. https://doi.org/10.7717/peerj.8164

Bowers, E. K. (2023). Coming of age in your local mating market: Just a numbers game? *Journal of Animal Ecology*, 92(5), 953–956. https://doi.org/10.1111/1365-2656.13923

Brazil, M. A. (1981). *The Behavioural Ecology of the Whooper Swan* (*Cygnus cygnus cygnus*). Doctoral dissertation, University of Stirling. https://dspace.stir.ac.uk/handle/1893/2395

Brazil, M. A. (2002). An addition to the diet of the whooper swan *Cygnus cygnus* from Japan. *Journal of The Yamashina Institute for Ornithology*, 33(2), 210–212.

Brennan, P. L. R., Clark, C. J., & Prum, R. O. (2010). Explosive eversion and functional morphology of the duck penis supports sexual conflict in waterfowl genitalia. *Proceedings of the Royal Society B: Biological Sciences*, 277(1686), 1309–1314.

Brennan, P. L. R., Prum, R. O., McCracken, K. G., Sorenson, M. D., Wilson, R. E., & Birkhead, T. R. (2007). Coevolution of male and female genital morphology in waterfowl. *PLOS ONE*, 2(5), e418. https://doi.org/10.1371/journal.pone.0000418

Briffa, M., & Elwood, R. W. (2001). Decision rules, energy metabolism and vigour of hermit-crab fights. *Proceedings of the Royal Society B: Biological Sciences*, 268(1476), 1841–1848. https://doi.org/10.1098/rspb.2001.1752

Brugger, C., & Taborsky, M. (1994). Male incubation and its effect on reproductive success in the black swan, *Cygnus atratus*. *Ethology*, 96(2), 138–146. https://doi.org/10.1111/j.1439-0310.1994.tb00889.x

Bruggers, R. L., & Jackson, W. B. (1977). Time budgets of mandarin ducks under semi-natural conditions. *Wildfowl*, 28, 7–14.

Bruun, M., & Smith, H. G. (2003). Landscape composition affects habitat use and foraging flight distances in breeding European starlings. *Biological Conservation*, 114(2), 179–187.

Calabuig, C. P., Green, A. J., Ferrer, M., Muriel, R., & Moreira, H. (2011). Sexual size dimorphism and sex determination by morphometric measurements in the Coscoroba Swan. *Studies on Neotropical Fauna and Environment*, 46(3), 177–184.

Cantlay, J. C., Martin, G. R., McClelland, S. C., Potier, S., O'Brien, M. F., Fernández-Juricic, E. et al. (2023). Binocular vision and foraging in ducks, geese and swans (Anatidae). *Proceedings of the Royal Society B: Biological Sciences*, 290, 20231213. https://doi.org/10.1098/rspb.2023.1213

Cao, L., Zhang, Y., Barter, M., & Lei, G. (2010). Anatidae in eastern China during the non-breeding season: Geographical distributions and protection status. *Biological Conservation*, 143(3), 650–659.

Carboneras, C. (1992). Family Anatidae (ducks, geese and swans). *In:* J. del Hoyo, A. Elliott, & J. Sargatal (Eds.), *Handbook of the Birds of the World: Vol. 1. Ostrich to Ducks* (pp. 536–628). Lynx Edicions.

Clayton, D. A. (1978). Socially facilitated behavior. *The Quarterly Review of Biology*, 53(4), 373–392.

Cooper, J. J., & Appleby, M. C. (1996). Individual variation in prelaying behaviour and the incidence of floor eggs. *British Poultry Science*, 37(2), 245–253.

Davis, E. S. (2002). Male reproductive tactics in the mallard, *Anas platyrhynchos*: Social and hormonal mechanisms. *Behavioral Ecology and Sociobiology*, 52(3), 224–231.

de Vries, A. C., Glasper, E. R., & Detillion, C. E. (2003). Social modulation of stress responses. *Physiology & Behavior*, 79(3), 399–407. https://doi.org/10.1016/S0031-9384(03)00152-5

Devries, J. H., Armstrong, L. M., Howerter, D. W., & Emery, R. B. (2023). Waterfowl distribution and productivity in the Prairie Pothole Region of Canada: Tools for conservation planning. *Wildlife Monographs*, 211(1). https://doi.org/10.1002/wmon.1074

Dirksen, S., Beekman, J. H., & Slagboom, T. H. (1991). Bewick's swans *Cygnus columbianus bewickii* in the Netherlands: Numbers, distribution and food choice during the wintering season. *Wildfowl*, 42, 228–237. https://wildfowl.wwt.org.uk/index.php/wildfowl/article/download/1398/pdf_44

DuBour, A. J. (2019). *Feeding ecology of scaup ducklings across a heterogeneous boreal wetland landscape.* Master's thesis, University of Alaska Fairbanks. https://scholarworks.alaska.edu/handle/11122/10617

Dunn, P. O., Whittingham, L. A., & Pitcher, T. E. (2001). Mating systems, sperm competition, and the evolution of sexual dimorphism in birds. *Evolution*, 55(1), 161–175. https://doi.org/10.1554/0014-3820(2001)055[0161:msscat]2.0.co;2

Eadie, J. M., & Lyon, B. E. (2011). The relative role of relatives in conspecific brood parasitism. *Molecular Ecology, 20*(24), 5114–5118. https://doi.org/10.1111/j.1365-294X.2011.05320.x

Fan, Y., Zhou, L., Cheng, L., Song, Y., & Xu, W. (2020). Foraging behavior of the Greater White-fronted Goose (*Anser albifrons*) wintering at Shengjin Lake: Diet shifts and habitat use. *Avian Research*, 11(1), 1–9. https://doi.org/10.1186/s40657-020-0189-y

Felsenstein J. (1985). Confidence limits on phylogenies: An approach using the bootstrap. *Evolution*, 39, 783–791.

Folliot, B., Caizergues, A., Barbotin, A., & Guillemain, M. (2017). Environmental and individual correlates of common pochard (*Aythya ferina*) nesting success. *European Journal of Wildlife Research*, 63(4), 69. https://doi.org/10.1007/s10344-017-1126-1

Forkman, B., & Haskell, M. J. (2004). The maintenance of stable dominance hierarchies and the pattern of aggression: Support for the suppression hypothesis. *Ethology,* 110(9), 737–744. https://doi.org/10.1111/j.1439-0310.2004.01009.x

Fox, A. D. (1993). Pre-nesting feeding selectivity of Pink-footed Geese *Anser brachyrhynchus* in artificial grasslands. *Ibis*, 135(4), 417–423. https://doi.org/10.1111/j.1474-919x.1993.tb02114.x

Fox, T., Eide, N. E., Bergersen, E., & Madsen, J. (2009). Resource partitioning in sympatric arctic-breeding geese: Summer habitat use, spatial and dietary overlap of Barnacle and Pink-footed Geese in Svalbard. *Ibis*, 151(1), 122–133. https://doi.org/10.1111/j.1474-919X.2008.00892.x

Fox, T., Francis, I. S., & Bergersen, E. (2006). Diet and habitat use of Svalbard Pink-footed Geese *Anser brachyrhynchus* during arrival and pre-breeding periods in Adventdalen. *Ardea*, 94(3), 691–699.

Frank, T., Probst, A., Konto, H. E., & Walter, I. (2007). The syrinx of the male mallard (*Anas platyrhynchos*): Special anatomical features. *Anatomia, Histologia, Embryologia*, 36(2), 121–126.

Freire, R., Appleby, M. C., & Hughes, B. O. (1997). Assessment of pre-laying motivation in the domestic hen using social interaction. *Animal Behaviour*, 54(2), 313–319.

Friedman, N. R., & Remeš, V. (2016). Global geographic patterns of sexual size dimorphism in birds: Support for a latitudinal trend? *Ecography*, 39(1), 17–25. https://doi.org/10.1111/ecog.01531

Frigerio, D., Weiss, B., & Kotrschal, K. (2001). Spatial proximity among adult siblings in greylag geese (*Anser anser*): Evidence for female bonding? *Acta Ethologica*, 3(2), 121–125. https://doi.org/10.1007/s102110000028

Gauthier, G. (1988). Territorial behaviour, forced copulations and mixed reproductive strategy in ducks. *Wildfowl*, 39, 102–114.

Giraldeau, L.-A., Caraco, T., & Valone, T. J. (1994). Social foraging: Individual learning and cultural transmission of innovations. *Behavioral Ecology*, 5(1), 35–43. https://doi.org/10.1093/beheco/5.1.35

Gluckman, T.-L. (2014). Pathways to elaboration of sexual dimorphism in bird plumage patterns. *Biological Journal of the Linnean Society*, 111(2), 262–273. https://doi.org/10.1111/bij.12211

Green, A. J. (1996). Analyses of globally threatened Anatidae in relation to threats, distribution, migration patterns, and habitat use. *Conservation Biology*, 10(5), 1435–1445. https://doi.org/10.1046/j.1523-1739.1996.10051435.x

Green, A. J. (2000). The scaling and selection of sexually dimorphic characters: An example using the marbled teal. *Journal of Avian Biology*, 31(3), 345–350. https://doi.org/10.1034/j.1600-048X.2000.310310.x

Greenberg, R., Etterson, M. A., & Danner, R. M. (2013). Seasonal dimorphism in the horny bills of sparrows. *Ecology and Evolution*, 3(2), 389–398. https://doi.org/10.1002/ece3.474

Gumułka, M., & Rozenboim, I. (2013). Mating activity of domestic geese ganders (*Anser anser* f. *domesticus*) during breeding period in relation to age, testosterone and thyroid hormones. *Animal Reproduction Science*, 142(3), 183–190. https://doi.org/10.1016/j.anireprosci.2013.09.021

Guo, X., Hasenjager, M. J., Fefferman, N. H., & Pinter-Wollman, N. (2024). Social interactions are impacted by food availability and group size. *Biology Open*, 13, bio060422. https://doi.org/10.1242/bio.060422

Hammell, G., Singer, H. V., & Armstrong, L. M. (2022). Comparative reproductive parameters of sympatric lesser scaup (*Aythya affinis*) and ring-necked duck (*Aythya collaris*) in parkland Manitoba. *Canadian Field-Naturalist*, 135(3), 278–292. https://doi.org/10.22621/cfn.v135i3.2507

Hario, M., Koljonen, M.-L., & Rintala, J. (2012). Kin structure and choice of brood care in a common eider (*Somateria m. mollissima*) population. *Journal of Ornithology*, 153(3), 963–973. https://doi.org/10.1007/s10336-012-0825-3

Holekamp, K. E., & Strauss, E. D. (2016). Aggression and dominance: An interdisciplinary overview. *Current Opinion in Behavioral Sciences*, 12, 44–51. https://doi.org/10.1016/j.cobeha.2016.09.005

Holopainen, S., Arzel, C., Dessborn, L., Elmberg, J., Gunnarsson, G., Nummi, P. et al. (2015). Habitat use in ducks breeding in boreal freshwater wetlands: A review. *European Journal of Wildlife Research*, 61(3), 339–363. https://doi.org/10.1007/s10344-015-0921-9

Holopainen, S., Miettinen, E., Väänänen, V.-M., Nummi, P., & Pöysä, H. (2024). Balancing between predation risk and food by boreal breeding ducks. *Ecology and Evolution*, 14, e11011. https://doi.org/10.1002/ece3.11011

Horrocks, N., Perrins, C., & Charmantier, A. (2009). Seasonal changes in male and female bill knob size in the mute swan *Cygnus olor*. *Journal of Avian Biology*, 40(5), 511–519. https://doi.org/10.1111/j.1600-048X.2009.04668.x

Hosseini Tayefeh, F., Ghasemi, S., Izadian, M., & Ghorbanzadeh, S. G. (2023). Wintering population fluctuation of the Anatidae family in the Bakhtegan Basin Wetlands, Iran. *Environmental Sciences*, 21(2), 205–228.

Johns, D. W., Devries, J. H., & Clark, R. W. (2024). Survival of northern pintail ducklings in grasslands and croplands: Implications for fall-seeded cereals. *Journal of Wildlife Management*, 88(4). https://doi.org/10.1002/jwmg.22558

Johnsgard, P. (2020). *The North American Swans: Their Biology and Conservation*. Lulu.com.

Johnsgard, P. A. (1960). Comparative behaviour of the Anatidae and its evolutionary implications. *Wildfowl*, 11, 31–45. https://wildfowl.wwt.org.uk/index.php/wildfowl/article/download/117/117

Johnsgard, P. A. (1961). Tracheal anatomy of the Anatidae and its taxonomic significance. *Wildfowl*, 12, 58–69.

Johnsgard, P. A. (1962). Evolutionary trends in the behaviour and morphology of the Anatidae. *Wildfowl*, 13, 130–148. https://wildfowl.wwt.org.uk/index.php/wildfowl/article/download/185/185

Johnsgard, P. A. (1963). Behavioral isolating mechanisms in the family Anatidae. https://digitalcommons.unl.edu/cgi/viewcontent.cgi?article=1023&context=johnsgard

Johnsgard, P. A. (1971). Observations on sound production in the Anatidae. *Wildfowl*, 22, 46–59. https://wildfowl.wwt.org.uk/index.php/wildfowl/article/download/420/420

Johnsgard, P. A. (2016). *Swans: Their Biology and Natural History*. https://digitalcommons.unl.edu/zeabook/38/

Johnson, W. C., & Collinge, S. K. (2004). Landscape effects on black-tailed prairie dog colonies. *Biological Conservation*, 115(4), 487–497. https://doi.org/10.1016/S0006-3207(03)00172-6

Johnson, W. P., Schmidt, P. M., & Taylor, D. P. (2014). Foraging flight distances of wintering ducks and geese: A review. *Avian Conservation and Ecology*, 9(2), 2. https://doi.org/10.5751/ACE-00663-090202

Kalmbach, E. (2006). Why do goose parents adopt unrelated goslings? A review of hypotheses and empirical evidence, and new research questions. *Ibis*, 148(1), 66–78. https://doi.org/10.1111/j.1474-919x.2006.00496.x

Kalmbach, E., van der Aa, P., & Komdeur, J. (2005). Adoption as a gosling strategy to obtain better parental care? Experimental evidence for gosling choice and age-dependency of adoption in greylag geese. *Behaviour*, 142(11), 1515–1533. https://doi.org/10.1163/156853905774831909

Kissane, Z. (2019). *Temporal and spatial use of resources by black swans (Cygnus atratus) on the Vasse-Wonnerup Wetlands*. Master's thesis, Murdoch University. Murdoch University Research Repository. https://researchrepository.murdoch.edu.au/id/eprint/54501/

Klaassen, R. H. G., Nolet, B. A., & Bankert, D. (2006). The influence of social interactions on the foraging path of Bewick's swans (*Cygnus columbianus bewickii*). *Ardea*, 94(3), 477–484. https://pure.knaw.nl/portal/en/publications/the-influence-of-social-interactions-on-the-foraging-path-of-bewi

Kleindorfer, S., Krupka, M. A., Katsis, A. C., Frigerio, D., & Common, L. K. (2024). Aggressiveness predicts dominance rank in greylag geese: Mirror tests and agonistic interactions. *Royal Society Open Science*, 11, 231686. https://doi.org/10.1098/rsos.231686

Kotrschal, K., Hemetsberger, J., & Dittami, J. (1993). Food exploitation by a winter flock of greylag geese: Behavioral dynamics, competition and social status. *Behavioral Ecology and Sociobiology*, 33(5), 289–295. https://doi.org/10.1007/BF00172926

Kouzov, S. A., Kravchuk, A. V., Koptseva, E. M., Gubelit, Y. I., Zaynagutdinova, E., & Abakumov, E. V. (2024). Ecological and phylogenetic aspects of the spring diet of three palaearctic species of swans. *BMC Ecology and Evolution*, 24, 22. https://doi.org/10.1186/s12862-024-02204-7

Kraaijeveld, K. (2005). Black swans *Cygnus atratus* adopt related cygnets. *Ardea*, 93(2), 163–169.

Kraaijeveld, K., Carew, P. J., Billing, T., Adcock, G. J., & Mulder, R. A. (2004). Extra-pair paternity does not result in differential sexual selection in the mutually ornamented black swan (*Cygnus atratus*). *Molecular Ecology*, 13(6), 1625–1633. https://doi.org/10.1111/j.1365-294x.2004.02158.x

Kumar S., Stecher G., Suleski M., Sanderford M., Sharma S., & Tamura K. (2024). Molecular Evolutionary Genetics Analysis Version 12 for adaptive and green computing. *Molecular Biology and Evolution*, 41, 1–9.

Kurk, C. D. (2008). *The Bill of Evolution: Trophic Adaptations in Anseriform Birds.* Doctoral dissertation, Leiden University. Leiden University Repository. https://scholarlypublications.universiteitleiden.nl/access/item%3A2923002/view

Lank, D. B., & Rockwell, R. F. (1995). Pairing, mating, and parental care. *In:* D. B. Lank & R. F. Rockwell (Eds.), *Avian Biology* (Vol. 6, pp. 81–98). Oxford University Press.

Lee, J.-Y., Nam, H.-K., Park, J.-Y., Kang, S.-G., Batbayar, N., Kim, D.-W. et al. (2023). Migration routes and differences in migration strategies of whooper swans between spring and autumn. *Avian Research*, 14, 100113. https://doi.org/10.1016/j.avrs.2023.100113

Legagneux, P., Inchausti, P., Bourguemestre, F., Latraube, F., & Bretagnolle, V. (2009). Effect of predation risk, body size, and habitat characteristics on emigration decisions in mallards. *Behavioral Ecology*, 20(1), 186–194. https://doi.org/10.1093/beheco/arn133

Li, Z., & Clarke, J. A. (2016). The craniolingual morphology of waterfowl (Aves, Anseriformes) and its relationship with feeding mode revealed through contrast-enhanced X-ray computed tomography and 2D morphometrics. *Evolutionary Biology*, 43(1), 12–25. https://doi.org/10.1007/S11692-015-9345-4

Liu, H., Yang, Z., He, Y., Yang, Q., Tang, Q., Yang, Z. H. et al. (2022). Metabolic profiling reveals that the olfactory cues in the duck uropygial gland potentially act as sex pheromones. *Animals*, 12(4), 413. https://doi.org/10.3390/ani12040413

Liu, L., Du, C., Sun, Y., Li, W., Zhang, J., Cao, L. et al. (2022). Spring diet and energy intake of tundra swan (*Cygnus columbianus*) at the Yellow River National Wetland in Baotou, China. *PeerJ*, 10, e13113. https://doi.org/10.7717/peerj.13113

Loba-Messet, L. E. G. A., Kakou, I. B., Yeo, D. M., Soro, D., & Adepo-Gourene, B. A. (2023). Sexual dimorphism of ducks (*Cairina moschata*, Linnaeus, 1758) reared in the southern forest of Côte d'Ivoire. *GSC Biological and Pharmaceutical Sciences*, 22(2), 213–221. https://doi.org/10.30574/gscbps.2023.22.2.0058

Lurz, P. W. W., Shirley, M. D. F., Rushton, S. P., & Sanderson, R. A. (2002). Modelling the consequences of duck migration patterns on the genetic diversity of aquatic organisms: A first step towards a predictive tool for wetland management. *Acta Oecologica*, 23(3), 205–212. https://doi.org/10.1016/S1146-609X(02)01151-7

MacFarlane, G. R., Blomberg, S. P., & Vasey, P. L. (2010). Homosexual behaviour in birds: Frequency of expression is related to parental care disparity between the sexes. *Animal Behaviour*, 80(2), 375–390. https://doi.org/10.1016/j.anbehav.2010.05.006

Mathiasson, S. (1973). A moulting population of non-breeding mute swans with special reference to flight-feather moult, feeding ecology and habitat selection. *Wildfowl*, 24, 43–53. https://wildfowl.wwt.org.uk/index.php/wildfowl/article/download/452/452

Maxson, S. J., & Riggs, M. R. (1996). Habitat use and nest success of overwater nesting ducks in west-central Minnesota. *Journal of Wildlife Management*, 60(1), 108–119. https://doi.org/10.2307/3802045

McKelvey, R. W. (1981). Some aspects of the winter feeding ecology of trumpeter swans at Port Alberni and Comox Harbour, British Columbia. Master's thesis, Simon Fraser University. http://summit.sfu.ca/item/3945

McKinney, F. (1965). The comfort movements of Anatidae. *Behaviour*, 25(1–2), 120–220. https://doi.org/10.1163/156853965X00138

McKinney, F., Derrickson, S. R., & Mineau, P. (1983). Forced copulation in waterfowl. *Behaviour*, 86(3), 250–293. https://doi.org/10.1163/156853983X00676

McNamara, J. M., Welham, R. K., & Houston, A. I. (1998). The timing of migration within the context of an annual routine. *Journal of Avian Biology*, 29(4), 416–423. https://doi.org/10.2307/3677160

Michot, T. C. (1996). Marsh loss in coastal Louisiana: Implications for management of North American Anatidae. *Environmental Management*, 13(3), 941–957.

Middleton, B. A., & van der Valk, A. G. (1987). The food habits of greylag and barheaded geese in the Keoladeo National Park, India. *Wildfowl*, 38, 94–102. https://lib.dr.iastate.edu/cgi/viewcontent.cgi?article=1088&context=bot_pubs

Mishkind, D., Lester, M., Tabin, C. J., & Goller, F. (2025). Courtship vocalizations in male ducks: Spectral composition and resonance of the syringeal bulla. *bioRxiv*, 2025-01. https://doi.org/10.1101/2025.01.13.632779

Monier, S. A. (2024). Social interactions and information use by foraging seabirds. *Biological Reviews*. https://doi.org/10.1111/brv.13089

Mukherjee, A., Bandyopadhyay, A., Pal, S., & Mukhopadhyay, S. K. (2021). Foraging habitats and foraging techniques of five wintering Anatidae waterfowl in light of genetic distances. *Russian Journal of Ecology*, 52, 567–577. https://doi.org/10.1134/S1067413621060103

Mulder, R. S., Williams, T. D., & Cooke, F. (1995). Dominance, brood size and foraging behavior during brood-rearing in the lesser snow goose: An experimental study. *The Condor*, 97(1), 99–106. https://doi.org/10.2307/1368987

Nielsen, C. L. R., Semel, B., Sherman, P. W., Westneat, D. F., & Parker, P. G. (2006). Host–parasite relatedness in wood ducks: Patterns of kinship and parasite success. *Behavioral Ecology*, 17(3), 491–496. https://doi.org/10.1093/beheco/arj057

Nolet, B. A., Gyimesi, A., & van Lith, B. (2014). Lower foraging efficiency of offspring constrains use of optimal habitat in birds with extended parental care. *Ibis*, 156(2), 387–394. https://doi.org/10.1111/ibi.12137

Norambuena, C., Jélvez, M., Mena, M., & Ratto, M. (2023). Seasonality in the feeding ecology of black-necked swans (*Cygnus melancoryphus*) in a temperate wetland of southern Chile. *Gayana*, 87(1), 10–22. https://doi.org/10.4067/s0717-65382023000100010

Nummi, P., & Pöysä, H. (1995). Habitat use by different-aged duck broods and juvenile ducks. *Wildlife Biology*, 1(3), 181–187. https://doi.org/10.2981/wlb.1995.017

O'Neal, B. J., Stafford, J. D., Larkin, R. P., & Michel, E. S. (2018). The effect of weather on the decision to migrate from stopover sites by autumn-migrating ducks. *Movement Ecology*, 6(1), 23. https://doi.org/10.1186/s40462-018-0141-5

Oguntunji, A. O., & Ayorinde, K. L. (2014). Sexual size dimorphism and sex determination by morphometric measurements in locally adapted Muscovy duck (*Cairina moschata*) in Nigeria. *Acta Agriculturae Slovenica*, 104(1), 15–24.

Ota, N., Gahr, M., & Soma, M. (2015). Tap dancing birds: The multimodal mutual courtship display of males and females in a socially monogamous songbird. *Scientific Reports*, 5(1), 16614. https://doi.org/10.1038/srep16614

Owen, M. (1996). Review of the migration strategies of the Anatidae: Challenges for conservation. *Gibier Faune Sauvage*, 13(2), 123–139.

Pagel, M., & Dawkins, M. (1997). Peck orders and group size in laying hens: Future contracts for non-aggression. *Behavioural Processes*, 40(1), 13–25. https://doi.org/10.1016/S0376-6357(96)00768-5

Peake, T. M. (2005). Eavesdropping in communication networks. *In:* P. K. McGregor (Ed.), *Animal Communication Networks* (pp. 13–37). Cambridge University Press.

Pecsics, T., Laczi, M., Nagy, G., & Csörgő, T. (2017). The cranial morphometrics of the wildfowl (Anatidae). *Ornis Hungarica*, 25(1), 44–57. https://doi.org/10.1515/orhu-2017-0004

Peiman, K. S., & Robinson, B. W. (2010). Ecology and evolution of resource-related heterospecific aggression. *The Quarterly Review of Biology*, 85(2), 133–158. https://doi.org/10.1086/652374

Pezzanite, B. M., Rockwell, R. F., Davies, J. C., Loonen, M. J. J. E., & Seguin, R. J. (2005). Has habitat degradation affected foraging behaviour and reproductive success of lesser snow geese (*Chen caerulescens caerulescens*)? *Ecoscience*, 12(4), 439–446. https://doi.org/10.2980/i1195-6860-12-4-439.1

Pfennig, K. S., & Hurlbert, A. H. (2012). Heterospecific interactions and the proliferation of sexually dimorphic traits. *Current Zoology*, 58(3), 450–459. https://doi.org/10.1093/czoolo/58.3.453

Poisbleau, M., Fritz, H., Guillemain, M., & Lacroix, A. (2005). Testosterone and linear social dominance status in captive male dabbling ducks in winter. *Ethology*, 111(5), 493–509. https://doi.org/10.1111/j.1439-0310.2005.01092.x

Post, D. J. van der, Franz, M., & Laland, K. N. (2017). The evolution of social learning mechanisms and cultural phenomena in group foragers. *BMC Evolutionary Biology*, 17(1), 49. https://doi.org/10.1186/s12862-017-0889-z

Price, T. D. (1998). An analysis of sexual size dimorphism in goose populations across geographic gradients. *Biological Journal of the Linnean Society*, 64(4), 425–445. https://doi.org/10.1111/j.1095-8312.1998.tb00343.x

Rees, E. C., Kirby, J. S., & Gilburn, A. S. (1997). Site selection by swans wintering in Britain and Ireland: The importance of habitat and geographic location. *Ibis*, 139(2), 337–352. https://doi.org/10.1111/j.1474-919x.1997.tb04633.x

Reeve, H. K., & Pfennig, D. W. (2003). Evolution of sexual dimorphism: The roles of selection and constraint. *Trends in Ecology & Evolution*, 18(9), 493–498. https://doi.org/10.1016/S0169-5347(03)00218-8

Rockwell, R. F., Witte, C. R., Jefferies, R. L., & Weatherhead, P. J. (2003). Response of nesting Savannah sparrows to 25 years of habitat change in a snow goose colony. *Écoscience*, 10(1), 33–37. https://doi.org/10.2980/1195-6860(2003)010<0033:ronssp>2.0.co;2

Rüppell, V. W. (1933). Physiologie und Akustik der Vogelstimme. *Journal für Ornithologie*, 81, 433–542.

Saitou N., & Nei M. (1987). The neighbor-joining method: A new method for reconstructing phylogenetic trees. *Molecular Biology and Evolution*, 4, 406–425.

Scheiber, I. B. R., Kralj, S., & Kotrschal, K. (2005). Sampling effort/frequency necessary to infer individual acute stress responses from fecal analysis in greylag geese (*Anser*

anser). *Annals of the New York Academy of Sciences*, 1046(1), 154–167. https://doi.org/10.1196/annals.1343.012

Scheiber, I. B. R., Weiß, B. M., Frigerio, D., & Kotrschal, K. (2005). Active and passive social support in families of greylag geese (*Anser anser*). *Behaviour*, 142(11), 1535–1557. https://doi.org/10.1163/156853905774831873

Scheiber, I. B. R., Weiß, B. M., Hirschenhauser, K., Nedelcu, I. T., Wascher, C. A. F., & Kotrschal, K. (2008). Does 'relationship intelligence' make big brains in birds? *Open Biology Journal*, 1, 6–8.

Scherer, S., & Hilsberg, T. (1982). Hybridisierung und Verwandtschaftsgrade innerhalb der Anatidae—eine systematische und evolutionstheoretische Betrachtung. *Journal of Ornithology*, 123(4), 357–380.

Scott, D. K. (1980). Functional aspects of the pair bond in winter in Bewick's swans (*Cygnus columbianus bewickii*). *Behavioral Ecology and Sociobiology*, 7(4), 323–327. https://doi.org/10.1007/BF00300673

Sears, J. N. E. (1989). Feeding activity and body condition of mute swans *Cygnus olor* in rural and urban areas of a lowland river system. *Wildfowl*, 40, 88–98. https://wildfowl.wwt.org.uk/index.php/wildfowl/article/download/812/812

Selander, R. K. (2017). Sexual selection and dimorphism in birds. *In: Sexual Selection and Dimorphism in Birds* (pp. 180–230). Routledge. https://doi.org/10.4324/9781315129266-8

Shurtleff, L. L., & Savage, C. (1996). *The Wood Duck and the Mandarin: The Northern Wood Ducks*. University of California Press.

Shutler, D., Gloutney, M. L., & Clark, R. G. (1998). Body mass, energetic constraints, and duck nesting ecology. *Canadian Journal of Zoology*, 76(10), 1805–1814. https://doi.org/10.1139/Z98-126

Silk, M. J., Cant, M. A., Cafazzo, S., Natoli, E., & McDonald, R. A. (2019). Elevated aggression is associated with uncertainty in a network of dog dominance interactions. *Proceedings of the Royal Society B: Biological Sciences*, 286(1906), 20190536. https://doi.org/10.1098/rspb.2019.0536

Smith, F. M., West, N. H., & Jones, D. R. (2000). The cardiovascular system. *In:* G. C. Whittow (Ed.), *Sturkie's Avian Physiology* (5th ed., pp. 141–232). Academic Press.

Soni, G. (2022). Disentangling the causes of age-assortative mating in bird populations with contrasting life-history strategies. *Journal of Animal Ecology*. https://doi.org/10.1111/1365-2656.13851

Sorenson, L. G., & Derrickson, S. R. (1994). Sexual selection in the northern pintail (*Anas acuta*): The importance of female choice versus male–male competition in the evolution of sexually-selected traits. *Behavioral Ecology and Sociobiology*, 35(6), 389–400. https://doi.org/10.1007/BF00165841

Steen, K. A., Therkildsen, O. R., Karstoft, H., & Green, O. (2012). A vocal-based analytical method for goose behaviour recognition. *Sensors*, 12(3), 3773–3788. https://doi.org/10.3390/s120303773

Tamura K., Nei M., & Kumar S. (2004). Prospects for inferring very large phylogenies by using the neighbor-joining method. *Proceedings of the National Academy of Sciences* (USA), 101, 11030–11035.

Therres, G. D., & Brinkler, D. F. (2004). Mute Swan interactions with other birds in Chesapeake Bay. *In:* M. C. Perry (Ed.), *Mute Swans and Their Chesapeake Bay Habitats: Proceedings of a Symposium* (pp. 43–46). U.S. Geological Survey.

Tibbetts, E. A., Pardo-Sanchez, J., & Weise, C. (2022). The establishment and maintenance of dominance hierarchies. *Philosophical Transactions of the Royal Society B*, 377(1845). https://doi.org/10.1098/rstb.2020.0450

Traylor, J. J., Alisauskas, R. T., & Kehoe, F. P. (2008). Ecological correlates of duckling adoption among white-winged scoters (*Melanitta fusca*): Strategy, epiphenomenon, or combination? *Behavioral Ecology and Sociobiology*, 62(7), 1085–1097. https://doi.org/10.1007/s00265-007-0536-6

Tu, Y. G., Yu, C. H., Huang, X. F., Shan, J., Sun, Z., & Wang, Z. (2009). Distribution and population of the overwintering Anatidae waterfowl in the Poyang Lake. *Acta Agriculturae Universitis Jiangxiensis*, 31(4), 760–764.

Van Der Leeuw, A. H. J., Kurk, K., Snelderwaard, P. C., Bout, R. G., & Berkhoudt, H. (2003). Conflicting demands on the trophic system of Anseriformes and their evolutionary implications. *Animal Biology*, 53(3), 259–301. https://doi.org/10.1163/157075603322539453

Wascher, C. A. F., Arnold, W., & Kotrschal, K. (2008). Heart rate modulation by social contexts in greylag geese (*Anser anser*). *Journal of Comparative Psychology*, 122(1), 100–107. https://doi.org/10.1037/0735-7036.122.1.100

Wascher, C. A. F., Scheiber, I. B. R., & Kotrschal, K. (2008). Heart rate responses to agonistic encounters in greylag geese (*Anser anser*). *Animal Behaviour*, 75(1), 593–603. https://doi.org/10.1016/j.anbehav.2007.06.011

Wascher, C. A., Scheiber, I. B., & Kotrschal, K. (2008). Heart rate modulation in bystanding geese watching social and non-social events. *Proceedings of the Royal Society B: Biological Sciences*, 275(1643), 1653–1659.

Weidinger, K., & van Franeker, J. A. (1998). Applicability of external measurements to sexing of the Cape petrel *Daption capense* at within-pair, within-population and between-population scales. *Journal of Zoology*, 245(4), 473–482.

Williams, M. (1996). Conservation dilemmas with island wildfowl (Anatidae): A perspective based on the New Zealand experience. *Biodiversity and Conservation*, 13(2), 849–865.

Włodarczyk, R., & Minias, P. (2015). Division of parental duties confirms a need for bi-parental care in a precocial bird, the mute swan *Cygnus olor*. *Animal Biology*, 65(2), 163–176. https://doi.org/10.1163/15707563-00002468

Włodarczyk, R., Minias, P., & Janiszewski, T. (2016). Size does not matter: No evidence for assortative mating by body size in the mute swan *Cygnus olor*. *Polish Journal of Ecology*, 64(2), 296–301. https://doi.org/10.3161/15052249PJE2016.64.2.013

Wood, K. A., Ham, P., Scales, J., Wyeth, E., & Rose, P. E. (2020). Aggressive behavioural interactions between swans (*Cygnus* spp.) and other waterbirds during winter: A webcam-based study. *Avian Research*, 11(1), 30. https://doi.org/10.1186/s40657-020-00216-7

Wood, K. A., Stillman, R. A., & Goss-Custard, J. D. (2015). The effect of kleptoparasite and host numbers on the risk of food-stealing in an avian assemblage. *Journal of Avian Biology*, 46(6), 589–596.

Xia, S., Duan, H., Wu, Y., Su, Y., Wang, B., Sun, G. et al. (2024). Study on the impact of reclamation and development of Yancheng coastal wetlands on the spatio-temporal evolution of wintering Anatidae habitat. *Ecological Indicators*, 161, 111892. https://doi.org/10.1016/j.ecolind.2024.111892

Yakubu, A., Ogah, D. M., & Idahor, K. O. (2009). Discriminant analysis of sexual dimorphism in morphological traits of African Muscovy ducks. *International Journal of Morphology*, 27(3), 821–825.

Yetter, A. P., Stafford, J. D., Hine, C. S., Bowyer, M. W., Havera, S. P., & Horath, M. M. (2009). Nesting biology of mallards in west-central Illinois. *Illinois Natural History Survey Bulletin*, 39(1), 1–38. https://doi.org/10.21900/J.INHS.V39.95

Yu, C., Zhou, L., Mahtab, N., Fan, S., & Song, Y. (2020). Microhabitat variables explain patch switching by wintering Bewick's swans through giving-up net energy intake rates. *Environmental Science and Pollution Research*, 27(15), 18843–18852. https://doi.org/10.1007/s11356-020-08343-w

Zelenkov, N. V. (2018). The earliest Asian duck (Anseriformes: *Romainvillia*) and the origin of Anatidae. *Doklady Biological Sciences*, 483(1), 225–227.

Zelenkov, N. V. (2020). Cenozoic evolution of Eurasian anatids (Aves: Anatidae s. l.). *Biology Bulletin Reviews*, 10(5), 417–426.

Zeng, Q., Reid, J., Saintilan, N., Colloff, M. J., Lei, G., Wen, L. et al. (2019). Contrasting diversity patterns of breeding Anatidae in the Northern and Southern Hemispheres. *Ecology and Evolution*, 9(17), 9990–10003.

Zhang, Y., Yao, Y., Wang, M. M., Yang, Y. Z., Gu, T. T., Cao, Z. F. et al. (2019). Comparisons in geese of the courtship, mating behaviors and fertility of the Carlos and Sichuan breeds and the breed crosses. *Animal Reproduction Science*, 204, 86–94. https://doi.org/10.1016/j.anireprosci.2019.03.008

Zulandt, T., Zulandt-Schneider, R. A., & Moore, P. A. (2008). Observing agonistic interactions alters subsequent fighting dynamics in the crayfish, *Orconectes rusticus*. *Animal Behaviour*, 75(1), 13–20. https://doi.org/10.1016/j.anbehav.2007.04.017

CHAPTER
5

Foraging Ecology of Anatidae Family

Poorva Meghwal, Utsav Jain and Vijay Kumar Koli*
Wildlife Research Laboratory, Department of Zoology, Mohanlal Sukhadia University, Udaipur – 313001, Rajasthan, India

1. Introduction

Food is the basic need of all living beings, and for waterbirds, it often determines their distribution, aggregation, and diversity in community structure. Food availability has a substantial impact on physiology, reproductive success, and survival (Beerens et al., 2011; Pearse et al., 2011). According to optimal foraging theory, the most efficient and advantageous habitat patches for foraging are always chosen in order to maximize net energy gain by minimizing energy costs and increasing energy intake (Páez et al., 2018). Wetlands provide a wide range of food, including fish, insects, mollusks, crustaceans, aquatic plants, and seeds, and a number of waterbird species have evolved different foraging strategies to obtain them (Hafner et al., 1986; Weller, 1999); however, these strategies can vary depending on prey abundance and spatiotemporal availability (Holm, 2002; Kuwae et al., 2010). In some situations, such as semi-arid and monsoonal regions, where water levels change with season and rainfall, the relationship between food availability and waterbirds becomes visible. When wetlands are full, invertebrates and fish thrive, attracting species that feed them. However, as the water recedes and shallow mudflats emerge, waders and dabbling ducks that graze on benthic invertebrates and vegetation have plenty of food. Seasonal fluctuations in water and food availability often shift pattern of waterbirds diversity and abundance (Shi et al., 2024). Food availability becomes

*Corresponding author: vijaykoli87@yahoo.in

DOI: 10.1201/9781003669784-5

increasingly important for migratory birds that rely on wetlands as stopover sites or wintering grounds. Wetlands with rich food provide the energy required for their long journey, whereas poor habitat poses a threat to their survival (Arzel et al., 2006; Sundar & Kittur, 2013; Donnelly et al., 2019).

The Anatidae (Aves: Anseriformes) is a family of waterbirds that includes ducks, geese, and swans. The family comprises diverse groups of waterfowl distributed all over the world excluding Antarctica (Johnsgard, 2010). They rely extensively on wetlands for food supplies such as plants, grasses, seeds, small fish, and invertebrates, which influence their diversity and abundance. Anatidae are extremely adaptive foragers with unique specializations to exploit different dietary niches. Members of this family are mostly herbivorous (such as geese and some species of ducks), but others can be classified as carnivores (some mergansers) or omnivores (dabbling ducks and mallards) during specific stages of development. Dabbling ducks feed seeds, aquatic vegetation, and small vertebrates in shallow waters, but diving ducks hunt submerged plants, mollusks, and fish in deeper areas. Geese and swans frequently graze on grasses and agricultural crops near wetlands (Barnes & Thomas, 1987; Johnsgard, 2010; Mukherjee et al., 2023).

2. Morphological Adaptations to Foraging

Anatidae members are masters of wetland life, with skulls, bills, tongues, legs, and digestive systems that evolved to match unique dietary requirements. These adaptations have also specialized them in various diets and reduced competition among species, allowing them to cohabit in similar wetland habits (Barnes & Thomas, 1987; Mukherjee et al., 2023).

2.1. Adaptations in Skull

Foraging efficiency is functionally related to cranial morphology. Herbivorous species (such as grazers—geese and swans) have enormous skulls with an enlarged quadrato-mandibular surface, which maximizes the efficacy of jaw-closure muscles required for grazing. Filter-feeding species such as lesser scaup and northern shoveler have somewhat elongated skulls (Li & Clarke, 2016). Omnivorous dabbling ducks have short and high skulls, but piscivorous species have elongated and gracile skulls. Cranial morphology in mergansers is influenced not only by dietary specialization, but also by hydrodynamic needs and drag forces. Furthermore, the occipital condyle's orientation varies by feeding guild, being caudally oriented in grazers and ventrally directed in filter-feeders. Skeletal variances are also visible in the palate. Grazers have an elongated pterygoid bone and shorter palatine bone than filter-feeders (Pecsics et al., 2017).

Anatidae exhibit advanced cranial kinesis—movements involving the upper jaw region including premaxilla and maxilla—which allows them to exploit a wide range of feeding niches, from filter feeding to grazing (Pecsics et al., 2017). These are evolutionary adaptations that enhance foraging versatility. Three major skull joints are involved in these movements: (1) Nasofrontal hinge—it is located between the nasal and frontal bones and allows for the elevation of the upper bill. (2) Quadrate bone articulation—the quadrate connects the lower jaw to the cranium, and its mobility facilitates movement of both mandible and upper bill. (3) Palate joints—flexible joints in the palate allow shearing and sliding movements significant for processing plant materials (e.g., geese and swans) (Bout & Zweers, 2001). Histological studies of mallards cranial joints (such as those connecting the quadrate, squamosal, and palatal bones) reveal highly dynamic articulations throughout development, allowing for the rapid bill motions required for surface feeding and filtering (Bailleul et al., 2017). During this movement, mallard repeatedly opens and closes its bill to expel water while capturing food particles (Zweers, 1973).

Kinematic studies employing X-ray reconstruction of moving morphology (XROMM) have indicated a more dynamic role for the quadrate bone in Anatidae skull mechanics. During feeding, the quadrate rotates in three dimensions, transmitting motion to the upper bill through a kinetic chain of articulations (Bout & Zweers, 2001). This coupling mechanism allows ducks to rapidly alter their bill position and gape, boosting efficiency in filtering aquatic invertebrates, seeds, and plant material (Dawson et al., 2011).

2.2. Adaptations in Bill

Foraging ecology is highly influenced by bill morphological diversification, which is linked to diverse feeding behaviors (Olsen, 2017). Bill morphology varies from highly specialized in mergansers and shovelers to more generalist in scoters (Pecsics et al., 2017). Differences in bill morphology indicate the type of food consumed and how it is acquired (Kehoe & Thomas, 1987). Anatids bills are generally moderate in length, flattened, blunt-tipped, and covered by a thin integumentary covering. A nail is present at the maxillary tip, and fine lamellae line the maxillary and mandibular margins (Pecsics et al., 2017). Bill lamellae filter water to trap food, and the distance between lamellae varies between species within the range of feeding guilds (Champagnon et al., 2010). Lamellae evolved for filter feeding but were later adapted or reduced in species that moved to grazing (Olsen, 2015). Interspecific variation in bill morphology, such as changes in lamellar density, corresponds to differences in diet composition and the size spectrum of food particles consumed (Mott, 1994).

The size of the bill is directly proportional to the amount of food consumed, and variations in the size of the bill may influence how biting occurs, as gape angle increases with length of the bill. Hence, species that feed on hard food items (such as rhizomes and tubers) benefit from taller bills (Jónsson & Afton, 2016). Furthermore, the amplitudes and gapes of bill rotations during the filter-feeding process are directly related to seed size (Zweers, 1999).

All members of the Anatidae family have a keratinized "nail" (unguis) on the front that can bend over the end like a hook or be shaped like spatula. It is a functionally and evolutionarily relevant trait that increases foraging efficiency. It helps in the capture of hard-shelled prey such as mussels, as well as the digging and sorting of food. The tip of the bill has dense clusters of mechanoreceptors from the Herbst and Grandry corpuscles, making a unique sensory region known as the bill-tip organ (Berkhoudt, 1979). These receptors aid ducks to detect food in muddy or murky water when it isn't visible. The size of nails varies across Anatidae, as do their feeding pattern. Geese and swans have relatively large, broad nails that help crop grasses and uprooting aquatic plants, but dabbling ducks have tiny nails designed for sorting seeds and invertebrates during filter feeding. In piscivorous species such as mergansers, the nails work with serrated bill margins to assist seize slippery fish (Zweers et al., 1994).

Anatidae possess following categories of bills:

1. **Sieve-like or lamellate beak:** It is broad, flattened bill with well-developed lamellae (comb-like structures) present along the margins. They filter-feed on soft plant materials, planktons, aquatic invertebrates, or mud. Ducks may vary the space between lamellae, allowing them to retain preferred prey while expelling others (Gurd, 2006). Its exaggeratedly wide and spoon-shaped design maximizes the surface area for filtering. Example: mallard, northern shoveler.
2. **Grazing or herbivorous beak:** It is shorter, deeper, and stronger with powerful jaw muscles and fewer lamellae. They evolved to graze on grasses, sedges, and agricultural crops. This type of bill has a larger bone surface for muscle attachment than filter feeders (Pecsics et al., 2017). Its structure also includes dorsal-arched culmen and tomium, which are more closely associated to grazing on aquatic plant material (such as leaves and roots) (Olsen, 2017). Example: greylag goose (*Anser anser*), Canada goose (*Branta canadensis*), swan (*Cygnus* spp.).
3. **Pointed or stout bill:** Few species that dive for mollusks, crustaceans, or aquatic plants have compact and pointed bills. This type of bill is suitable for opening shells and grasping prey underwater. Example: pochards, eiders.
4. **Serrated or piscivorous bill:** It is long, narrow, and serrated type bill, design to catch slippery fish (Olsen, 2017). This form of beak lacks lamellae, however the edge is lined with sharp tooth-like serrations that ensure a firm grip on aquatic prey. Example: mergansers.
5. **Intermediate or generalist type beak:** It is moderately sized broad bill with intermediate lamella, which allows birds to forage on a variety of foods, including plants and seeds to invertebrates. Example: gadwall, wigeon.

2.3. Visual Cues Adapted for Foraging

In Anatids, feeding ecology seems to play a major role in determining the position and extent of the binocular visual field (Martin & Katzir, 1999). Visual forager typically exhibits a broader region of binocular overlap and a more centrally aligned bill, which enhance visual precision in controlling bill placement during feeding (Guillemain et al., 2002; Martin et al., 2007). Ducks that forage primarily using visual cues tend to have eyes positioned more anteriorly compared with species that rely on tactile feeding strategies (Guillemain et al., 2002; Martin et al., 2007). Furthermore, the accommodative can also effectively differentiate dividing ducks from non-diving species (Sivak et al., 1985). As cornea loses refractive power in the water, diving ducks (e.g., hooded merganser (*Lophodytes cucullatus*) and the common goldeneye (*Bucephala clangula*)) compensate by changing the shape of their lenses. This change considerably enhances optical power of their eyes, allowing them to visually locate prey beneath water's surface (Sivak et al., 1985). Non-diving or dabbling ducks, such as mallards (*Anas platyrhynchos*) and wood ducks (*Aix sponsa*), have an order of magnitude lower accommodative mechanisms (Sivak et al., 1985).

Eye morphology is closely linked to the ambient light conditions experienced by a species (Kirk, 2006). Generally, species active in low-light environments have larger corneas than species foraging in high-light environments (Pettigrew et al., 1988; Kirk, 2006). A larger corneal diameter enhances visual sensitivity, as corneal size constrains the amount of light entering the eye when the pupil is fully dilated (Kirk, 2006). Hence, species that forage in low light environments have larger corneal diameters than species that forage in high light environments (Lisney et al., 2013). Several species of waterfowl, such as diving ducks, exhibit crepuscular or nighttime feeding behavior, while pursuit-diving ducks (such as mergansers) and dabbling ducks seem to be restricted to foraging only under bright light conditions (McNeil et al., 1992; Sjöberg, 1985). On the other hand, Canada goose (*Branta canadensis*) does not have such preference. Despite this, they can feed at any time of day (opportunistically) (McNeil et al., 1992).

The capacity to constrict the size of pupil region assists effective amphibious vision (seeing underwater and in air) (Gislén et al., 2003). Furthermore, because the width of the cornea is directly proportional to the size of the pupil (Kirk, 2006), diving and pursuit diving ducks may have an advantage over non-diving ducks due to their smaller corneal diameters (Lisney et al., 2013). Diving and pursuit-diving ducks attain visual accommodation under water by tightly constricting the pupil (to 1 mm diameter) and displacing the lens against iris disc, causing the central part of the lens to protrude. This mechanism significantly enhances the eye's optical power, thereby improving vision underwater (Levy & Sivak, 1980).

Blue ducks (*Hymenolaimus malacorhynchos*) have frontally situated eyes that provide a broad binocular visual field for the slender and tapering bill. In addition, there is a substantial blind area at the back of the head. This visual field morphology is quite similar to that of other visually guided foragers (such as penguins; family: Spheniscidae), who get their food from the water column (Martin et al., 2007). Pink-eared ducks (*Malacorhynchus membranaceus*) have tactile-feeding anatid-like visual traits, such as a limited narrow frontal binocular field with the bill on the periphery and holistic visual coverage. Thus, while blue ducks can obtain food from rock surfaces, they are largely visual feeders in the water column. Hence, variations in water clarity are expected to have a major impact on their foraging efficiency (Martin et al., 2007).

2.4. Adaptations in Feet

Members of Anatidae family exhibit significant variety in feet morphology based on habitat utilization and foraging techniques. The majority of ducks, geese, and swans have *palmate feet*, with the front three toes fully webbed, which provide strong propulsion for swimming and effective surface locomotion (Livezey & Humphrey, 1986; Kear, 2005). Diving species, such as stiff-tailed ducks, have *lobed feet* with expanded lobes on each toe. This modification increases thrust during the power stroke and reduces drag on recovery. Thus, it allows for effective foraging in water (Raikow, 1973; Olsen, 2017). Grazing geese and other land-based birds have less webbing. It enhances terrestrial movement and allows for effective swimming when required (Johnsgard, 2010).

Dabbling ducks have small, webbed feet that are ideal for foraging in shallow waters (Özkan et al., 2024; Cherry & Morris, 2008). The West Indian whistling ducks (*Dendrocygna arborea*) are characterized by their strong, quick-witted, and fairly webbed feet which enable them to navigate through dense vegetation and shallow aquatic water bodies while foraging on aquatic plants, aquatic insects, and smaller invertebrates. They also have elongated and thin toes that provide them support and enhance propulsion in water (Madge & Burn, 1988). On the other hand, mandarin ducks (*Aix galericulata*) possess highly developed webbed feet with more systematic webbing that almost reaches the tip of their toes. Unlike West Indian whistling ducks (*Dendrocygna arborea*), mandarin ducks (*Aix galericulata*) rely less on terrestrial locomotion and predominantly to semi-aquatic lifestyle, spending the most of their time in or near water bodies (Kear, 2005).

The paddle size in ducks represents an evolutionary trade-off, balancing the relatively smaller paddle feet size suitable for terrestrial locomotion with the larger paddle feet size advantageous for movement in aquatic environments. The American wigeon (*Mareca americana*, a dabbling duck) is one of the terrestrial species. They have relatively smaller paddle-like feet, which are thought to be an adaptation that enhances their agility for terrestrial locomotion (mostly through grazing). Diving ducks, on other hand, have larger paddle-like feet than dabbling ducks, reflecting that they rely more on aquatic locomotion (Raikow, 1973).

Common eiders (*Somateria*; tribe: Mergini) lack larger paddles than other duck species and roughly comparable with *Anas* (common dabbling ducks) species. As a result, it is concluded that they are not much efficient divers. They are ground-nesting species, and their *Anas*-like paddle size is better suited to terrestrial locomotion (Raikow, 1973).

Ruddy ducks (*Oxyura jamaicensis*) feet are well adapted for swimming both underwater and above the water's surface, but they are inept and clumsy on land (unable to walk more than a few steps) (Kortright, 1942). This is owing to their posteriorly placed feet and specialized bone and muscle structure on their bodies (Raikow, 1970), and their large paddles make land movement difficult (Raikow, 1973).

Swans have robust, lengthened, and narrower toes with well-developed webbing, allowing them to generate effective push while swimming (Özkan et al., 2024; Johnsgard, 2010). The shape and size of geese feet vary with species, having webbed toes (Özkan et al., 2024; Kear, 2005).

3. Foraging Strategies

Anatidae birds exhibit remarkable diversity in their foraging strategies for meeting their everyday energy requirements, which is most likely due to their ecology, morphology, and evolutionary adaptations (Lisney et al., 2013). A variety of adaptations aid them in niche partitioning, reducing competition for food within coexisting species (Mukherjee et al., 2021). For example, inside a single wetland, a few dabbling ducks filter small seeds and invertebrates from shallow water, while diving ducks feed on food at depth, and geese graze on grasses, reducing conflict (Nudds & Bowlby, 1984; Nudds et al., 1994; Khan & Khan, 2024). Second, morphological adaptation has a significant impact on foraging strategies such as bill form, lamellar density, and nail size, as mentioned in the earlier text (Kear, 2005). Third, habitat availability, habitat utilization, and seasonal availability all play an additional role. On the other side, energetic requirements influence feeding choices. For example, migratory species require energy-rich food (e.g., seeds, invertebrates) for long-distance flights, whereas herbivorous geese rely on high biomass and low-quality vegetation to maintain group grazing (Sedinger, 1997). Finally, social behavior and predation pressure also shape strategies (Lazarus & Inglis, 1978).

Table 1 depicts foraging mode and representative diet of the Anatidae. The most frequently adapted foraging strategies include:

3.1. Dabbling

Dabbling, which implies feeding by tipping up, is one of the most common strategies used by the Anatini tribe to feed at or near the water's surface (Fig. 1A). However, swans and geese occasionally use dabbling as a means of feeding. Thus, they are collectively known as dabbling ducks (Guillemain et al., 2002).

Dabbling ducks use a wide range of foraging strategies (Fig 1). Dabbling ducks feed by picking or skimming food from surface water or "tipping-up"; they may tip up in deep water, drowning their neck and head (Pöysä, 1986). They may also grub in mud or shallow water, keeping their eyes above the surface to monitor their surrounding while foraging. In comparison to deep foraging, it is a safer practice with a lower risk of predation (Guillemain et al., 2000). Mallard and pintail prefer to feed under the surface, with their feet and bodies elevated and their heads in the water reaching for underwater plants, a technique known as "upending" (Kear, 2005; Pecsics et al., 2017).

Dabbling ducks sometimes have a tendency to be vigilant while foraging while their eyes are above water, however this is dependent on their vision

Table 1. Foraging modes, diet, and dietary guilds of Anatidae subfamilies and tribes

Subfamily	Tribe	Foraging Mode	Feeding Guild	Representative Diet Composition	Key References
Anseranatinae	*Anseranatini* (magpie geese)	Grazing	Herbivorous	Seeds of wild rice (*Oryza*), grasses (*Cynodon*, *Echinochloa*, *Paspalum*)	Frith & Davies, 1961; Johnsgard, 2010
Anserinae	*Dendrocygnini* (whistling or wood ducks)	Grazing and diving	Primarily herbivorous, occasionally omnivorous	Seeds, buds, and leaves of aquatic/terrestrial plants (e.g., *Echinochloa*, *Eleocharis*, *Fimbristylis*), fruits (e.g., *Roystonea*), plus occasional invertebrates	Hohman et al., 1996; Johnsgard, 2010
	Anserini (swans and true geese)	Grazing (mainly)	Largely herbivorous, some omnivory	Aquatic macrophytes (pondweeds, *Sagittaria*, *Vallisneria*), grasses (*Zostera*, *Eriophorum*), herbs (*Salicornia*), crops (wheat, rice, barley), and occasional animal matter (invertebrates, amphibians, fish spawn)	Bailey et al., 2008; Johnsgard, 2010; Chen et al., 2019
	Cereopsini (Cape Barren goose)	Grazing	Herbivorous	Tussock grass (*Poa*), sedges, clovers (*Trifolium*)	Guiler, 1967; Johnsgard, 2010
	Stictonettini (freckled duck)	Dabbling and filter-feeding	Omnivorous	Seeds of *Polygonum*, *Rumex*, algae, small insects	Briggs, 1982; Johnsgard, 2010
Anatinae	*Tadornini* (shelducks, sheldgeese)	Dabbling and grazing	Omnivorous to carnivorous	Grasses, clovers, algae, grains, invertebrates (mollusks, worms, midges)	Walmsley & Moser, 1981; Viain et al., 2013; Johnsgard, 2010
	Tachyerini (steamer ducks)	Diving	Carnivorous	Crabs, mollusks, crustaceans, small fish	Johnsgard, 2010

(*Contd.*)

(*Contd.*)

Subfamily	Tribe	Foraging Mode	Feeding Guild	Representative Diet Composition	Key References
	Cairinini (perching ducks)	Dabbling and grazing	Omnivorous to carnivorous	Aquatic vegetation (*Nymphaea*, *Oryza*), grains, acorns, and nuts, invertebrates, small reptiles	Johnsgard, 2010; Brochet et al., 2012; Soons et al., 2016
	Merganettini (torrent ducks)	Diving	Carnivorous	Aquatic insect larvae (caddisflies, mayflies, stoneflies), small mollusks	Johnsgard, 2010; Cerón et al., 2010; Cerón, 2015
	Anatini (dabbling ducks)	Dabbling (mainly), some grazing and diving	Omnivorous	Aquatic plants (*Najas*, *Myriophyllum*, *Ruppia*), seeds, grains, algae (*Chara*), invertebrates (mollusks, insects, shrimps, ostracods)	Swanson, 1984; Johnsgard, 2010
	Aythyini (pochards)	Diving (mainly) and some dabbling	Omnivorous	Submerged vegetation (*Nymphaea*, *Ceratophyllum, Chara, Nitella*), plus mollusks (*Mytilus*, *Littorina*), crustaceans, insect larvae	Olney, 1968; Johnsgard, 2010; Sekiya et al., 2000
	Mergini (sea ducks)	Diving	Carnivorous (mainly)	Crustaceans, mollusks, echinoderms, fish, insect larvae; some species also consume aquatic plants	Johnsgard, 2010
	Oxyurini (stiff-tailed ducks)	Diving and dabbling	Omnivorous	Aquatic plants (*Azolla*, *Polygonum*), plus invertebrates (crustaceans, mollusks, midges); Musk ducks also eat fish and frogs	Johnsgard, 2010

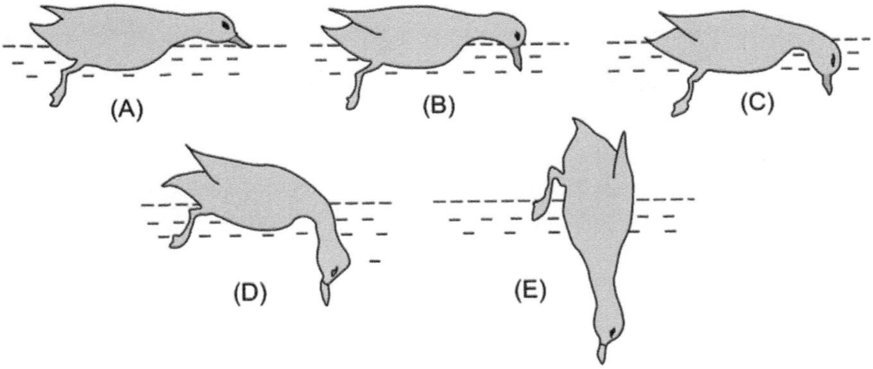

Figure 1. Outline diagrams depicting feeding methods of dabbling ducks. (A) Straining or picking from surface, (B) submerged bill, (C) submerged head, (D) submerged neck, and (E) upending (adapted from Pöysä, 1986).

capabilities. Despite their close association, shovelers and wigeons have distinct dietary preferences and foraging strategies. Shovelers are *filter feeders* and *plankivorous*; they filter the upper layers of the water column to remove zooplanktons (DuBowy, 1996; Guillemain, Fritz, & Guillon, 2000). Wigeons are herbivorous and have been recorded pecking plant leaves from swampy places (Owen & Thomas, 1979; Jacobsen, 1992). Both of these species require lengthy eating intervals, thus they spend less time being vigilant in order to maximize food intake (Guillemain et al., 2001).

3.2. Diving

Diving birds' foraging strategies and diet preferences vary across tribes Aythyini (pochards) and Mergini (mergansers) (Lisney et al., 2013). Pochards typically dive shallow and eat a variety of aquatic vegetation and invertebrates. In contrast, pursuit-divers mergansers dive deeper in search of invertebrates and fishes (Lisney et al., 2013; Cantlay et al., 2023). Similarly, sea ducks in Mergini tribe may dive to great depths to forage. These birds have strong legs and adaptations for deep diving, allowing them to reach mollusks, echinoderms, and fish (Žydelis & Richman, 2015).

Lesser scaup (Aythya affinis) is a diving duck that has evolved several adaptations for efficient foraging beneath water. Active swimming allows for better oxygen usage, forceful hindlimb thrusts, and buoyancy management. These adaptations enable them to dive to higher depths despite high energy demands (Stephenson, 1994). Common pochard, tufted duck, and greater scaup are well-known diving ducks. Zebra mussels are mostly used as food and are collected in clusters from the bottoms of estuaries (Marchowski et al., 2016).

According to research, tidal conditions influence birds' diving methods as well as their food preferences. Ancient murrelets are planktivorous sea birds that dive into deep, rough water during big tides. Piscivorous birds require moderate flow at different depths; pigeon guillemots use swift currents and dive while floating downstream. The diving ducks that feed on invertebrates like shallow, quiet water. They feed on mussels and other benthic organisms. They are seen avoiding strong water currents. Some animals demonstrate adaptive feeding strategies based on tides to avoid competition (Holm & Burger, 2002). Canvasbacks, lesser scaups, and ruddy ducks all have different underwater diving tactics. Canvasbacks forage by piercing the substrate while keeping their bodies right angles to the foundation; this form of behavior is good for probing in soft remnants. Lesser scaups and ruddy ducks open and reopen their lips swiftly by sliding their beaks in lateral sections of curves near the substrate surface. Their bodies remain at various angles. Scaups use their vision to find prey in the water column, whereas ruddy ducks do not. Underwater posture and mobility also vary among species, indicating adaptations to specific prey and environmental conditions (Tome & Wrubleski, 1988).

3.3. Grazing

Birds graze the grasses and sedges that grow alongside wetland margins and near agricultural fields. This behavior is particularly noticeable in geese, whose beak structures are specialized for cutting vegetation that grows closer to land (Lorenzen & Madsen, 1986). These birds have a narrower bill and a wider bone surface. Grazers have a large skull with a large quadrate-mandibular, which improves the efficiency of jaw-closure muscles, which are essential for grazing. They have longer pterygoid and shorter palatine bones than filter feeders (Pecsics et al., 2017). Greylag goose (*Anser anser*) and bar-headed geese (*Anser indicus*) are known to graze in such areas (Kumar, 2024). Swans also feed on vegetables, which they acquire through grazing and dipping. Greylag geese dig for underground plant portions. They dig *Spartina anglica* rhizomes and *Scirpus maritimus* tubers (Esselink et al., 1997).

Some ducks, including ruddy shelducks (*Tadorna ferruginea*), have adapted this tactic, foraging both in grasslands and wetlands. They represent grazing as a periodic foraging strategy, rather than a permanent one. During migration, as energy requirements increase and aquatic food becomes scarce, these birds adapt to terrestrial grazing to meet their needs. Agricultural fields and pasture land are critical for population survival during these times. As a result, Anatidae broaden their ecological niche beyond water-based feeding by integrating grazing in their foraging strategy (Kumar, 2024).

Geese (true geese) are known to rely heavily on land plants and terrestrial vegetation. They are seen cropping grasses in meadows and farmland (Vickery & Gill, 1999). Swans are near relatives, however geese differ from swans in that they graze terrestrial vegetation, whereas swans eat on submerged water plants via neck-dipping and upending. Sheldgeese, a duck subfamily, are ecologically similar to geese, living on short grasses and plants (Johnsgard, 2010).

3.4. Filter Feeding

Filter feeders capture small food items from the water column (Cantlay et al., 2023). Filter feeding in Anatidae consists of two components: a water pumping system and a filtration system (Kooloos et al., 1989; Kooloos & Zweers, 1991). Pumping mechanisms operate as suction-pressure pumps. The lingual bulges retract and rise, pulling water into the mouth when the beak opens. The bulges then descend and the tongue protrudes, allowing water to flow over the lingual bulges due to inertial forces. In contrast, two mechanisms are engaged in the filtering process. In shovelers and mallards, particles are filtered by direct impact when they are larger than the mesh width between the maxillary lamellae. Another mechanism used by mallard and tufted ducks is the lingual cushion, which divides water flow into two streams that pass through the bill's lamellae to separate tiny grains. As the flow swirls, tiny seeds become caught against the lamellar walls. Tiny tongue hairs brush these seeds into grooves, which lingual scrapers subsequently transport backward into the throat (Kooloos et al., 1989).

During filter feeding, the amplitudes and gapes of the beaks of mallards (*Anas platyrhynchos*), shovelers (*Anas clypeata*), and tufted ducks (*Aythya fuligula*) were observed to be positively linked with seed size. Filter-feeding animals have smaller and weaker pterygoid bones and adductor muscles because stronger jaw and neck muscles are unnecessary. Shovelers can filter a wide range of planktonic species with their highly specialized bill. Filter-feeding anatids have a long, gracile skull (Pecsics et al., 2017).

Filter feeders' bills include a fine comb-like structure known as lamellae; when water and sediment permeate the bill, the lamellae filter away non-edible elements and trap food items such as invertebrates, seeds, and so on (Kooloos & Zweers, 1991). Most dabbling ducks have between 50 and 70 lamellae. Northern shovelers have highly developed lamellae and are employed to collect microscopic prey from aquatic environments (Kooloos et al., 1989; Skieresz-Szewczyk & Jackowiak, 2014).

Shovelers that feed by filtering the top few centimeters of water (dabbling) are predominantly planktivorous (Guillemain et al., 2002). Mallards (*Anas platyrhynchos*), northern shovelers (*Spatula clypeata*), and pink-eared ducks (*Malacorhynchus membranaceus*) are filter feeders with dorsal eyes and a small binocular field that allows for a wide visual range (Cantlay et al., 2023).

4. Factors Influencing Foraging

Anatidae foraging is linked to a variety of morphological and behavioral adaptations, seasonal migration patterns, and niche utilization (Rizzo & Battisti, 2009). These waterfowls occupy a variety of aquatic environments, including lakes, rivers, estuaries, and wetlands. Their dominant foraging methods (such as dabbling, diving, and grazing) are critical in maximizing energy flow or consumption, assessing reproductive production, and determining temporal and spatial distribution patterns (Dessborn et al., 2011). Furthermore, physical characteristics adapted for foraging in anatids have evolved alongside dietary niche and habitat selection, establishing foraging as a crucial mechanism driving trophic niche partitioning and permitting interspecific cohabitation within the family. Thus, foraging behavior serves as a foundation for completely examining the ecological dynamics of the Anatidae family.

4.1. Habitat

Habitat selection is one of the most powerful influences on foraging. Because of their high productivity, wetlands typically sustain the coexistence of many diverse waterbird species that use shared food resources (Hafner et al., 1986; Weller, 1999). Various species like to eat at varying water depths. Dabblers and divers use different microhabitats and foraging methods, however dabblers prefer shallow water (<30 cm) for feeding (Nudds et al., 2000). Divers use deeper habitats. Sometimes, the presence of submerged macrophytes and emergent plants can influence diet choices (Elmberg et al., 1994). Productive wetlands also promote increased food supply, which benefits omnivorous species (Guillemain et al., 2002).

Mallards, being generalists, adopt a variety of foraging tactics, but shallow foraging is preferred since it is safer and less costly than foraging in deep water. Ducks prefer shallow foraging when food is abundant (Guillemain et al., 2000). When food in shallow water became scarce, only deep foraging was increased, and only after emptying the shallow area did they totally switch (Guillemain et al., 2000). Similarly, mallards feed throughout all depth gradients. Wigeon, on the other hand, is largely a "terrestrial grazer," while the long-necked pintail is a "deep-water specialist." Despite these distinctions, there is significant interspecific overlap in foraging strategies, even when the sexes are considered independent. Statistically, wigeon has significantly different foraging strategies than any other species. Pintail differs significantly from mallard, while mallard and teal have no significant variations in foraging behavior in either sex (Arzel & Elmberg, 2015). Despite sharing a similar proportion of deeper foraging behaviors, teal and mallard occupy distinct functional feeding niches because of differences in depth reach, neck length, and bill lamellar density (Nudds et al., 2000).

Thus, the literature suggests that foraging strategies in ducks are dependent on the varied costs of foraging approaches rather than the rate of food intake; the shift was induced by food depletion (Guillemain et al., 2000). Variations in foraging habitat result in divergent foraging behavior. Resource partitioning reduces competition by allowing different waterfowl species to cohabit in wetland ecosystems. However, species' food preferences differ not just among themselves, but also between foraging and breeding sites. Thus, resource partitioning typically entails differences in habitat selection, food preference, and feeding practices (Pöysä, 1983; Pöysä, 1984).

Swans, geese, and some ducks are obligate herbivores, and their dietary habits in cropland, rotational grassland, and pasture are extensively documented. Food availability and nutritional composition are important factors in habitat selection (Fox et al., 2017). However, larger fields tend to attract more species, while smaller fields with the same crops are often avoided (Jensen et al., 2008).

4.2. Morphological and Physiological Adaptations

Ecomorphological studies show that physical differences across species have adaptive value (Pecsics et al., 2017). Differences in eating habitats and diet types are intimately related to differences in feeding mechanisms, which are then linked to differences in morphological features. Studies have also shown that interspecific competition diminishes resource availability and promotes physical divergence between sexes (Nudds & Kaminski, 1984).

Functional and morphological adaptations of the beak have developed to increase energy efficiency in exploiting preferred food items (Kooloos et al., 1989). The filter-feeding method in Anseriformes was determined by combining beak anatomy, seed size, and kinematics. Variations in these parameters indicated dietary specialization. The volume and gape movements of the beak increased in proportion to the size of the seed. The filtration and pumping procedures influence the kinematics of the mouth, limiting lingual engagements and water flow. *Anas clypeata*'s filtering effectiveness was heavily influenced by lamellae structure, whereas the same of the *Anas platyrhynchos* and *Aythya fuligula* were not. This demonstrates interspecific variability in dependency on morphological traits. Thus, filtering mechanisms have enough modulations to manage a wide variety of seed sizes, but morphological and functional adaptations are tuned to maximize efficiency for staple food resources (Kooloos et al., 1989).

Variation in gizzard structure has also been documented among species, possibly due to feeding patterns (Pecsics et al., 2017). Herbivorous species such as Canada goose (*Branta canadensis*), brant (*Branta bernicla*), ring-necked duck (*Aythya collaris*), redhead (*Aythya americana*), and gadwall (*Mareca strepera*) have larger gizzards than carnivorous species such as long-tailed duck (*Clangula hyemalis*), common goldeneye (*Bucephala clangula*), and bufflehead (*Bucephala albeola*), which primarily consume soft-bodied invertebrates. Because terrestrial plants contain high levels of structural carbohydrates (such as cellulose), herbivores that feed primarily on them have proportionally larger gizzards than those that feed on aquatic vegetation, which receives much of its support from water rather than cellulose. Brant (*Branta bernicla*) had the largest gizzard of any herbivore, while Canada geese (*Branta canadensis*) had considerably lighter gizzards than other species grazing on water vegetation (Barnes & Thomas, 1987).

Aside from beak and gizzard changes, binocular field vision is another morphological adaptation. The eyes are positioned laterally, allowing for a wide field of vision. Forward overlapping degree (binocular field) varies between species based on their foraging strategy. Wider binocular fields benefit visually foraging species, such as dabbling ducks, who see foliage or seeds near the water's surface. Species with narrower fields and higher lateral vision use it to remain vigilant and protect themselves from predators (Cantlay et al., 2023).

4.3. Evolutionary Relationships and Genetics

Genetic distance influences foraging strategies (Mukherjee et al., 2021). Early branching species in the evolutionary tree, such as the lesser whistling duck, exhibited generalist foraging, but pintails and pochards (later-diverging species) developed specialized methods (upending, diving, etc.). Red-crested pochard demonstrated the transitional behavior of dabblers and divers from the Anatinae and Aythyinae groups. It moved between dipping, dabbling, and diving, depending on the habitat. This brings together intermediate foraging strategies and adaptive radiation in Anatidae (Mukherjee et al., 2021).

4.4. Energetic Efficiency

To survive, birds must balance energy costs and gains. The rate of net energy intake has a significant impact on foraging in Anatidae, which is influenced by a variety of foraging (e.g., searching, consuming, and processing food) and non-foraging activities such as diving, flying, swimming, growth and production, and waste excretion (Lovvorn et al., 2009) (Fig. 2). A study of tundra swans found

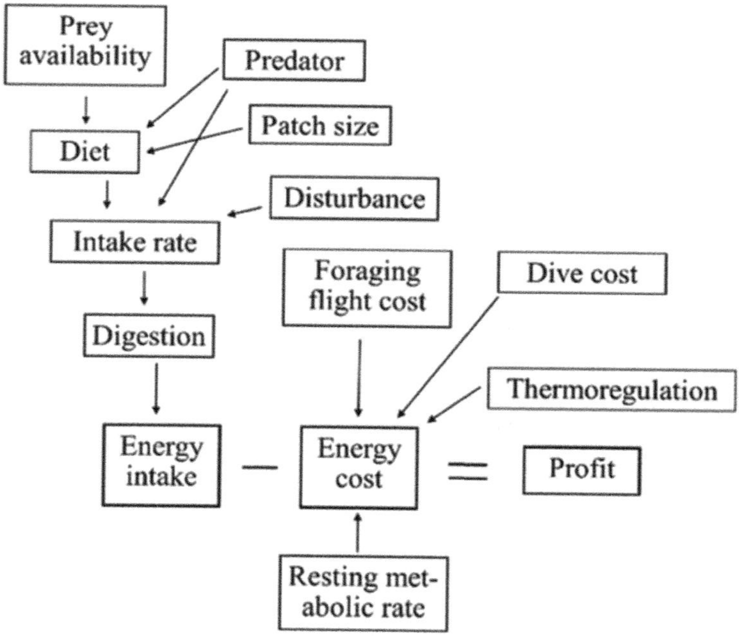

Figure 2. Flow chart depicting energy balance in Anatidae during feeding and foraging (adapted from Lovvorn et al., 2009).

that their dispersal was determined by net energy intake rate rather than total food biomass. The net energy intake rate takes into consideration both foraging costs and accessibility (Gyimesi et al., 2012a, 2012b). This demonstrates that members of the Anatidae improve foraging by optimizing energy gain rather than merely chasing bigger food biomass (Miller & Eadie, 2006). Herbivorous waterfowl are known to choose food with high protein, soluble carbohydrate, and digestibility, as well as low fiber and phenolic chemical with water content (Fox et al., 2017).

Human disturbance can also influence foraging behavior in Anatidae, however not always through food exhaustion, but rather by modifying eating energetic efficiency (Taylor et al., 2019). In a study of Bewick's swans, it was discovered that giving-up densities did not differ between undisturbed and disturbed creeks, but when calculating giving-up net energy intake rates (for food accessibility and foraging costs) between undisturbed and disturbed areas, it was found to be higher in disturbed areas (Gyimesi et al., 2012a, 2012b). It demonstrates how disruption affects foraging by changing the energy cost and gain balance (Fig. 2).

4.5. Environmental Conditions

Anatidae's foraging technique is heavily influenced by environmental conditions including day duration, temperature, and seasonal fluctuations. A study found that as winter approached and temperatures dropped and days became shorter, ducks (e.g., wintering gadwall, pintail, green-winged teal) increased their foraging speed (meters per minute) and rates (dips per minute) (Krapu, 1974). This behavioral modification allows them to gather the most food in the shortest amount of time while reducing cold exposure (Hepp, 1985; McKnight, 1998). Furthermore, environmental factors such as temperature have a direct impact on their foraging tactics and prey availability (Owen-Smith, 1998). Waterbirds frequently congregate in huge flocks within wetland habitats to meet their energy demands and receive adequate nutrients, particularly in lakes with high productivity (Li et al., 2013). Waterbirds display foraging plasticity as the carrying capacity of their natural habitats declines, using new sites and changing their foraging behavior (Zheng et al., 2015).

Ducks are primarily visual foragers, limiting their foraging to daylight hours. However, due to shortened daylight hours at high latitudes and a cold climate, some species, such as Steller's eiders in North Norway, have been observed feeding after sunset (Systad & Bustnes, 2001). Similarly, common eiders in southwest Greenland have two feeding techniques. They graze during the day around the outer coast, but mostly at night in fjords to avoid predators (Merke & Mosbech, 2008).

4.6. Food Abundance

When food abundance was reduced, wintering gadwalls and American coots adjusted their foraging behavior to maintain energy intake. This emphasizes how food abundance has a significant influence on foraging. When food availability is reduced, waterbirds may increase their foraging rates and change their foraging schedules (McKnight, 1998; Nolet et al., 2002). Gadwalls, for example, have enhanced their foraging pace and frequently steal food from coots. Coots conducted greater dives, resulting in fewer surface pecks. They boosted their foraging rates to maximize food intake while avoiding cold exposure. Food supply interacts with environmental forces, resulting in behavioral modifications in foraging (McKnight, 1998).

4.7. Behavioral Strategies

Foraging habit in Anatidae is inextricably linked to kleptoparasitism (one species stealing food from another). Bewick's swan (*Cygnus bewickii*), along with the closely related whooper swan (*Cygnus cygnus*) and whistling swan (*Cygnus columbianus*), frequently forages alongside dabbling ducks and pochards (Sherwood, 1960). These species frequently adopt a foraging method that involves stomping the sediment to find buried food items. Although such a foraging approach has the potential to harm ducks' food resources, it may also increase food availability by bringing previously unavailable foods to the surface. Ducks have also been documented to engage in kleptoparasitism, taking food supplies (such as tubers) from swans (Beekman et al., 1991).

5. Ecological Roles

During feeding and foraging throughout habitats, Anatidae species play a variety of ecological roles, making them an essential component of their ecosystems.

5.1. Role in Seed Dispersal

A wide range of duck species have also been identified as key agents of internal (endozoochorous) and external (epizoochorous) seed dispersion (Figuerola et al., 2002; Green et al., 2008; van Leeuwen et al., 2012). They regularly consume diaspores from marsh sediment seed banks (Green et al., 2002). Research on the Pacific black duck (*Anas superciliosa*), grey teal (*Anas gracilis*), and chestnut teal (*Anas castanea*) in Australia has confirmed that both nomadic and seasonal migratory ducks serve as important vectors of endozoochorous and epizoochorous seed dispersal (Green et al., 2008; Raulings et al., 2011). In the Mediterranean region, the nomadic and globally vulnerable marbled teal (*Marmaronetta angustirostris*) (subfamily: Aythyinae) has been identified as a key seed distribution vector (Green et al., 2002; Fuentes et al., 2005). Piscivorous migrants, such as the red-breasted merganser (*Mergus serrator*) and the bufflehead (*Bucephala albeola*), have been observed carrying seeds externally in New Jersey (Vivian-Smith & Stiles, 1994).

True geese (Anatidae, subfamily Anserinae) are mostly terrestrial waterbirds, as opposed to ducks. They are mostly grazers (surface feeders) that consume vegetative plant parts and grasses. They contribute significantly to the ecology by dispersing plants with ripe fruits (Janzen, 1984). Hawaiian goose (*Branta sandvicensis*) is well-known for dispersing berries (e.g., *Vaccinium reticulatum*, *Fragaria chiloensis*, *Coprosma ernodeoides*, etc.) and terrestrial plant seeds such as alien sow-thistle (*Sonchus asper*) (Guppy, 1906). Furthermore, seeds of smooth cordgrass (*Spartina alterniflora*), seashore saltgrass (*Distichlis spicata*), and many other species have been found on the feathers and feet of brant goose (*Branta bernicla*) (Vivian-Smith & Stiles, 1994).

Investigations into diaspore dispersal by true swans (family: Anatidae; subfamily: Anserinae) have been sparse, however they may be more important aquatic plant vectors than geese (Green et al., 2008). Ridley (1930) found that both Bewick's swans (*Cygnus bewickii*) and mute swans (*Cygnus olor*) ate aquatic plant seeds (such as *Potamogeton natans*).

Members of the Tadorninae (shelduck–sheldgoose; subfamily of Anatidae), including the Magellan Goose (*Chloephaga picta*), eat the berries of the crowberry or diddle-dee shrub (*Empetrum rubrum*) (Ridley, 1930). Common fruit eaters include the Magellan goose (*Chloephaga picta*) and the ashy-headed sheldgoose (*Chloephaga poliocephala*) (Willson et al., 1997). Furthermore, the common shelduck (*Tadorna tadorna*) disperses crop plant seeds (such as Salicornia) (Viain et al., 2013). In South Africa, the Egyptian goose (*Alopochen aegyptiaca*) is believed to be a major seed distribution vector (Halse, 1984). Investigations from fossilized feces of *Thambetochen chauliodous* (an extinct and flightless anatid species from Hawaii), a near ancestor of dabbling ducks, revealed a high quantity of pteridophyte spores (James & Burney, 1997).

5.2. Role in Nutrient Cycling

Foraging behavior in Anatidae contributes to nutrient cycling by allocating nutrients between land-based and water-based habitats, hence maintaining ecosystem productivity and biodiversity. Members of the Anatidae family can efficiently modulate the nutrient composition of waterbodies in a variety of ways, including (a) modifying nutrient forms through nutrient cycling or internal loading, (b) adding external nutrients from another waterbody through external nutrient loading, and (c) transferring nutrients from aquatic to terrestrial ecosystems (Adhurya et al., 2018).

6. Conclusion

The present chapter highlight the foraging diversity in Anatidae family associated with their morphological, habitat, and behavioral adaptations. These birds have evolved a number of ways to find and consume food, from dabbling at the water's surface and grazing on grasses to diving deep underwater or filtering tiny prey. Their bills, skulls, eyes, and even digestive systems have been shaped by their feeding requirements, allowing them to share wetlands without intense competition. Seasons and temperature also play an important role: during breeding, many species switch to protein-rich animal food, while in winter or migration, they rely more on seeds and crops for energy. On other hand, they also help wetlands thrive—spreading seeds, cycling nutrients, and supporting overall biodiversity. Hence, protecting these habitats is therefore vital, not only for Anatidae but for the many ecological roles they play.

References

Adhurya, S., Das, S., & Ray, S. (2018). Guanotrophication by waterbirds in freshwater lakes: A review on ecosystem perspective. *In: International Workshop of Mathematical Analysis and Applications in Modeling* (pp. 253–269). Singapore: Springer Singapore. https://doi.org/10.1007/978-981-15-0422-8_22

Arzel, C., & Elmberg, J. (2015). Time use and foraging behaviour in pre-breeding dabbling ducks Anas spp. in sub-arctic Norway. *Journal of Ornithology*, 156(2), 499–513. https://doi.org/10.1007/s10336-014-1151-8

Arzel, C., Elmberg, J., & Guillemain, M. (2006). Ecology of spring-migrating Anatidae: A review. *Journal of Ornithology*, 147(2), 167–184. https://doi.org/10.1007/s10336-006-0054-8

Bailey, M., Petrie, S. A., & Badzinski, S. S. (2008). Diet of mute swans in lower Great Lakes coastal marshes. *The Journal of Wildlife Management*, 72(3), 726–732. https://doi.org/10.2193/2007-133

Bailleul, A. M., Witmer, L. M., & Holliday, C. M. (2017). Cranial joint histology in the mallard duck (*Anas platyrhynchos*): New insights on avian cranial kinesis. *Journal of Anatomy*, 230(3), 444–460. https://doi.org/10.1111/joa.12562

Barnes, G. G., & Thomas, V. G. (1987). Digestive organ morphology, diet, and guild structure of North American Anatidae. *Canadian Journal of Zoology*, 65(7), 1812–1817. https://doi.org/10.1139/z87-274

Beekman, J. H., van Eerden, M. R., & Dirksen, S. (1991). Bewick's swan *Cygnus columbianus bewickii* utilising the changing resource of *Potamogeton pectinatus* during autumn in The Netherlands. *Wildfowl*, 238–248.

Beerens, J. M., Gawlik, D. E., Herring, G., & Cook, M. I. (2011). Dynamic habitat selection by two wading bird species with divergent foraging strategies in a seasonally fluctuating wetland. *The Auk*, 128(4), 651–662. https://doi.org/10.1525/auk.2011.10165

Berkhoudt, H. (1979). The morphology and distribution of cutaneous mechanoreceptors (Herbst and Grandry corpuscles) in bill and tongue of the mallard (*Anas platyrhynchos* L.). *Netherlands Journal of Zoology*, 30(1), 1–34. https://doi.org/10.1163/002829680X00014

Bout, R. G., & Zweers, G. A. (2001). The role of cranial kinesis in birds. *Comparative Biochemistry and Physiology Part A: Molecular & Integrative Physiology*, 131(1), 197–205. https://doi.org/10.1016/S1095-6433(01)00470-6

Briggs, S. V. (1982). Food habits of the freckled duck and associated waterfowl in north-western New South Wales. *Wildfowl*, 33(33), 88–93.

Brochet, A. L., Mouronval, J. B., Aubry, P., Gauthier-Clerc, M., Green, A. J., Fritz, H. et al. (2012). Diet and feeding habitats of Camargue dabbling ducks: What has changed since the 1960s? *Waterbirds*, 35(4), 555–576. https://doi.org/10.1675/063.035.0406

Cantlay, J. C., Martin, G. R., McClelland, S. C., Potier, S., O'Brien, M. F., Fernández-Juricic, E. et al. (2023). Binocular vision and foraging in ducks, geese and swans (Anatidae). *Proceedings of the Royal Society B*, 290(2006), 20231213. https://doi.org/10.1098/rspb.2023.1213

Cerón, G. (2015). Torrent ducks (*Merganetta armata*) diving and feeding in hot springs. *Waterbirds*, 38(2), 214–216. https://doi.org/10.1675/063.038.0212

Cerón, G., Trejo, A., & Kun, M. (2010). Feeding habits of torrent ducks (*Merganetta armata armata*) in Nahuel Huapi National Park, Argentina. *Waterbirds*, 33(2), 228–235. https://doi.org/10.1675/063.033.0211

Champagnon, J., Guillemain, M., Elmberg, J., Folkesson, K., & Gauthier-Clerc, M. (2010). Changes in mallard *Anas platyrhynchos* bill morphology after 30 years of supplemental stocking. *Bird Study*, 57(3), 344–351. https://doi.org/10.1080/00063657.2010.486028

Chen, Y., Zhang, Y., Cao, L., De Boer, W. F., & Fox, A. D. (2019). Wintering swan geese maximize energy intake through substrate foraging depth when feeding on buried *Vallisneria natans* tubers. *Avian Research*, 10(1), 6. https://doi.org/10.1186/s40657-019-0145-x

Cherry, P., & Morris, T. R. (2008). *Domestic Duck Production: Science and Practice*. CABi, Digital Library.

Dawson, M. M., Metzger, K. A., Baier, D. B., & Brainerd, E. L. (2011). Kinematics of the quadrate bone during feeding in mallard ducks. *Journal of Experimental Biology*, 214(12), 2036–2046. https://doi.org/10.1242/jeb.047159

Dessborn, L., Brochet, A. L., Elmberg, J., Legagneux, P., Gauthier-Clerc, M., & Guillemain, M. (2011). Geographical and temporal patterns in the diet of Pintail *Anas acuta*, Wigeon *Anas penelope*, Mallard *Anas platyrhynchos* and Teal *Anas crecca* in the Western Palearctic. *European Journal of Wildlife Research*, 57(6), 1119–1129. https://doi.org/10.1007/s10344-011-0577-z

Donnelly, J. P., Naugle, D. E., Collins, D. P., Dugger, B. D., Allred, B. W., Tack, J. D. et al. (2019). Synchronizing conservation to seasonal wetland hydrology and waterbird migration in semi-arid landscapes. *Ecosphere*, 10(6), e02758. https://doi.org/10.1002/ecs2.2758

DuBowy, P. J. (1996). *Northern Shoveler: Anas Clypeata*. American Ornithologists' Union.

Elmberg, J., Nummi, P., Pöysä, H., & Sjöberg, K. (1994). Relationships between species number, lake size and resource diversity in assemblages of breeding waterfowl. *Journal of Biogeography*, 75–84. https://doi.org/10.2307/2845605

Esselink, P., Helder, G. J., Aerts, B. A., & Gerdes, K. (1997). The impact of grubbing by Greylag Geese (*Anser anser*) on the vegetation dynamics of a tidal marsh. *Aquatic Botany*, 55(4), 261–279. https://doi.org/10.1016/S0304-3770(96)01076-5

Figuerola, J., Green, A. J., & Santamaría, L. (2002). Comparative dispersal effectiveness of wigeongrass seeds by waterfowl wintering in south-west Spain: Quantitative and qualitative aspects. *Journal of Ecology*, 90(6), 989–1001. https://doi.org/10.1046/j.1365-2745.2002.00734.x

Fox, A. D., Elmberg, J., Tombre, I. M., & Hessel, R. (2017). Agriculture and herbivorous waterfowl: A review of the scientific basis for improved management. *Biological Reviews*, 92(2), 854–877. https://doi.org/10.1111/brv.12258

Frith, H. J., & Davies, S. J. J. F. (1961). Ecology of the magpie goose, *Anseranas semipalmata* Latham (Anatidae). *CSIRO Wildlife Research*, 6(2), 91–141. https://doi.org/10.1071/CWR9610091

Fuentes, C., Green, A. J., Orr, J., & Ólafsson, J. S. (2005). Seasonal variation in species composition and larval size of the benthic chironomid communities in brackish wetlands in Southern Alicante, Spain. *Wetlands*, 25(2), 289–296. https://doi.org/10.1672/5

Gislén, A., Dacke, M., Kröger, R. H., Abrahamsson, M., Nilsson, D. E., & Warrant, E. J. (2003). Superior underwater vision in a human population of sea gypsies. *Current Biology*, 13(10), 833–836.

Green, A. J., Figuerola, J., & Sánchez, M. I. (2002). Implications of waterbird ecology for the dispersal of aquatic organisms. *Acta Oecologica*, 23(3), 177–189. https://doi.org/10.1016/S1146-609X(02)01149-9

Green, A. J., Jenkins, K. M., Bell, D., Morris, P. J., & Kingsford, R. T. (2008). The potential role of waterbirds in dispersing invertebrates and plants in arid Australia. *Freshwater Biology*, 53(2), 380–392. https://doi.org/10.1111/j.1365-2427.2007.01901.x

Guiler, E. R. (1967). The Cape Barren goose, its environment, numbers and breeding. *Emu-Austral Ornithology*, 66(3), 211–235. https://doi.org/10.1071/MU966211

Guillemain, M., Duncan, P., & Fritz, H. (2001). Switching to a feeding method that obstructs vision increases head-up vigilance in dabbling ducks. *Journal of Avian Biology*, 32(4), 345–350. https://doi.org/10.1111/j.0908-8857.2001.320409.x

Guillemain, M., Fritz, H., & Blais, S. (2000). Foraging methods can affect patch choice: An experimental study in Mallard (*Anas platyrhynchos*). *Behavioural Processes*, 50(2–3), 123–129. https://doi.org/10.1016/S0376-6357(00)00095-4

Guillemain, M., Fritz, H., & Duncan, P. (2002). The importance of protected areas as nocturnal feeding grounds for dabbling ducks wintering in western France. *Biological Conservation*, 103(2), 183–198. https://doi.org/10.1016/S0006-3207(01)00120-3

Guillemain, M., Fritz, H., & Guillon, N. (2000). Foraging behavior and habitat choice of wintering Northern Shoveler in a major wintering quarter in France. *Waterbirds*, 353–363. https://doi.org/10.2307/1522173

Guillemain, M., Martin, G. R., & Fritz, H. (2002). Feeding methods, visual fields and vigilance in dabbling ducks (Anatidae). *Functional Ecology*, 16(4), 522–529. https://doi.org/10.1046/j.1365-2435.2002.00652.x

Guppy, H. B. (1906). *Observations of a Naturalist in the Pacific Between 1896 and 1899: Plant-Dispersal* (Vol. 2). Macmillan and Company Limited.

Gurd, D. B. (2006). Filter-feeding dabbling ducks (Anas spp.) can actively select particles by size. *Zoology*, 109(2), 120–126. https://doi.org/10.1016/j.zool.2005.10.002

Gyimesi, A., Franken, M. S., Feige, N., & Nolet, B. A. (2012a). Human disturbance of Bewick's swans is reflected in giving-up net energy intake rate, but not in giving-up food density. *Ibis*, 154(4), 781–790. https://doi.org/10.1111/j.1474-919X.2012.01253.x

Gyimesi, A., Varghese, S., De Leeuw, J., & Nolet, B. A. (2012b). Net energy intake rate as a common currency to explain swan spatial distribution in a shallow lake. *Wetlands*, 32(1), 119–127. https://doi.org/10.1007/s13157-011-0256-6

Hafner, H., Dugan, P. J., & Boy, V. (1986). Use of artificial and natural wetlands as feeding sites by Little Egrets (*Egretta garzetta* L.) in the Camarge Southern France. *Colonial Waterbirds*, 149–154. https://doi.org/10.2307/1521207

Halse, S. A. (1984). Diet, body condition, and gut size of Egyptian geese. *The Journal of Wildlife Management*, 48(2), 569–573. https://doi.org/10.2307/3801191

Hepp, G. R. (1985). Effects of environmental parameters on the foraging behavior of three species of wintering dabbling ducks (Anatini). *Canadian Journal of Zoology*, 63(2), 289–294. https://doi.org/10.1139/z85-044

Hohman, W. L., Stark, T. M., & Moore, J. L. (1996). Food availability and feeding preferences of breeding fulvous whistling-ducks in Louisiana ricefields. *The Wilson Bulletin*, 137–150.

Holm, K. J., & Burger, A. E. (2002). Foraging behavior and resource partitioning by diving birds during winter in areas of strong tidal currents. *Waterbirds*, 25(3), 312–325. https://doi.org/10.1675/1524-4695(2002)025[0312:FBARPB]2.0.CO;2

Holm, T. E. (2002). Habitat use and activity patterns of mute swans at a molting and a wintering site in Denmark. *Waterbirds*, 25, 183–191.

Jacobsen, O. W. (1992). Factors affecting selection of nitrogen-fertilized grassland areas by breeding Wigeon *Anas penelope*. *Ornis Scandinavica*, 23(2), 121–131. https://doi.org/10.2307/3676439

James, H. F., & Burney, D. A. (1997). The diet and ecology of Hawaii's extinct flightless waterfowl: Evidence from coprolites. *Biological Journal of the Linnean Society*, 62(2), 279–297. https://doi.org/10.1111/j.1095-8312.1997.tb01627.x

Janzen, D. H. (1984). Dispersal of small seeds by big herbivores: Foliage is the fruit. *The American Naturalist*, 123(3), 338–353.

Jensen, R. A., Wisz, M. S., & Madsen, J. (2008). Prioritizing refuge sites for migratory geese to alleviate conflicts with agriculture. *Biological Conservation*, 141(7), 1806–1818. https://doi.org/10.1016/j.biocon.2008.04.027

Johnsgard, P. A. (2010). *Ducks, Geese, and Swans of the World*. Revised Edition. University of Nebraska Press.

Jónsson, J. E., & Afton, A. D. (2016). Do foraging methods in winter affect morphology during growth in juvenile snow geese? *Ecology and Evolution*, 6(21), 7656–7670. https://doi.org/10.1002/ece3.2481

Kear, J. (2005). *Ducks, Geese, and Swans: Species Accounts* (*Cairina to Mergus*). Vol. 2. Oxford University Press.

Kehoe, F. P., & Thomas, V. G. (1987). A comparison of interspecific differences in the morphology of external and internal feeding apparatus among North American Anatidae. *Canadian Journal of Zoology*, 65(7), 1818–1822. https://doi.org/10.1139/z87-275

Khan, B., & Khan, M. M. H. (2024). Comparative feeding behavior of ducks and geese (family: Anatidae) in Rajshahi division of Bangladesh. *Bangladesh Journal of Zoology*, 52(3), 341–348.

Kirk, E. C. (2006). Eye morphology in cathemeral lemurids and other mammals. *Folia Primatologica*, 77(1–2), 27–49. https://doi.org/10.1159/000089694

Kooloos, J. G. M., & Zweers, G. A. (1991). Integration of pecking, filter feeding and drinking mechanisms in waterfowl. *Acta Biotheoretica*, 39(2), 107–140. https://doi.org/10.1007/BF00046595

Kooloos, J. G. M., Kraaijeveld, A. R., Langenbach, G. E. J., & Zweers, G. A. (1989). Comparative mechanics of filter feeding in *Anas platyrhynchos*, *Anas clypeata* and *Aythya fuligula* (Aves, Anseriformes). *Zoomorphology*, 108(5), 269–290. https://doi.org/10.1007/BF00312160

Kortright, F. H. (1942). *The Ducks, Geese and Swans of North America*. Stackpole.

Krapu, G. L. (1974). *Feeding Ecology of Pintail Hens During Reproduction*. US Fish and Wildlife Service.

Kumar, P. (2024). Foraging ecology of wetland birds in heterogeneous habitats: A case study of the Dighal wetlands, Haryana, India. *Journal of Applied and Natural Science*, 16(4), 1627. https://doi.org/10.31018/jans.v16i4.5976

Kuwae, T., Miyoshi, E., Sassa, S., & Watabe, Y. (2010). Foraging mode shift in varying environmental conditions by dunlin *Calidris alpina*. *Marine Ecology Progress Series*, 406, 281–289. https://doi.org/10.3354/meps08519

Lazarus, J., & Inglis, I. R. (1978). The breeding behaviour of the pink-footed goose: Parental care and vigilant behaviour during the fledging period. *Behaviour*, 65(1–2), 62–87. https://doi.org/10.1163/156853978X00198

Levy, B., & Sivak, J. G. (1980). Mechanisms of accommodation in the bird eye. *Journal of Comparative Physiology*, 137(3), 267–272. https://doi.org/10.1007/BF00657122

Li, D., Chen, S., Lloyd, H. U. W., Zhu, S., Shan, K. A. I., & Zhang, Z. (2013). The importance of artificial habitats to migratory waterbirds within a natural/artificial wetland mosaic, Yellow River Delta, China. *Bird Conservation International*, 23(2), 184–198. https://doi.org/10.1017/S0959270913000099

Li, Z., & Clarke, J. A. (2016). The craniolingual morphology of waterfowl (Aves, Anseriformes) and its relationship with feeding mode revealed through contrast-enhanced X-ray computed tomography and 2D morphometrics. *Evolutionary Biology*, 43(1), 12–25. https://doi.org/10.1007/s11692-015-9345-4

Lisney, T. J., Stecyk, K., Kolominsky, J., Schmidt, B. K., Corfield, J. R., Iwaniuk, A. N. et al. (2013). Ecomorphology of eye shape and retinal topography in waterfowl (Aves: Anseriformes: Anatidae) with different foraging modes. *Journal of Comparative Physiology A*, 199(5), 385–402. https://doi.org/10.1007/s00359-013-0802-1

Livezey, B. C., & Humphrey, P. S. (1986). Flightlessness in steamer-ducks (Anatidae: Tachyeres): Its morphological bases and probable evolution. *Evolution*, 40(3), 540–558. https://doi.org/10.1111/j.1558-5646.1986.tb00506.x

Lorenzen, B., & Madsen, J. (1986). Feeding by geese on the Filsø farmland, Denmark, and the effect of grazing on yield structure of spring barley. *Ecography*, 9(4), 305–311. https://doi.org/10.1111/j.1600-0587.1986.tb01224.x

Lovvorn, J. R., Grebmeier, J. M., Cooper, L. W., Bump, J. K., & Richman, S. E. (2009). Modeling marine protected areas for threatened eiders in a climatically changing Bering Sea. *Ecological Applications*, 19(6), 1596–1613. https://doi.org/10.1890/08-1193.1

Madge, S., & Burn, H. (1988). *Waterfowl: An Identification Guide to the Ducks, Geese, and Swans of the World*. Houghton Mifflin, Boston.

Marchowski, D., Jankowiak, Ł., & Wysocki, D. (2016). Newly demonstrated foraging method of herring gulls and mew gulls with benthivorous diving ducks during the nonbreeding period. *The Auk*, 133(1), 31–40. https://doi.org/10.1642/AUK-15-62.1

Martin, G. R., & Katzir, G. (1999). Visual fields in short-toed eagles, *Circaetus gallicus* (Accipitridae), and the function of binocularity in birds. *Brain Behavior and Evolution*, 53(2), 55–66. https://doi.org/10.1159/000006582

Martin, G. R., Jarrett, N., & Williams, M. (2007). Visual fields in blue ducks *Hymenolaimus malacorhynchos* and pink-eared ducks *Malacorhynchus membranaceus*: Visual and tactile foraging. *Ibis*, 149(1), 112–120. https://doi.org/10.1111/j.1474-919X.2006.00611.x

McKnight, S. K. (1998). Effects of food abundance and environmental parameters on foraging behavior of Gadwalls and American coots in winter. *Canadian Journal of Zoology*, 76(11), 1993–1998. https://doi.org/10.1139/z98-167

McNeil, R., Drapeau, P., & Goss-Custard, J. D. (1992). The occurrence and adaptive significance of nocturnal habits in waterfowl. *Biological Reviews*, 67(4), 381–419. https://doi.org/10.1111/j.1469-185X.1992.tb01188.x

Merke, F. R., & Mosbech, A. (2008). Diurnal and nocturnal feeding strategies in common eiders. *Waterbirds*, 31(4), 580–586. https://doi.org/10.1675/1524-4695-31.4.580

Miller, M. R., & McA. Eadie, J. (2006). The allometric relationship between resting metabolic rate and body mass in wild waterfowl (Anatidae) and an application to estimation of winter habitat requirements. *The Condor*, 108(1), 166–177. https://doi.org/10.1093/condor/108.1.166

Mott, K. M. (1994). *A Functional Morphological Approach to the Study of Filter-Feeding in Dabbling Ducks, Anas* spp. Guelph, Ontario, Canada: University of Guelph.

Mukherjee, A., Bandyopadhyay, A., Pal, S., & Mukhopadhyay, S. K. (2021). Foraging habitats and foraging techniques of five wintering Anatidae waterfowl in light of genetic distances. *Russian Journal of Ecology*, 52(6), 567–577. https://doi.org/10.1134/S1067413622010088

Mukherjee, A., Pal, S., Das, P., & Mukhopadhyay, S. K. (2023). Time activity budget and foraging behavior: Important determinants of resource sharing and guild structure in wintering waterbirds. *European Journal of Wildlife Research*, 69(2), 30. https://doi.org/10.1007/s10344-023-01648-4

Nolet, B. A., Bevan, R. M., Klaassen, M., Langevoord, O., & Van der Heijden, Y. G. J. T. (2002). Habitat switching by Bewick's swans: Maximization of average long-term energy gain? *Journal of Animal Ecology*, 71(6), 979–993. https://doi.org/10.1046/j.1365-2656.2002.00662.x

Nudds, T. D., & Bowlby, J. N. (1984). Predator–prey size relationships in North American dabbling ducks. *Canadian Journal of Zoology*, 62(10), 2002–2008. https://doi.org/10.1139/z84-293

Nudds, T. D., & Kaminski, R. M. (1984). Sexual size dimorphism in relation to resource partitioning in North American dabbling ducks. *Canadian Journal of Zoology*, 62(10), 2009–2012. https://doi.org/10.1139/z84-294

Nudds, T. D., Sjöberg, K., & Lundberg, P. (1994). Ecomorphological relationships among Palearctic dabbling ducks on Baltic coastal wetlands and a comparison with the Nearctic. *Oikos*, 295–303. https://doi.org/10.2307/3546150

Nudds, T. D., Elmberg, J., Pöysä, H., Sjöberg, K., & Nummi, P. (2000). Ecomorphology in breeding Holarctic dabbling ducks: The importance of lamellar density and body length varies with habitat type. *Oikos*, 91(3), 583–588. https://doi.org/10.1034/j.1600-0706.2000.910321.x

Olney, P. J. S. (1968). The food and feeding-habits of the pochard, Aythya ferina. *Biological Conservation*, 1(1), 71–76. https://doi.org/10.1016/0006-3207(68)90026-8

Olsen, A. M. (2015). Exceptional avian herbivores: Multiple transitions toward herbivory in the bird order Anseriformes and its correlation with body mass. *Ecology and Evolution*, 5(21), 5016–5032. https://doi.org/10.1002/ece3.1787

Olsen, A. M. (2017). Feeding ecology is the primary driver of beak shape diversification in waterfowl. *Functional Ecology*, 31(10), 1985–1995. https://doi.org/10.1111/1365-2435.12890

Owen, M., & Thomas, G. J. (1979). The feeding ecology and conservation of wigeon wintering at the Ouse Washes, England. *Journal of Applied Ecology*, 795–809. https://doi.org/10.2307/2402854

Owen-Smith, N. (1998). How high ambient temperature affects the daily activity and foraging time of a subtropical ungulate, the greater kudu (*Tragelaphus strepsiceros*). *Journal of Zoology*, 246(2), 183–192. https://doi.org/10.1111/j.1469-7998.1998.tb00147.x

Özkan, E., Mücaviroğlu, E., Nicoleta, M., & Günay, E. (2024). Exploring shape variance in waterbirds' pad feet: A geometric morphometric analysis. *Harran University Journal of the Faculty of Veterinary Medicine*, 13(2), 141–147.

Páez, D. J., Restif, O., Eby, P., & Plowright, R. K. (2018). Optimal foraging in seasonal environments: Implications for residency of Australian flying foxes in food-subsidized urban landscapes. *Philosophical Transactions of the Royal Society B: Biological Sciences*, 373(1745), 20170097. https://doi.org/10.1098/rstb.2017.0097

Pearse, A. T., Alisauskas, R. T., Krapu, G. L., & Cox Jr, R. R. (2011). Changes in nutrient dynamics of midcontinent greater white-fronted geese during spring migration. *The Journal of Wildlife Management*, 75(8), 1716–1723. https://doi.org/10.1002/jwmg.223

Pecsics, T., Laczi, M., Nagy, G., & Csörgő, T. (2017). The cranial morphometrics of the wildfowl (Anatidae). *Ornis Hungarica*, 25(1), 44–57. https://doi.org/10.1515/orhu-2017-0004

Pettigrew, J. D., Dreher, B., Hopkins, C. S., McCall, M. J., & Brown, M. (1988). Peak density and distribution of ganglion cells in the retinae of microchiropteran bats: Implications for visual acuity (Part 2 of 2). *Brain, Behavior and Evolution*, 32(1), 48–56. https://doi.org/10.1159/000316042

Pöysä, H. (1983). Morphology-mediated niche organization in a guild of dabbling ducks. *Ornis Scandinavica*, 317–326. https://doi.org/10.2307/3676325

Pöysä, H. (1983). Resource utilization pattern and guild structure in a waterfowl community. *Oikos*, 295–307. https://doi.org/10.2307/3544594

Pöysä, H. (1984). Temporal and spatial dynamics of waterfowl population in a wetland area – A community ecological approach. *Ornis Fennica*, 61(4), 99–108.

Pöysä, H. (1986). Species composition and size of dabbling duck (Anas spp.). *Ornis Fennica*, 63(2), 33–41.

Raikow, R. J. (1970). *Evolution of Diving Adaptations in the Stiff-Tail Ducks*. University of California Press, Berkeley.

Raikow, R. J. (1973). Locomotor mechanisms in North American ducks. *The Wilson Bulletin*, 295–307.

Raulings, E., Morris, K. A. Y., Thompson, R., & Nally, R. M. (2011). Do birds of a feather disperse plants together? *Freshwater Biology*, 56(7), 1390–1402. https://doi.org/10.1111/j.1365-2427.2011.02576.x

Ridley, H. N. (1930). *The Dispersal of Plants Throughout the World*. L. Reeve & Co., Ltd.

Rizzo, E. & Battisti, C. (2009). Habitat preferences of Anatidae (Aves, Anseriformes) in a Mediterranean patchy wetland (central Italy). *Ekologia*, 28(1), 66–73.

Sedinger, J. S. (1997). Adaptations to and consequences of an herbivorous diet in grouse and waterfowl. *The Condor*, 99(2), 314–326. https://doi.org/10.2307/1369937

Sekiya, Y., Hiratsuka, J. I., Yamamuro, M., Oka, N., & Abe, M. (2000). Diet selectivity and shift of wintering common pochards and tufted ducks in a eutrophic coastal lagoon. *Journal of Marine Systems*, 26(2), 233–238. https://doi.org/10.1016/S0924-7963(00)00057-9

Sherwood, G. A. (1960). The whistling swan in the west with particular reference to Great Salt Lake Valley, Utah. *The Condor*, 62(5), 370–377. https://doi.org/10.2307/1365166

Shi, J., Meng, L., Xia, S., Liu, S., & Zhou, L. (2024). Habitat suitability and determinants for Anatidae in multi-watershed composite wetlands in Anhui, China. *Animals*, 14(7), 1010. https://doi.org/10.3390/ani14071010

Sivak, J. G., Hildebrand, T., & Lebert, C. (1985). Magnitude and rate of accommodation in diving and nondiving birds. *Vision Research*, 25(7), 925–933. https://doi.org/10.1016/0042-6989(85)90203-2

Sjöberg, K. (1985). Foraging activity patterns in the Goosander (*Mergus merganser*) and the red-breasted merganser (*M. serrator*) in relation to patterns of activity in their major prey species. *Oecologia*, 67(1), 35–39. https://doi.org/10.1007/BF00378448

Skieresz-Szewczyk, K., & Jackowiak, H. (2014). Scanning electron microscopy investigation of the filter-feeding apparatus in the domestic goose (*Anser anser f. domestica*) and the domestic duck (*Anas platyrhynchos f. domestica*). *Microscopy: Advances in Scientific Research and Education* (A. Méndez-Vilas, Ed.), 1, 2014.

Soons, M. B., Brochet, A. L., Kleyheeg, E., & Green, A. J. (2016). Seed dispersal by dabbling ducks: an overlooked dispersal pathway for a broad spectrum of plant species. *Journal of Ecology*, 104(2), 443-455. https://doi.org/10.1111/1365-2745.12531

Stephenson, R. (1994). Diving energetics in lesser scaup (*Aythyta affinis*, Eyton). *Journal of Experimental Biology*, 190(1), 155–178. https://doi.org/10.1242/jeb.190.1.155

Sundar, K. G., & Kittur, S. (2013). Can wetlands maintained for human use also help conserve biodiversity? Landscape-scale patterns of bird use of wetlands in an agricultural landscape in north India. *Biological Conservation*, 168, 49–56. https://doi.org/10.1016/j.biocon.2013.09.016

Swanson, G. A. (1984). Invertebrates consumed by dabbling ducks (Anatinae) on the breeding grounds. *Journal of the Minnesota Academy of Science*, 50(3), 37–40.

Systad, G. H., & Bustnes, J. O. (2001). Coping with darkness and low temperatures: Foraging strategies in Steller's eiders, *Polysticta stelleri*, wintering at high latitudes. *Canadian Journal of Zoology*, 79(3), 402–406. https://doi.org/10.1139/z00-213

Taylor, D. P., Dvorett, D. A., Vradenburg, J. N., & Smith, L. M. (2019). Influence of environmental stress and anthropogenic disturbance on the energy expenditure of wintering northern pintails (*Anas acuta*). *Waterbirds*, 42(3), 294–303. https://doi.org/10.1675/063.042.0305

Tome, M. W., & Wrubleski, D. A. (1988). Underwater foraging behavior of canvasbacks, lesser scaups, and ruddy ducks. *The Condor*, 90(1), 168–172. https://doi.org/10.2307/1368445

van Leeuwen, C. H., Van der Velde, G., van Groenendael, J. M., & Klaassen, M. (2012). Gut travellers: Internal dispersal of aquatic organisms by waterfowl. *Journal of Biogeography*, 39(11), 2031–2040.

Viain, A., Corre, F., Delaporte, P., Joyeux, E., & Bocher, P. (2013). Numbers, diet and feeding methods of common shelduck *Tadornatadorna* wintering in the estuarine bays of Aiguillon and Marennes-Ol. *Wildfowl*, 61(61), 121–141.

Vickery, J., & Gill, J. (1999). Managing grassland for wild geese in Britain: A review. *Biological Conservation*, 89(1), 93–106. https://doi.org/10.1016/S0006-3207(98)00134-7

Vivian-Smith, G., & Stiles, E. W. (1994). Dispersal of salt marsh seeds on the feet and feathers of waterfowl. *Wetlands*, 14(4), 316–319. https://doi.org/10.1007/BF03160638

Walmsley, J. G., & Moser, M. E. (1981). The winter food and feeding habits of shelduck in the Camargue, France. *Wildfowl*, 32(32), 99–106.

Weller, M. W. (1999). *Wetland Birds: Habitat Resources and Conservation Implications*. Cambridge University Press.

Willson, M. F., Traveset, A., and Sabag, C. (1997). Geese as frugivores and probable seed-dispersal mutualists. *Journal of Field Ornithology*, 68, 144–146.

Zheng, M., Zhou, L., Zhao, N., & Xu, W. (2015). Effects of variation in food resources on foraging habitat use by wintering Hooded Cranes (*Grus monacha*). *Avian Research*, 6(1), 11. https://doi.org/10.1186/s40657-015-0020-3

Zweers, G. A. (1973). Structure, movement, and myography of the feeding apparatus of the mallard (*Anas platyrhynchos* L.): A study in functional anatomy. *Netherlands Journal of Zoology*, 24(4), 323–467.

Zweers, G. A. (1999). Cranial kinesis in birds: Consequences for the evolution of the jaw apparatus. *Proceedings of the 22nd International Ornithological Congress* (Durban, South Africa), 82–93.

Zweers, G. A., Berkhoudt, H., & Vanden Berge, J. C. (1994). Behavioral mechanisms of avian feeding. *In: Biomechanics of Feeding in Vertebrates* (pp. 241–279). Berlin, Heidelberg: Springer Berlin Heidelberg. https://doi.org/10.1007/978-3-642-57906-6_9

Žydelis, R., & Richman, S. E. (2015). Foraging behavior, ecology, and energetics of sea ducks. *Ecology and Conservation of North American Sea Ducks. In: Studies in Avian Biology*, 46, 241–266.

CHAPTER
6

Migration of the Members of Anatidae Family

Anupama Yadav[1,2]* **and Khushboo Chaturvedi**[1]

[1] Department of Zoology, University of Lucknow, Lucknow – 226007, Uttar Pradesh, India

[2] CSIR-Central Drug Research Institute, Lucknow, Uttar Pradesh, India

1. Overview of the Anatidae Family and the Importance of Migration for Their Survival and Ecological Roles

The order Anseriformes of class Aves includes Anatidae family, the largest family of waterfowl encompassing ducks, geese, and swans which are small to large sized birds having webbed-feet and broad bills. These birds are omnipresent except Antarctica (Livezey, 1997), show widespread distributions, and are abundant in number. Primarily, they live around the water bodies and are adapted to swim, float, and dive. Members of the family Anatidae are widely distributed across diverse inland aquatic ecosystems, including lentic habitats such as lakes and ponds, as well as lotic systems such as rivers and streams. While these freshwater sites are critical for nesting, certain taxa exhibit a seasonal shift in habitat, migrating to marine environments during the non-breeding period. Anatidae is a diverse family of birds with 53 genera, including 173 species. Their conservation is a matter of concern as, out of 173 species, several are extinct, and many are critically endangered. The 53 genera of Anatidae are categorized into 5 subfamilies: Anatinae, Anserinae, Dendrocygninae, Stictonettinae, and Tadorninae. The subfamily Anatinae (largest) includes eight groups: Tadornini

*Corresponding author: anupamayadav737@gmail.com

DOI: 10.1201/9781003669784-6

(shelducks and allies: 5 genera, 14 species); Tachyerini (steamer ducks: 1 genus, 4 species); Cairini (perching ducks and allies: 9 genera, 13 species); Merganettini (torrent duck (*Merganetta armata*)); Anatini (dabbling ducks: 4 genera, 40 species); Aythyini (pochards: 2 genera, 15 species); Mergini (mergansers and allies: 7 genera, 18 species); and Oxyurini (stifftails: 3 genera, 8 species) (Howard, 2003). Within the family Anatidae, members of the subfamilies Anseranatinae and Anserinae typically exhibit sexual monomorphism regarding their plumage. Conversely, taxa within the subfamily Anatinae are characterized by sexual dimorphism, often displaying pronounced visual differences between sexes. The chromatic range of these feathers encompasses a spectrum of brown, gray, and white, as well as various black-and-white configurations. For many Anatidae species, the young birds mostly feed on invertebrates. Whereas, in general, they have herbivorous diet and are monogamous birds during adulthood. However, some adults also relish upon fish, mollusks, or aquatic arthropods. The members of Anatidae are strongly gregarious. These birds frequently assemble in flocks of diverse sizes, from small clusters (a dozen) or fewer to vast congregations that can reach several hundred thousand individuals and largely depend on time of year, species composition, and area. When in group, they usually feed, roost, and loaf together either on water or on the land, or they move about together, such as during migration (Carboneras, 1992). Several members of the Anatidae family undertake annual migrations (Newton, 2023); however, during the non-breeding season, tropical and subtropical birds remain in proximity of their breeding grounds.

Migration is a ubiquitous trait among Anatidae, with species exhibiting diverse strategies, from short-distance movements to transcontinental journeys (Alerstam et al., 2003). The migratory flights of Anatids are supported by pointed and short wings that are strengthened by strong flight muscles that produce quick beats during migratory flight. Anatidae species exhibit V-formation flight, which reduces energy expenditure through aerodynamic drafting (Portugal et al., 2014). This adaptation is principally crucial for species such as tundra swans (*Cygnus columbianus*), which are long-distance migratory birds. Waterfowl use multiple cues for navigation, including geomagnetic fields, celestial patterns, and landscape features (Moayedi et al., 2025). Eurasian teals (*Anas crecca*) follow consistent migratory routes across years, suggesting a strong endogenous component to their navigation. These birds typically migrate along the Black Sea–Mediterranean flyway, with stopovers in Southeastern Europe, and show individual consistency in timing and route selection (Giunchi et al., 2019). Migratory Anatidae utilize several primary migratory routes, often referred to as flyways, to migrate between non-breeding and breeding grounds. Two major flyways in Asia are the Central–South Asian flyway and the East Asian flyway, according to Wetlands International. The East Asia–Australasian flyway is another significant route, extending from the Arctic to Australia and New Zealand. Within the Western Palearctic region, major flyways include the North Sea, Black Sea, and Caspian Sea flyways.

Seasonal migration becomes a necessity, particularly in the temperate region, owing to extreme cold during winters. In temperate zones across the Northern and Southern Hemispheres, their breeding grounds, which provide abundant food and water in spring and summer, often freeze in winter, and the limited areas that stay unfrozen cannot sustain the larger populations post breeding season (Carboneras, 1992). Several members of the Anatidae family are economically and culturally significant species; some species such as mallards (*Anas platyrhynchos*) and Muscovy ducks (*Cairina moschata*) are domesticated for food and other products, while others such as wood duck (*Aix sponsa*) and Canada goose (*Branta canadensis*) are hunted for sport and admired for their beauty. Habitat loss and pollution pose ongoing threats to waterfowl populations, among many other challenges.

Migration is a crucial phenomenon for the survival of Anatidae birds as it enables them to access resources such as food and breeding grounds, reducing competition. It allows Anatidae to move to areas where food sources are abundant and breeding conditions are suitable, ensuring their survival and reproductive success. Migration also provides them the opportunity to adapt to seasonal changes in their environment and maximize their chances of survival. Anatidae can influence ecosystem dynamics by feeding on aquatic plants and insects, impacting plant communities and the food web. Long-distance migratory birds such as the greater white-fronted goose (*Anser albifrons*) migrate from breeding grounds in Arctic regions to wintering sites in the temperate regions, to exploit seasonal food abundance (Ely & Takekawa, 1996). Their journey is driven not only by the need to access abundant food and avoid harsh climatic conditions, but also to reduce predation pressures from animals such as the polar bear (*Ursus maritimus*) (Alerstam et al., 2003; Prop et al., 2013). Migration allows anatids to capitalize particularly on seasonal peaks in food availability. Breeding sites in the Arctic and temperate zones provide abundant insects and vegetation during summer, while wintering sites in milder climates offer open water and plant resources (Dingle, 2014). Without migration, these birds would face starvation or reproductive failure in harsh winter conditions.

The movements of Anatidae can modify habitats through their foraging and nesting activities, creating and maintaining certain habitat structures. By moving to different areas, Anatidae can reduce competition with other species, including other Anatidae, for resources. Migratory behavior reduces predation pressure by enabling synchronized breeding and rapid fledging. Species such as Arctic tern (*Sterna paradisaea*) have been reported to nest in Arctic regions (Fijn et al., 2013), probably because predator densities are lower during the short summer. Additionally, migrating in large flocks perhaps dilutes individual predation risk. For example, brant geese (*Branta bernicla*) migrate in tightly coordinated flocks to minimize exposure to predators such as Arctic foxes. In fact, flocking evolved among species due to ecological factors such as food availability, its distribution, and predation pressure experienced while foraging (Beauchamp, 2022). Moreover, these birds may undertake short as well as long-distance migratory journeys, frequently following the transcontinental migratory flyways (Owen & Black, 1990), which facilitates outbreeding and genetic diversity and thereby reduces the chances of inbreeding. Particularly, long-distance migration promotes genetic mixing across populations, reducing inbreeding and enhancing adaptability (Webster et al., 2002). Studies on the greater white-fronted goose show that continental scale migratory connectivity facilitates phenotypic flexibility (Fox & Abraham, 2017) and perhaps maintains genetic diversity, which is crucial for resilience against diseases and environmental changes. Thus, this migratory behavior not only ensures their survival but also maintains genetic diversity by connecting distant populations, which is crucial for long-term species suppleness (Webster et al., 2002). Without migration, Anatidae would face severe challenges, including food shortages, increased predation, and reproductive failure, highlighting its fundamental role in their life cycle (Dingle, 2014).

Migration is ecologically important too, as it influences ecosystem dynamics. Beyond their own survival, migratory Anatidae also play crucial ecological roles in their ecosystems (Green & Elmberg, 2014). For instance, the seasonal migration of the members of Anatidae family can contribute to nutrient cycling (Boros, 2021; de Arruda Almeida et al., 2019; Hahn et al., 2007) via guano deposition through fecal matter which enriches wetlands with nitrogen and phosphorus, enhancing primary productivity and supporting other aquatic organisms (Post et al., 1998), and their seasonal migration also shapes landscapes (especially in areas where they are concentrated primarily through nutrient dispersal and seed dispersal). Many species act as key pollinators and seed dispersers; they ingest and transport plant seeds across vast distances, which helps maintain wetland vegetation and biodiversity (Figuerola & Green, 2002). For example, seed dispersal by mallards and other dabbling ducks facilitates aquatic plant regeneration across continents (Figuerola & Green, 2002; Soons et al., 2016). Research in the East Asian–Australasian flyway shows that teal species significantly disperse aquatic plant seeds in their wintering grounds (Soons et al., 2016). Such transport of nutrients and seed dispersion is often widespread ranging from breeding sites to overwintering sites, and vice versa. Therefore, migration is a critical survival strategy for Anatidae birds, enabling them to exploit seasonal resources across vast geographical ranges (Newton, 2023).

Anatidae birds also play a critical role in forming food webs by being prey for predators and modifying habitats through their foraging behaviors (New World Encyclopedia, 2023). Such ecological functions underscore their importance in sustaining healthy wetland ecosystems, one of the most productive and ecologically rich habitats on this planet. Not only does the Anatidae family contributes for faunal diversity and richness, its ecological services for enhancing floral abundance are also no less. For example, grazing of Arctic stoloniferous grass by lesser snow geese (*Anser caerulescens*) has been shown to stimulate their growth due to overcompensation (Bazely & Jefferies, 1989), creating a feedback loop that likely benefits other herbivores too.

2. Climate and Habitat Change as Drivers of Migration in the Anatidae Family

Climate shapes the intricate migratory pattern, which plays a pivotal role in the life-history of Anatids. These birds rely on finely tuned environmental cues to time their movements between breeding and wintering grounds, but anthropogenic climate change and habitat destruction are increasingly disrupting these patterns. Rising global temperatures, shifting precipitation regimes, rise in sea levels, and the destruction of wetlands are altering the suitability of critical stopover and breeding sites, forcing anatids to adapt alternative migratory behaviors or face population declines. Studies document how these changes affect migration timing, route selection, and reproductive success, with cascading consequences for ecosystems and biodiversity.

Environmental changes are resulting in altered migratory behaviors in some anatids. For instance, Canada goose and mallard showed significant advancement in their arrival to their breeding ground among the many species of Anseriformes studied by Murphy-Klassen et al. (2005) in response to rising global temperature. Some birds retort to rising air temperatures of spring by migrating early (Žalakevičius, 1990). For instance, the birds initiated migration following the second day of high ambient temperature, which helped in optimizing the flight conditions (Žalakevičius, 1993). The relationship between migration and daily weather as reviewed by Richardson (1990) suggested the best conditions for migration to be calm winds in the direction of the migratory pathways, no precipitation, low pressure, and warm temperatures. Eventually, weather at the time of take-off, along the route, and at the final destination, influences the timing of bird migration (Richardson, 1978). As per Murphy-Klassen et al. (2005), approximately 40% waterbirds during the study showed significant advancement in spring arrival dates. Waterfowls use open water as stopover sites, thus early appearance of open water sources along the migratory route might result in early arrival of these birds to their breeding grounds. Thus, a better predictor of their arrival dates could be the dates of ice breakup. Spring and autumn migration time and prominent behavioral and habitat features of some key Anatidae birds are shown in Table 1.

Coastal wetlands, essential for wintering and migrating Anatidae, are under threat from rising sea levels and saltwater intrusion. Studies suggest that water-level changes affect food availability and accessibility for migratory waterbirds, with sustained high water levels potentially reducing suitable habitats and foraging opportunities (Aharon-Rotman et al., 2017). A study from Anhui Province, China, reported that large floodplains of the Yangtze River were the most suitable habitats for migratory waterbirds (Shi et al., 2024). Unambiguously, flat terrains near farmlands were associated with higher bird richness, whereas increased elevation and human activities resulted in lower bird richness (Shi et al., 2024). With the help of ensemble prediction model, Shi et al. (2024) predicted a significant

Table 1. Spring and autumn migration timings, and key behavioral and habitat features of selected Anatidae species

Species	Spring Departure	Autumn Arrival	Behavior and Habitat
Mallard (*Anas platyrhynchos*)	Feb–Mar	Sep–Oct	Early migrant; flexible staging behavior; income breeder.
Northern pintail (*Anas acuta*)	Feb–Mar	Sep–Oct	Sensitive to wetland conditions; relies on staging site quality.
Eurasian wigeon (*Mareca penelope*)	Mar–Apr	Oct–Nov	Later spring migrant; prefers shallow wetlands.
Eurasian teal (*Anas crecca*)	Feb–Mar	Sep–Oct	Small-bodied; highly dependent on stopover food availability.
Bar-headed goose (*Anser indicus*)	Mar–Apr	Oct–Nov	High-altitude migrant; timing influenced by Himalayan snowmelt.
Ruddy shelduck (*Tadorna ferruginea*)	Mar	Oct	Uses Central Asian flyways; timing linked to monsoon onset.
Gadwall (*Mareca strepera*)	Mar–Apr	Oct–Nov	Prefers freshwater marshes; moderate-distance migrant.
Northern shoveler (*Spatula clypeata*)	Feb–Mar	Sep–Oct	Filter-feeding duck; migrates early to exploit ephemeral wetlands.
Ferruginous duck (*Aythya nyroca*)	Mar	Oct	Near-threatened; prefers deeper wetlands; sensitive to disturbance.
Tufted duck (*Aythya fuligula*)	Mar–Apr	Oct–Nov	Diving duck; uses large water bodies; late migrant.

loss in the coastal marsh habitats and overall decline in quantity and quality of habitat for ducks. Studies suggest coastal marshes are particularly vulnerable to a rise in sea-level (Moon et al., 2021). Conversion of wetlands to croplands accounted for the loss of 3.4 million km^2 of inland wetlands since 1700 as per the previous literature (Fluet-Chouinard et al., 2023). Wetland loss has been localized in Europe, the United States, and China, and projections indicate that by 2100, up to 30% of coastal wetlands could be lost, forcing ducks and geese to seek alternative habitats or face population bottlenecks (Galbraith et al., 2002; Fluet-Chouinard et al., 2023). The Yellow Sea, a major stopover for millions of waterbirds and ducks, has seen a 65% reduction in tidal flats (Studds et al., 2017). In North America, the draining of prairie potholes, a vital breeding habitat for mallards and blue-winged teal (*Spatula discors*), has reduced nesting success by fragmenting habitats and increasing predation (Ballard et al., 2013).

Of the several impacts of global climate change on migratory Anatidae, the disruption of phenological synchrony, that is, the alignment between migration timing and resource availability, is one of the most significant impacts. Many species, particularly Arctic-nesting geese such as the greater white-fronted goose and the swan goose (*Anser cygnoides*), rely on seasonal peaks in vegetation growth to fuel reproduction (Nolet et al., 2020). However, Arctic warming has accelerated snowmelt and plant emergence, leading to a trophic mismatch between geese arrival and peak forage timing (Doiron et al., 2015). This desynchronization reduces gosling survival and overall reproductive success of these arctic nesting geese (Doiron et al., 2015).

Breeding and nesting requirements serve as primary drivers of seasonal movements across continents for migrants. In order to optimize reproductive success, migration patterns are intricately linked to the availability of safe nesting sites, abundant food resources, and favorable climatic conditions during critical life-history stages. Breeding site selection is influenced by multiple factors, including predator avoidance, habitat structure, resource availability, and social dynamics, all of which interact to shape migratory strategies of anatids (Deng et al., 2021; Sun et al., 2023). As per previous literature, geese depend largely on shoals for their nutrition and protection from predators (Deng et al., 2021). However, dabbling ducks and diving ducks prefer more open waters compared to vegetation cover (Zhu et al., 2022).

Many Anatidae species (such as lesser snow goose, bufflehead, etc.) exhibit strong philopatry, that is, the tendency to return to natal breeding sites, which plays a crucial role in maintaining migratory connectivity (Gauthier, 1990; Robertson & Cook, 1999). This behavior ensures access to familiar habitats with predictable food supplies, nevertheless this makes the populations vulnerable to localized environmental changes. For example, the degradation of coastal nesting sites due to permafrost thaw and increased predation pressure from expanding fox populations has forced some snow goose colonies to shift their breeding locations (Samelius et al., 2011; Tape et al., 2013). Correspondingly, ducks such as the common eider (*Somateria mollissima*) rely on specific island habitats for nesting, where isolation from terrestrial predators enhances reproductive success (Tomlik, 2019). However, rising sea levels and human disturbance threaten these critical sites, potentially disrupting long-established migration routes.

Predation pressure is also one of the major selective forces shaping the breeding migrations of Anatidae (Green, 1996). Many species migrate to remote or inaccessible areas to minimize nest predation, while others build nests in proximity to more aggressive species. For instance, pochard (*Aythya ferina*) and tufted duck (*Aythya fuligula*) often nest in association with small larids, such as the black-headed gull (*Larus ridibundus*), to avoid nest predation (Väänänen, 2000). Similarly, sea ducks such as the long-tailed duck (*Clangula hyemalis*) and common merganser (*Mergus merganser*) breed near oligotrophic lakes or coastal areas with low predator densities (Harvey & Connors, 2002). However, anthropogenic changes, including the expansion of invasive predators (e.g., foxes and raccoons) and human disturbance, have eroded these refuges, leading to increased nest failure rates (Harvey & Connors, 2002). In response, some species, such as the wood duck, have adapted to use artificial nest boxes, which has helped in stabilizing their population in fragmented habitats (Semel & Sherman, 1995).

Social learning and genetic predispositions also play a role in shaping migratory behavior related to breeding (Newton, 2007). Many geese and swans migrate in family groups, with juveniles learning migration routes from their parents (Newton, 2007). This cultural transmission ensures the perpetuation of traditional stopover and breeding sites but can limit adaptability when environmental conditions change. For example, the decline of bulrush marshes has disrupted the migration patterns of the declining long-tailed duck population, which relies on these habitats for both breeding and molting (Meehan et al., 2009). In contrast, species with more flexible genetic programing, such as the mallard, can adjust migration routes more readily in response to habitat alterations (van Toor et al., 2013).

Breeding and nesting requirements are fundamental drivers of migration in Anatidae, shaping their movements and life-history strategies in profound ways. The loss of critical habitats, climate-induced phenological mismatches, and increasing predation pressure threaten the reproductive success of many species, with cascading effects on population dynamics. Effective conservation requires international cooperation, habitat restoration, and measures to address climate change. Protecting the ecological integrity of breeding sites is vital to ensure the persistence of these migratory birds and their ecosystems.

3. Anatomical and Metabolic Adaptations for Migration in the Anatidae Family

Migration represents one of the most energetically demanding feats in the avian world, requiring specialized physiological adaptations that enable these birds to undertake long-distance flights, endure extreme environmental conditions, and efficiently utilize energy reserves. The anatomical and metabolic modifications observed in migratory Anatidae reflect millions of years of evolutionary refinement, allowing them to optimize flight performance, thermoregulation, and fuel utilization during their arduous journeys. These adaptations include musculoskeletal enhancements for sustained flight, cardiovascular and respiratory modifications for high-altitude performance, digestive system plasticity to accommodate rapid fuel deposition, and endocrine mechanisms that precisely regulate migratory physiology. Understanding these adaptations provides crucial insights into how Anatidae cope with the physiological challenges of migration and how environmental changes may impact their migratory capabilities.

The flight apparatus of migratory Anatidae demonstrates numerous anatomical specializations that maximize aerodynamic efficiency and endurance. Compared to non-migratory waterfowl, migratory species exhibit proportionally larger pectoral muscles anchored to an enlarged sternum with a deeper keel, providing greater mechanical advantage for the powerful wing strokes needed during long-distance flight (Norberg, 1990; Nyeland et al., 2003; Rayner, 1988). The humerus and ulna bones show increased cortical thickness and mineralization to withstand the mechanical stresses of prolonged flapping flight (Habib & Ruff, 2008; Osiak-Wicha et al., 2024). Wing morphology is particularly correlated with aerodynamic performance of birds while flying, with high aspect ratio wings (long and narrow) facilitate soaring and gliding during long overwater crossings, whereas low aspect ratio exhibits (short and broad wings) better maneuverability in windy coastal environments (Norberg, 1990).

Migratory Anatidae employ several behavioral adaptations that complement their anatomical specializations. Many species such as white geese, Canadian geese, and cranes fly in V-formations or echelon flight to save energy and improve navigation (Ballerini et al., 2008). The two most frequently observed formations, that is, the V and echelon formations, offer reduced drag and increased lift significantly enhancing the aerodynamic benefits to the migrating birds, allowing them to fly farther and with greater ease (Lissaman & Shollenberger, 1970). Formation flight does not offer the similar benefits to all the migratory birds. As per the data, the lead bird and the last following bird exert more energy to fly (Kshatriya & Blake, 1992). The uplift generated by the leader bird improves the performance of the subsequently following birds in the line (Mirzaeinia et al., 2020). The lead bird switches its position in the wake of each other. Since the lead bird and the last bird spend more energy in their respective positions, they switch their position with the middle birds who have more available energy as a result of formation flight benefit (Mirzaeinia et al., 2020).

The metabolic transformations associated with migration in birds are among the most extreme in vertebrates. Premigratory hyperphagia increases body mass primarily through subcutaneous and visceral fat deposition (Blem, 1990). The liver undergoes hypertrophy and upregulates lipogenic enzymes to facilitate rapid fat synthesis (Jenni-Eiermann & Jenni, 1994; Pearson et al., 1983). Interestingly, corticosterone has also been reported to have permissive effects on food intake facilitating hyperphagia and henceforth fat deposition (Bauer & Watts, 2021). Besides corticosterone, certain appetite-regulating hormones show differential response with respect to avian feeding habit. Ghrelin, a potent orexigenic hormone, produced in the gut, stimulates food intake in mammals, but in contrast, birds' ghrelin typically inhibits food intake (Kaiya et al., 2013). Likewise, leptin produced by the adipose tissue is also known to stimulate food intake in mammals (Friedman & Halaas, 1998), but its role is questionable in birds (Boswell & Dunn, 2015). Beside appetite-stimulating hormones, anorexigenic hormones such as glucagon-like peptide 1 (GLP-1), cholecystokinin (CCK), and gastrin-releasing peptide (GRP) produced by the gut and adipose tissue control the appetite and fuel reserves in birds. Studies on mammals show inhibitory effect of these hormones on food intake (Dakin et al., 2004). Similarly, in migratory birds held on short photoperiods (imitating winter time), intraperitoneal injections of CCK reduced food intake for short term, but not for long term (Richardson et al., 1993). CCK has short half-life of 1–2 minutes, thus its inhibitory effects on food intake are short-lived (Eikenaar, 2017). Studies on chicken and turkey report the reduced food intake in response to bombesin, imitating gastrin-releasing peptide (Denbow, 1989; Savory, 1987). Thus, the endocrine regulation of hyperphagia in migratory birds is yet not disclosed. During flight, metabolic rates increase over resting levels, with fat providing the maximum energy through β-oxidation (McWilliams et al., 2004). Unique among vertebrates, migratory waterfowl maintain high rates of fatty acid transport across muscle membranes via specialized fatty-acid-binding proteins (FABPs) than in non-migratory species (Sheta et al., 2025). Thus, fats are the primary fuel for long-distance, energy-seeking flights in migratory waterfowls.

During migratory flights, the birds' performance depends on various environmental factors such as wind speed, air temperature, air density, oxygen availability, and humidity. Some waterfowls such as geese and ducks thrive effortlessly at high altitudes following certain respiratory and cardiovascular adaptations. For instance, bar-headed goose (*Anser indicus*), which migrates twice annually through the Himalayas (4,500–6,500 m), and Andean goose (*Chloephaga melanoptera*), a resident of the Andes (4,000–5,500 m), deal with the high altitudes differentially. While bar-headed geese vigorously increase total ventilation, Andean geese and crested ducks gradually increase lung oxygen extraction of lung oxygen, which is the difference in percentage of oxygen between inspired and expired air (Laguë, 2017). Hemoglobin and hematocrit values are relatively higher for migrants compared to non-migrants (Yap et al., 2019). Besides, there exists a positive correlation in hemoglobin concentration with the altitude, suggesting that high-altitude migrants boost their oxygen-carrying capacity by enhancing their hemoglobin (Yap et al., 2019). Both the abovementioned species double their cardiac output, but bar-headed geese attain this by increasing the heart rate, while Andean geese accomplish it mainly by enhanced stroke volume at the most severe levels of hypoxia (Laguë, 2017). The Andean goose is one of those species whose lungs are exceptionally well-specialized for increased pulmonary O_2 diffusion (Maina et al., 2017). A unique feature of its lungs is the presence of highly vascularized infundibula (the structure from where the air capillaries ascend), which subsequently increases the respiratory surface area, henceforth, the blood available for gaseous exchange. Andean goose has higher blood–gas barrier with respect to other birds, as it continuously lives in hypoxic environment. Since its heart rate and cardiac output are frequently elevated, thicker blood–gas barrier prevents any structural failure in this species (Maina et al., 2017).

Migratory flight generates substantial metabolic heat that must be efficiently dissipated. The bill serves as a crucial thermoregulatory organ during flight to facilitate heat loss (Hagan & Heath, 1980). A study on mallards suggests that counter-current heat exchange is important for brain cooling with help of *rete ophthalmicum*, that is, the connection of different heat loss areas. According to Midtgård (1984), the head's vascular architecture indicates that sphincteric veins play a critical role in modulating venous blood flow from the evaporative regions of the palate and nasal cavity. However, the unprotected eyes of mallards also serve as an efficient source of effective cooling during strenuous flights.

4. Distinct Flyways and Migration Patterns of Anatidae Family

The migratory journeys of Anatidae are organized along well-defined flyways that represent some of the most spectacular and ecologically significant movement patterns in the avian world (Fig. 1).

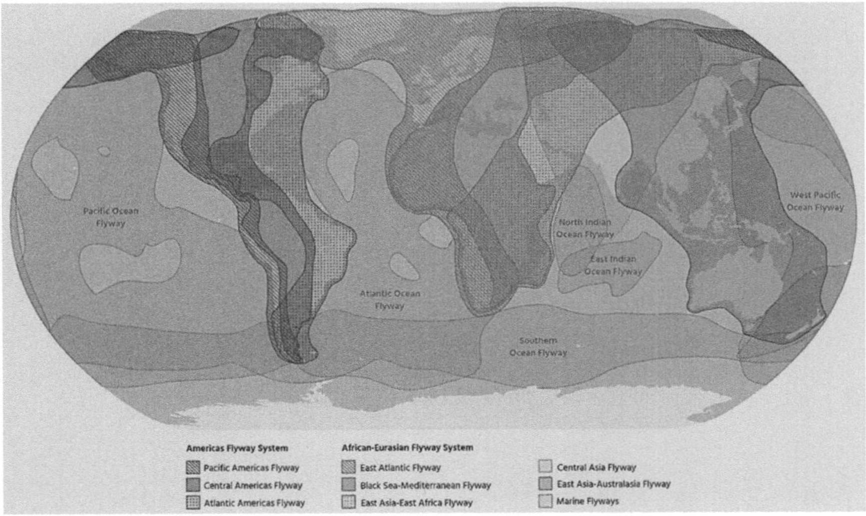

Figure 1. Different flyways adopted by the waterfowls during migration across the globe (picture is taken and modified from datazone.birdlife.org).

These flyways, broadly defined as "migration corridors" that connect breeding and non-breeding ranges, are shaped by an intricate relationship of geographic barriers, wind patterns, habitat availability, and evolutionary history. Globally, Anatidae migration is concentrated along eight major flyways: *East Asian–Australasian flyway, Central Asian flyway, African–Eurasian flyway, Americas flyway, Pacific Americas flyway, Atlantic Americas flyway, East Atlantic flyway, and West Pacific flyway* (Boere & Stroud, 2006) as tabulated in Table 2. Each flyway supports different species based on its ecological environment and the species' evolutionary adaptations. For example, the East Asian–Australasian flyway is utilized by species such as the northern pintail (*Anas acuta*) and the Baikal teal (*Anas formosa*), which cover large distances to take advantage of specific breeding and wintering resources (Xu et al., 2019; BirdLife International, 2025). The dabbling ducks (Anas species) are known to migrate across Atlantic and Mississippi flyways due to their adaptability to diverse habitats (Aagaard et al., 2015). In contrast, the whooper swan (*Cygnus cygnus*), commonly found in the East Atlantic flyway, follows a more north–south route, taking advantage of

Table 2. Different major flyways with their geographic locations and representative species

Flyway Name	Geographical Coverage	Prominent Anatidae Species
East Asian–Australasian Flyway (EAAF)	Alaska, Russian Far East, East Asia, Southeast Asia, Australia, New Zealand	Baikal teal (*Anas formosa*), northern pintail (*Anas acuta*), swan goose (*Anser cygnoides*)
Central Asian Flyway (CAF)	Siberia, Central Asia, Himalayas, Indian Subcontinent, Maldives	Bar-headed goose (*Anser indicus*), ruddy shelduck (*Tadorna ferruginea*), Eurasian teal (*Anas crecca*)
African–Eurasian Flyway	Arctic and temperate Europe, Middle East, Central Asia to sub-Saharan Africa	Greater white-fronted goose (*Anser albifrons*), Eurasian wigeon (*Mareca penelope*), garganey (*Spatula querquedula*)
Americas Flyway	Arctic Canada and Alaska through North, Central, and South America	Blue-winged teal (*Spatula discors*), American black duck (*Anas rubripes*), snow goose (*Anser caerulescens*)
Pacific Americas Flyway	Western North America from Alaska to Chile via Coastal Wetlands	Brant goose (*Branta bernicla*), northern shoveler (*Spatula clypeata*), canvasback (*Aythya valisineria*)
Atlantic Americas Flyway	Eastern North America through Caribbean to South America	American wigeon (*Mareca americana*), Canada goose (*Branta canadensis*), wood duck (*Aix sponsa*)
East Atlantic Flyway	Arctic and sub-Arctic Europe through Western Europe to West Africa	Eurasian teal (*Anas crecca*), greylag goose (*Anser anser*), northern pintail (*Anas acuta*)
West Pacific Flyway	Russian Far East, Kamchatka, Kuril Islands to Japan, Korea, and Southeast Asia	Falcated duck (*Mareca falcata*), greater scaup (*Aythya marila*), whooper swan (*Cygnus cygnus*)

the temperate climates for breeding in the arctic and subarctic regions (BirdLife International, 2025). The lesser white-fronted goose (*Anser erythropus*), another example, uses routes such as the East Atlantic and others, and is particularly sensitive to environmental changes along these paths (BirdLife International, 2025). Each flyway supports unique assemblages of species and populations, with migration strategies finely tuned to regional ecological conditions. The East Asian–Australasian flyway (EAAF), for instance, hosts more than 201 waterbird species, including Baer's pochard (*Aythya baeri*), which is critically endangered, and the declining swan goose, which undertake journeys of up to 6,000 km between Arctic breeding grounds and Southeast Asian wintering sites (Yong et al.,

2015). Understanding these flyway systems is critical for effective conservation, as Anatidae populations face increasing challenges as a result of habitat loss, climate change, and hunting pressure along their entire migratory routes.

The configuration of continental flyways reflects fundamental geographic constraints and habitat distributions. In North America, the four flyways align with major watershed systems: the Pacific flyway follows the Western Cordillera and Coastal Wetlands, the Mississippi flyway centers on the continent's largest river drainage, the Central flyway spans the Great Plains and Prairie Pothole Region, and the Atlantic flyway traces the Eastern Seaboard (Bellrose, 1968). These routes are not arbitrary but have evolved in response to Pleistocene glaciations and subsequent wetland distributions, with genetic evidence suggesting flyway fidelity has been maintained for millennia (Louchart, 2008; Milá et al., 2006). The Atlantic flyway's snow geese, for example, show distinct mitochondrial DNA haplotypes between populations separated during Pleistocene indicating long-term separation (Quinn, 1992). Mountain ranges such as the Himalayas and deserts such as the Gobi act as formidable barriers that funnel migrants into narrow corridors, creating migration bottlenecks where majority of the population may pass within narrow zone (Yong et al., 2015).

Different Anatidae species exhibit remarkable niche partitioning within flyways, reducing competition for resources. In the Central Asian flyway, the bar-headed goose carries out its migratory flight at extreme altitudes over the mighty Himalayan range, while the ruddy shelduck (*Tadorna ferruginea*) circumvents these mountains via lower-elevation valleys (Parr, 2020; Takekawa, 2009). The Mississippi flyway demonstrates trophic stratification, with diving ducks such as canvasbacks (*Aythya valisineria*) utilizing deepwater habitats while dabbling ducks such as mallards dominate shallow wetlands (Yetter et al., 2019). Phenological differences also reduce the overlap with northern pintails migrating 2–3 weeks earlier than other dabbling ducks in the Pacific flyway, likely to exploit ephemeral prairie wetlands (Zarzycki, 2017). Such specialization makes flyways function as multi-species migration highways, with satellite tracking revealing that Anatidae species show strong flyway fidelity but adjust stopover sites annually based on local conditions (Eren et al., 2024; Wong, 2021).

Flyway-scale conservation is imperative as threats vary dramatically across regions. In the Yellow Sea, where majority of migratory shorebirds and Anatidae stage, the East Asian–Australasian flyway has lost more than 65% of its intertidal habitats (Murray et al., 2014). The Atlantic Americas flyway faces wetland fragmentation from urban development with historical tidal marshes lost since European settlement (Creaseman et al., 1992; Mitsch & Hernandez, 2013). Climate change alters flyway dynamics differently across latitudes; Arctic-breeding species in the East Atlantic flyway now depart breeding grounds 15 days earlier than in 1990 due to advanced spring thaw (Clausen & Clausen, 2013), while drought in the Central flyway's Prairie Pothole Region has reduced wetland availability in dry years (Johnson et al., 2010). International conservation agendas such as the Ramsar Convention and regional initiatives such as the African–Eurasian

Waterbird Agreement (AEWA) have established flyway-scale protections for safeguard of critical Anatidae sites (Kirby et al., 2008).

Protecting Anatidae flyways requires integrated strategies addressing habitat loss (especially coastal wetlands), climate adaptation, and sustainable hunting. The EAAF partnership's work to safeguard the Yellow Sea demonstrates flyway-scale success (Murray et al., 2014), while the Western Hemisphere Shorebird Reserve Network preserves key sites across the Americas (Myers et al., 1987). Emerging threats such as wind energy development and avian influenza necessitate flyway-wide monitoring (Gilbert et al., 2006; Zhao et al., 2022). As Anatidae serve as indicators of wetland health, their flyways represent ecological lifelines that sustain biodiversity across hemispheres; their conservation is both an ecological imperative and a test of international environment governance.

5. Mechanisms for Navigation and Multisensory Integration in Anatidae Family

The navigational capabilities of migratory Anatidae represent one of the most sophisticated orientation systems in the animal kingdom, integrating multiple sensory modalities to guide birds across thousands of kilometers with remarkable precision. These birds employ a complex suite of innate and learned mechanisms that include geomagnetic sensing, celestial navigation, olfactory cues, visual landmarks, and even infrasound detection, all integrated through specialized neural pathways. The bar-headed goose, for instance, demonstrates sensitivity to both inclination and intensity of the magnetic field of Earth, allowing precise navigation during trans-Himalayan migrations (Spivey & Bishop, 2014; Wiltschko & Wiltschko, 2019). Birds possess magnetoreceptive cryptochrome proteins (Cry) in retinal ganglion cells that are responsive to magnetic fields through quantum radical pair mechanisms (Becker et al., 2009; Wiltschko & Wiltschko, 2019), while iron-rich magnetite clusters in the upper beak (likely in trigeminal nerve endings) provide a three-dimensional magnetic map sense (Falkenberg et al., 2010). The avian navigation system demonstrates extraordinary neural specialization for multisensory integration. The cluster N region of the forebrain (part of the visual wulst) in migratory birds shows greater activation during magnetic orientation tasks (Heyers et al., 2007). This region receives input from retinal cryptochromes and projects to the hippocampal formation, which contains "place cells" that encode spatial memory (Mouritsen et al., 2016). The trigeminal system transmits magnetite-based magnetic information to the trigeminal ganglion which is then passed to the trigeminal brainstem nuclei (Wiltschko & Wiltschko, 2019). Neuroplasticity in these pathways is seasonally regulated, with increased hippocampal neurogenesis (Barnea & Nottebohm, 1994). The anterior dorsal thalamus acts as a neural switch between the magnetic compass and celestial navigation systems, allowing rapid cue prioritization based on environmental conditions (Zapka et al., 2009). Celestial navigation complements magnetic sensing, with polarized light detection through specialized double cones in the retina allowing orientation at twilight during migratory flights (Muheim et al., 2006). In birds, otolith, found spread on otolithic layer, has significant amount of iron as in case of spot-billed duck, suggesting its putative role in geomagnetic sensing and in navigation using the magnetic compass (Harada et al., 2001).

Migratory Anatidae have evolved specialized adaptations to overcome navigational challenges. During trans-equatorial migrations, species such as the blue-winged teal recalibrate their magnetic compass to account for polarity reversal, likely using sunset polarization patterns as reference (Wolf, 1970). Over featureless oceans, wandering albatrosses (*Diomedea exulans*; family: Diomedeidae) use infrasound from waves to orient (Gillies et al., 2023), a capability suspected in sea ducks. In polar regions, where magnetic cues become unreliable, barnacle geese switch to celestial navigation based on stellar rotation (Wiltschko & Wiltschko, 2009). Atmospheric conditions influence cue weighting, for instance, an overcast night might obstruct the visual cues used for navigation by mallards (Weller et al., 2022), while smoke from wildfires triggers emergency orientation responses in migrating geese (Overton et al., 2021).

Protecting the navigational capacity of Anatidae requires addressing emerging threats. Electromagnetic noise from anthropogenic sources may disrupt magnetoreception (Engels et al., 2014), while wetland loss eliminates crucial orientation landmarks (Berthold, 2001). The navigational adaptations of Anatidae represent a pinnacle of evolutionary innovation; their preservation is essential for maintaining the ecological phenomenon of long-distance bird migration.

6. Challenges and Threats Concerning Survival of the Anatidae Family

Despite having significant ecological roles, migratory anatids face challenges such as habitat loss and climate change during migration, which can disrupt migratory routes as well as synchronization with seasonal events. In fact, the survival of migratory Anatidae and their ecological contributions are progressively vulnerable due to human activities and environmental variations (Fox & Madsen, 2017). Wetland destruction due to agriculture, urbanization, and pollution has eliminated critical stopover and wintering sites, leading to population declines of waterbirds (Buriro & Lund, 2025; Mengdi et al., 2025). Climate change further exacerbates these challenges by altering migration timing and disrupting the synchrony between bird arrivals and peak resource availability, as seen in barnacle geese arriving earlier in the Arctic to find insufficient grass growth for their young (Jonker et al., 2013). Overhunting and bycatch in fisheries add additional pressures, pushing some species toward vulnerability or endangerment (Mateo, 2009).

The waterfowl family faces significant threats due to natural as well as anthropogenic factors such as climate change, urbanization, and habitat loss. Many species arrive at breeding grounds either too early or too late to synchronize with peak food availability (Visser et al., 2012). Climate change disrupts migration timing due to shifting temperature regimes (Lehikoinen et al., 2013). Many Anatidae species rely on photoperiod and temperature cues for migration, but warmer winters cause mismatches in arrival times and peak food availability (Najafabadi et al., 2015). For example, the northern pintail now arrives earlier in Europe, but its breeding success declines if insect hatches are asynchronous (Doxa et al., 2012). Warmer springs cause earlier insect hatches, leaving late-arriving ducklings with reduced food (Lehikoinen et al., 2013). Also, snow geese (*Anser caerulescens*) now arrive too late to exploit peak Arctic forage, and research prediction suggested that climate change due to global warming should cause an earlier commencement of reproductive phenology in these geese. Besides, high spring temperatures may also decrease the fledging success (Dickey et al., 2008). Rising temperatures and altered precipitation reduce wetland availability. Prairie Pothole Regions, critical for North American ducks, are drying due to increased evaporation (Johnson et al., 2010). Climate change compounds these habitat losses by altering the migratory and breeding time (Knudsen et al., 2011). In the Arctic, where warming occurs at twice the global rate, rapid environmental changes are transforming breeding habitats in ways that may render them unsuitable for some species (Doiron et al., 2015). Additionally, increasing frequency of extreme weather events disrupts migration and reduces survival rates, particularly for juvenile birds who are making their first journeys (Lehikoinen et al., 2013). Also, increased frequency of storms and droughts destroys nests and reduces survival rates. Sea-level rise also threatens coastal habitats of species relying on estuarine marshes for breeding and stopover (Clausen et al., 2013). Further, non-native plants such as *Phragmites australis* outcompete native vegetation (Tozer & Beck, 2018; Wilcox et al., 2003), reducing food resources. Predatory mammals (e.g., raccoons, red fox, and striped skunk) around nests of Anatidae members are responsible for higher nest predation rates and limiting the survival of ducklings (Docken, 2011).

The threats to migratory Anatidae are numerous with habitat loss representing the most immediate danger (Davidson, 2014). The Sunda teal (*Anas gibberifrons*) faces habitat collapse in Indonesia due to mangrove deforestation (Iwamura et al., 2013). Global wetland destruction has been catastrophic, with more than 50% of the Earth's wetlands have disappeared over the last century (Gardner et al., 2015). The Yellow Sea region, a crucial stopover point for millions of waterfowl, has suffered particularly dramatic losses, with nearly 65% of its tidal flats disappearing due to coastal development (Murray et al., 2014). Drainage for agriculture and urban development reduces stopover sites critical for refueling (Ramsar Convention, 2018). Such threats disrupt migratory patterns, breeding success, and food availability, leading to population declines. Although, the Anatidae family plays crucial ecological roles in wetland ecosystems, yet destructive anthropogenic pressures and human negligence are severely affecting their populations dynamics.

Urban expansion fragments wetlands, isolating populations and reducing genetic diversity (Soulé et al., 2005). Industrial expansion in Tianjin's Nangang district has significantly degraded the Tangjiahe Estuary wetland, leading to observable reductions in avian biodiversity at this formerly critical nesting site for relict gulls (*Ichthyaetus relictus*), as documented in longitudinal studies (Liu et al., 2006). Urban runoff introduces heavy metals and pesticides into wetlands, bioaccumulating in birds' bodies (Biswas & Ch, 2025; Chastel et al., 2022). Additionally, crowded urban habitats facilitate avian botulism outbreaks (Rocke & Bollinger, 2007; Work et al., 2010). Feeding by humans alters natural foraging behaviors, increasing aggression and dependency. Wetland drainage for agriculture eliminates breeding grounds. In the Mekong Delta, rice intensification has reduced wintering habitats for the Indian spot-billed duck (*Anas poecilorhyncha*) (Buckton et al., 1999).

Unsustainable hunting, overharvesting, illegal poaching, habitat destruction, and pollution are major anthropogenic threats to Anatidae birds. These pressures are also driving population declines in many species of Anatidae family. Practices such as unsustainable hunting continue to threaten populations in certain regions, despite international agreements regulating harvests (Fox & Madsen, 2017; Wilcove & Wikelski, 2008). Hunting pressure exacerbates declines in vulnerable species such as the Madagascar pochard (*Aythya innotata*) that nearly went extinct due to hunting and habitat loss until few birds were rediscovered and released back in nature under the Durrell Project managed by Dr. Glyn Young. The practice of hunting is particularly observed in regions with weak enforcement of wildlife protection laws. In Europe and North America, regulated hunting is permitted for species such as the Canada goose and mallard, but overhunting still occurs (Fox & Madsen, 2017). Climate shifts disrupt migratory timing, making some species more vulnerable to hunting. Early arrivals of emperor geese due to warmer winters (Uher-Koch et al., 2021) expose them to extended hunting seasons. Additionally, droughts force waterfowl into smaller wetlands, increasing their concentration and susceptibility to overharvest (Johnson et al., 2005). Coastal species perhaps face habitat loss from rising sea levels which could be pushing them into suboptimal inland areas where hunting pressure is higher. From the 1900s up to the early 1970s, ornithological research pointed to hunting and deforestation as the key factors behind the decrease in bird numbers in the Republic of Korea (Moores et al., 2014). Moreover, hunting often results in unintended bycatch, where nontarget species are killed. The Kabul River wetlands, a critical habitat for migratory birds, have suffered significant damage due to unregulated hunting, harmful practices, and extensive water pollution. Experts and locals note an 80% decline in bird numbers, with illegal hunting being a major cause of recent overexploitation (Samuel & Hussain, 2014).

Wetlands, crucial for Anatidae breeding and feeding, are increasingly drained for agriculture. Population surveys indicate that the Southeast Asian eastern sarus crane (*Grus antigone sharpie*; family: Gruidae) subspecies has undergone dramatic decline, with only an estimated 1,000 individuals once remaining. Of these, the

Mekong River Delta in Southern Vietnam supports more than 30% of the global population during the dry season period (Buckton & Safford, 2004). Similar trends are observed in Europe's farmlands where intensive farming was observed to be correlated with reduced populations of farmland birds; 9 out of 28 bird species were analyzed, including 5 species facing larger declines across the Europe (Sanderson et al., 2013). Expanding cities and transportation networks fragment habitats, isolating populations. Land reclamation constructions such as artificial reservoirs and dams have resulted into rapid decline in the number of shorebirds, due to reduction in their habitat size and feeding sources (Kang et al., 2011; Yu et al., 2014). These man-made structures can impede the hydrological connectivity between water bodies, leading to altered water levels and potentially reducing the suitable habitats for waterbirds (Cui et al., 2024). Drewitt and Langston (2008) reviewed about bird-strikes and suggested infrastructural development, including wind farms and power lines, to cause deadly obstacles along flyways. In their study, they concluded that priority areas for future work may include advanced remote technology for monitoring bird collision, analyzing its cause, and designing innovative and effective strategies to mitigate bird collision. Further, pollution from agricultural runoff, plastic waste, and industrial contaminants accumulates in wetland habitats, affecting waterfowl health and reproduction. Ma et al. (2024) held mercury contamination responsible for declining the migratory shorebirds along the East Asian–Australasian flyway.

These multifaceted threats demand equally comprehensive solutions that address all facets of the problems across entire migratory ranges (Piersma & Lindström, 2004). In 2010, Fox et al. estimated the size of 30 defined populations of geese that winter in the Western Palearctic (this estimation also included five reintroduced or released populations of three species) and reported that Svalbard/Greenland light-bellied brant geese number being critically small within restricted range (numbered less than 10,000 birds). Disturbance from wind turbines and power lines also poses risks for waterbirds such as mallard and other shorebirds, such as gulls (Krijgsveld et al., 2009). Collisions with wind turbines kill approximately 100,000 waterfowl annually in North America (Drewitt & Langston, 2008). In a very recent research, Heim et al. (2025) claimed high microplastic pollution in bodies of waterbirds of urban freshwaters via analysis of fecal samples (up to 72 microplastic particles/mg feces). They reported microplastics in form of fibers in approximately 98% of the collected fecal samples of Anatidae and other families such as black-headed gull (*Chroicocephalus ridibundus*), mallard, jack snipe (*Lymnocryptes minimus*), and common moorhen (*Gallinula chloropus*). Further, they suggested the possibilities of migratory birds polluting the remote sites with microplastics, through their fecal matter. Similarly, a study done in South Africa confirmed microplastic contamination in wetlands (freshwater ecosystems) where 408 feather brushings and 283 fecal samples from African ducks were analyzed. The study reported microplastic fibers in 5% fecal and 10% feather samples. Reportedly, such plastic contamination in birds differed across duck species and was attributed to differential foraging behavior (Reynolds & Ryan, 2018).

Heavy metal pollution such as lead poisoning from ingested shotgun pellets also remained a major issue as per the study outcomes in the current decade. Kanstrup et al. (2020) resurveyed the 1970s survey of shotgun pellet bulk in Ringkøbing Fjord. In Denmark, lead poisoning was detected in water birds and lead gun shots used for hunting were banned on 26 Ramsar sites under legislation, back in 1986. They found lead shot in large amount, even after more than three decades of regulation because lead accumulation persisted in nature and it remained irreversibly hazardous to wildlife and ecosystems. Further, lead poisoning from ingested fishing weights remains a leading cause of death in mute swans (*Cygnus olor*) (Perrins et al., 2003). Lead toxicity from spent ammunition and industrial runoff continues to affect waterfowl. A study from Britain showed that 41.7% of dead mute swans had lethal lead levels (>20 μg/dL) (Newth et al., 2016). Long-term effects of consuming polluted diets lead to digestive blockages and toxic chemical absorption.

Habitat fragmentation, climate change, and invasive species have intensified threats to Anatidae population but predation and natural obstacles too have always shaped their population dynamics. To effectively conserve the declining Anatidae species, it is crucial to keep a tab on these threats. A major cause of unsuccessful reproduction in Anatidae family is nest predation. Mammalian predators such as raccoon (*Procyon lotor*), coyote (*Canis latrans*), red fox (*Vulpes vulpes*), and the American badgers (*Taxidea taxus*) are known major nest predators, particularly in fragmented habitats of the Southcentral Saskatchewan's Prairie Pothole Region (Larivière & Messier, 1998). The effects of human intervention due to human trail in the wetland edges were found to be associated with nest damage in a study carried out keeping in mind the waterfowl management in breeding locations (Olson & Rohwer, 1998). Juvenile waterfowl also face high predation from fish (e.g., pike (*Esox lucius*)), turtles, and birds of prey. Based on data from Suisun Marsh, California, the survival rate of mallard ducklings to the fledging stage was very low, with less than 3% surviving during two dry years compared to under 17% during two wet years (Peterson et al., 2024). To manage this issue, the authors suggested maintenance of low salinity in wetland, at least within 1 km of nesting sites during post-hatch period. Another group of researchers also showed that in Western Hudson Bay, early arrival of polar bear due to climate change reduced the reproductive output of snow geese as their eggs were predated (due to "trophic match" in absence of seals), which led to a decline in their nesting population; however, due to "stochastic mismatch" (when the geese egg-laying did not match with bear arrival), geese tend to periodically rebound, which prevents their complete collapse (Rockwell et al., 2011). Sovada et al. (2001) also examined the predator–prey dynamics affecting waterfowl nesting success in Arctic regions and Prairie regions, and found that in Arctic regions, elevated rates of egg and chick predation among waterfowl typically occur when previously secluded populations become accessible to predators. In contrast, Prairie environments face predation challenges primarily due to widespread habitat deterioration and shifts in predator composition.

Besides predation pressure, Anatidae birds also face natural obstacles that contribute toward decline in their numbers in real time. Storms and droughts disrupt feeding and breeding cycles of these birds. Masto et al. (2022) used citizen science and GPS-telemetry data to investigate how extreme climatic events (ECEs), particularly polar vortex disruptions, impact the distribution of wintering waterfowl across North America and reported that waterfowl reacted differently to ECEs depending on body size, foraging strategy, and habitat preference. Geese delayed northward migration and wetland obligate species tended to shift southward, while smaller-bodied ducks were less affected. Similarly, the effect of polar vortex disruption on the survival of mallard was analyzed across Arkansas, Louisiana, and Tennessee, and researchers found that mallards were more likely (5.4 times) to die during ECE winters (Blake-Bradshaw et al., 2024). Other than storms, prolonged droughts also pose problems to waterfowl population. Mid-summer wetland desiccation caused by drought severely threatens brood-rearing waterfowl, and this environmental stressor was a key factor behind the 40% decline in the breeding waterfowl populations of California, observed between 2011 and 2015 (Brady & Weaver, 2025). In areas dominated by saline or brackish wetlands, prolonged drought conditions can cause severe reproductive losses among waterfowl due to insufficient freshwater availability, which may have played a role in the observed declines in mallard populations within the Suisun Marsh as drought intensity escalated (Siegel et al., 2011; U.S. Fish and Wildlife Service, 2013). While migrating, most birds fly at 5,000 to 6,000 meters above the sea level, where oxygen levels are roughly half of those at sea level, with rare instances recorded up to 7,290 m (Hawkes et al., 2013). Further, natural disease outbreaks (e.g., avian cholera, botulism) cause mass mortality. In North America, botulism kills thousands of diving ducks annually (Russel et al., 2019). From pathological investigation of Anatidae, mute swan obliterative endophlebitis has been identified within the intestinal venous system, and has been attributed to adult schistosomes, most likely *Trichobilharzia filiformis* (van Bolhuis et al., 2004). This vascular pathology underscores the potential severity of host–parasite interactions in wild Anatidae and highlights the importance of monitoring parasitic threats across migratory ranges.

7. Conservation Efforts

For Anatidae birds and other water birds, wetlands are suitable stopover, breeding, and wintering habitats. Their protection is central to sustaining migratory populations. Such initiatives often include habitat conservation, international cooperation, and sustainable management of human interventions. Studies based on habitat suitability modelling, such as those in Anhui Province, China (Shi et al., 2024), used maximum entropy (MaxEnt) and generalized linear mixed models (GLMM) to identify high-suitability habitats based on terrain, farmland proximity, and human disturbance. They found that flat terrains adjacent to farmlands supported higher bird richness, whereas factors such as elevation and anthropogenic disturbance were associated with low species abundance. These are significant results in terms of biosecurity and prevention of waterbird epidemics in different areas. Similarly, to support Anatidae during early dry seasons, Poyang and Dongting lakes in China have been prioritized for restoration using landscape ecology and spatial analysis (Wang et al., 2023). They used spatial autocorrelation analysis and their study involved three landscape pattern indices for investigating the distribution pattern of crucial sites and evaluating the efficiency of habitat restoration for biodiversity conservation. Boere (2003) emphasized on coordinated conservation across entire migratory routes, not just isolated sites, and advocated for ecosystem-level management, recognizing wetlands as dynamic systems influenced by climate, land use, and human activity.

Other than this, migratory Anatidae traverse vast flyways, which requires coordinated international conservation efforts. The Central South Asian flyway and East Asian–Australasian flyway support the maximum diversity of migratory bird species among global flyway networks. Developed under the Asia-Pacific Migratory Waterbird Conservation Strategy (Miyabayashi & Mundkur, 1999), the Anatidae Site Network identifies and protects key sites across the East Asian flyway. Whereas, the WWF's Wetlands for Asian Flyways Initiative (WAF) (World Wildlife Fund, n.d.) aims to restore and manage wetland networks by 2030, ensuring ecological connectivity and community stewardship for stabilizing the declining population of migratory birds and conservation of species-abundant wetlands. Asian Waterbird Conservation Fund (AWCF) provides grants for site-level conservation projects, including those benefiting Anatidae birds. Another flyway protection initiative that was proposed in 2021, the Asian Development Bank (ADB) Regional Flyway Initiative aimed to raise $3 billion to protect priority wetlands along the East Asian–Australasian flyway. This program was initiated to support both biodiversity and local livelihoods with opportunities in recreation and tourism. According to research conducted by Bandivadekar et al. (2012), Veer Dam in Maharashtra's Satara district serves as a seasonal habitat for bar-headed geese, with recorded flock sizes reaching as high as 487 birds during the wintering period. The study confirmed flyway connectivity between Central Asia and India and suggested that site to be qualified as an Important Bird Area (IBA) under Criterion A4i, given that observed flocks represent nearly 1% of the global population. Similarly, Kumar et al. (2024) provided a detailed assessment of avian diversity in a Ramsar site Bhindawas Wildlife Sanctuary, an IBA in Haryana, and documented Central–South Asian flyway species such as Eurasian wigeon (*Mareca penelope*), gadwall (*Mareca strepera*), and northern pintail, all of which are winter migrants, and use Bhindawas for overwintering. Assessing the threat in that area, they suggested to integrate empirical evidence from this site into regional conservation strategies which can help prioritize protection efforts and inform flyway-level management policies specifically meant for Anatidae populations. As India's wetland networks continue to evolve under international frameworks such as Ramsar and AEWA, Bhindawas serves as a miniature for balancing biodiversity objectives with sustainable land-use practices.

Global conservation efforts for migratory Anatidae are strongly supported by international frameworks, with the AEWA standing out as a key legally enforceable agreement under the Convention on Migratory Species (CMS). This treaty encompasses 254 waterbird species and is implemented across 119 countries (Lewis, 2016; UNEP-AEWA, 2023). AEWA outlines targeted conservation actions such as species-specific plans, like those for the lesser white-fronted goose, and mandates protective measures including the prohibition of lead ammunition in wetland areas to stop toxic contamination (Convention on Migratory Species, 2025). The agreement fosters cross-border collaboration and encourages the use of resources such as the Critical Site Network tool to pinpoint vital breeding, stopover, and wintering habitats along migratory bird routes (Miyabayashi & Mundkur, 1999). Complementary agreements such as the Ramsar Convention and the World Heritage Convention strengthen wetland protection by designating ecologically significant sites, for example, the Wadden Sea and Bharatpur, as Ramsar wetlands and World Heritage sites (Ramsar Sites Information Service, 2017). These designations recognize wetlands' ecological value in sustaining Anatidae during their migratory cycles. AEWA also facilitates multinational monitoring through the International Waterbird Census (IWC), promoting harmonized research methodologies such as telemetry and habitat assessments (UNEP-AEWA, 2023). However, governance challenges persist, especially in regions with limited institutional capacity or fragmented domestic legislation (Runge et al., 2017). Studies from Iran and the Asia-Pacific region further underscore the importance of AEWA's flyway-level coordination, noting that conservation efforts are most effective when aligned with local ecological realities and supported by robust monitoring frameworks (Gallo-Cajiao et al., 2024; Nourani et al., 2015). It is believed that such treaties not only support multiscale conservation strategies but also offer policy justification for wetland restoration and ecosystem-level management across flyways.

Local community involvement is increasingly recognized as the basis of effective conservation strategies for migratory Anatidae species. Community-based wildlife management (CBWM) integrates traditional ecological knowledge with modern conservation science, enabling localized stewardship of wetlands and flyway habitats (Convention on Migratory Species, 2025; Owusu-Achiaw & Osei-Owusu, 2023). Sustainable hunting practices, when regulated and culturally embedded, can also contribute to conservation by aligning livelihood needs with species protection, particularly for Anatidae populations. Kamp et al. (2011) focused their study on wetland-dependent birds, including Anatidae species such as garganey (*Spatula querquedula*) and ruddy shelduck, and examined how community land-use practices and traditional irrigation systems in Kazakhstan and Central Asia influence breeding success and habitat availability. They alarmed that loss of traditional agro-pastoral systems may lead to future declines in Anatidae populations, despite recent recoveries. UNEP/CMS Resolution 12.25 (Convention on Migratory Species, 2017) prioritized the identification and protection of crucial habitats used for breeding, staging, molting, and wintering by migratory species and encourages transboundary cooperation and World Heritage Site nominations for intertidal wetlands (e.g., Wadden Sea, Banc d'Arguin National Park). Resolution 12.25 is significant for species such as lesser white-fronted goose, baer's pochard, and ferruginous duck which depend heavily on intertidal habitats and are enlisted in Appendix I and II of CMS. Case studies from Central Asia and India demonstrate that empowering local stakeholders through education, participatory monitoring, and benefit-sharing mechanisms fosters long-term commitment to conservation goals. Kumawat and Chadoo (2025) highlight how community-based wetland management in Keoladeo National Park has enhanced habitat resilience and supported wintering populations of bar-headed geese and other Anatidae species by integrating local stewardship with invasive species control and water management strategies. Similarly, Shoaib and Singh (2023) document the role of local engagement in Patna Bird Sanctuary, where restoration efforts and participatory monitoring have helped maintain high seasonal diversity of Anatidae, including northern pintail and lesser whistling duck. Moreover, initiatives such as the Kunming-Montreal Global Biodiversity Framework (GBF) highlight linking sustainable-use benefits directly to Indigenous peoples and local communities (IPLCs), reinforcing their role as custodians of critical flyway habitats. In the context of migratory Anatidae, such community-led approaches not only mitigate anthropogenic pressures but also enhance resilience against climate-induced habitat shifts, making them indispensable to ecosystem-level conservation planning. Further, these findings collectively affirm that empowering local communities through restoration projects, benefit-sharing models, and culturally sensitive hunting regulations not only enhances ecological outcomes but also fosters long-term commitment to flyway conservation goals.

The Ramsar Convention, formally known as the Convention on Wetlands of International Importance especially for waterfowl habitat, has been instrumental in global conservation efforts for migratory Anatidae. Adopted in 1971, it remains a basis of global conservation for migratory Anatidae species. It was the first multilateral environmental treaty to prioritize ecosystem-level protection, particularly wetlands critical for waterbirds (Gajjar & Bhagat, 2023). The principle of "wise use" under the Ramsar Convention has progressively shifted toward safeguarding the ecological character of wetlands, an integrated expression of their biological, chemical, and physical attributes that are vital for supporting migratory waterfowl (Finlayson et al., 2011). Ramsar's three pillars—"wise use," "designation of Ramsar sites," and "international cooperation"—form the backbone of its conservation strategy. Despite its foundational role, gaps remain in its successful implementation. Less than one-third (31%) of worldwide threatened Anatidae birds have ever been documented on Ramsar sites, and this further dropped to 10% for highly threatened species (21 taxa) (Green, 1996), underscoring the need for more targeted site designations. Despite ongoing conservation challenges, Indian Ramsar sites such as Keoladeo National Park,

Table 3. Conservation strategies, region/scope, and their implementation mechanism for Anatidae birds

Conservation Strategy/ Policy	Region/Scope	Implementation Mechanism
Ramsar Convention on Wetlands	Global	Wetland designation based on biodiversity criteria (Criterion 6: 1% population threshold)
AEWA (African–Eurasian Waterbird Agreement)	Africa–Eurasia	Binding treaty under Convention on Migratory Species (CMS); action plans; favorable reference values (FRVs); population trend monitoring
East Asian–Australasian Flyway Partnership (EAAFP)	East Asia to Australasia	Voluntary flyway partnership; flyway site network; Communication, Education, and Public Awareness (CEPA) and regional coordination
Asia-Pacific Migratory Waterbird Strategy (1996–2005)	Asia-Pacific	Thematic action plans for Anatidae, cranes, shorebirds; site inventories and capacity building
Yangtze Wetland Reserves (China)	China	Legal protection of lakes; hunting bans; seasonal water management linked to Normalized Difference Vegetation Index (NDVI) habitat optimization
North American Waterfowl Management Plan (NAWMP)	USA, Canada, Mexico	Joint ventures for habitat restoration and adaptive harvest management
Convention on Biological Diversity (CBD) and Global Species Action Plan (GSAP)	Global	National Biodiversity Strategies and Action Plans (NBSAPs); connectivity corridors; genetic diversity targets
Captive Breeding and Reintroduction	Global (species-specific)	Ex-situ breeding and monitored reintroduction of endangered Anatidae species
Invasive Species Control	Islands, nesting sites	Predator removal (rats, mink); restoration of native flora and hydrology
Climate Change Adaptation	Global	Artificial wetlands, corridor planning, hydrological buffering for migratory shifts
Monitoring Frameworks and Cross-Treaty Synergies	Global (Cross-jurisdictional)	Integration of datasets; standardized metrics; coordinated reporting via Ramsar, AEWA, CMS, and CBD platforms

Harike Lake, and Sambhar Lake remain vital seasonal habitats for significant congregations of Anatidae species, notably bar-headed geese and northern pintail (Chandra et al., 2021). These locations function as essential breeding grounds, stopover points, and winter refuges, with their conservation playing a pivotal role in ensuring the continued survival of migratory bird species.

Recent evaluations show that Ramsar-designated wetlands support higher waterbird abundance and diversity than non-designated sites, especially when backed by strong governance and community engagement (Gaget et al., 2020). However, the convention's non-binding nature and reliance on national implementation lead to wide variation in conservation outcomes. For instance, while some countries have developed robust management plans and monitoring frameworks, others struggle with enforcement, funding, and ecological degradation due to invasive species and water scarcity. The convention has also adapted to emerging challenges in recent times. Satellite tracking and ecological modelling are increasingly used to identify critical stopover sites and assess habitat quality along flyways (Lei et al., 2019). These tools help bridge knowledge gaps in migratory connectivity, allowing Ramsar sites to be integrated into broader flyway-level conservation strategies. Moreover, Ramsar's alignment with the GBF and the Sustainable Development Goals (SDGs) position it as a key player in climate-resilient conservation planning. While the Ramsar Convention has laid a strong foundation for wetland and Anatidae conservation, its future effectiveness hinges on adaptive management, transboundary cooperation, and integration with other treaties such as AEWA and CMS. Strengthening site-level implementation, expanding coverage to underrepresented flyways, and leveraging new technologies will be essential to safeguard migratory Anatidae during this period of accelerating environmental adjustments. Below is a table showing various conservation strategies, their region of scope, and implementation mechanism being followed during their execution for the conservation of waterfowls (Table 3).

8. Conclusion

Migration is indispensable for anatid survival, ensuring access to resources, genetic diversity, and reproductive success. Ecologically, these birds sustain wetland health through nutrient cycling, seed dispersal, and trophic interactions. However, anthropogenic threats demand urgent conservation measures, including habitat protection, climate adaptation, and international cooperation. Preserving migratory anatids is not just about saving species, it is about maintaining the ecological integrity of the wetlands they inhabit.

As a critical ecological and evolutionary strategy for waterfowls, migration facilitates access to optimal breeding and wintering habitats, enhances survival rates, and maintains genetic diversity while reducing competition. The phenomenon is particularly pronounced in Northern Hemisphere species, where breeding grounds become inhospitable during winter (Alerstam et al., 2003). Understanding migration in Anatidae is essential for their conservation, as many species face threats from habitat degradation, hunting, and climate change (Newton, 2023). Migration patterns are shaped by environmental factors, including climate, food availability, and predation pressures. Anatidae migration between breeding and wintering grounds helps in transport of nutrients and seed dispersal, and influences vegetation dynamics through different feeding habits. Further, long-distance migrations facilitate gene flow between distant populations, and therefore migratory movements of distant species aid in genetic diversity in environment due to outbreeding.

Due to migratory behavior, Anatidae species utilize geographically distinct breeding and wintering sites. Breeding grounds in the Arctic regions and temperate regions of the globe provide abundant food while reducing predation during summer, whereas wintering sites in milder climates offer open water and food resources (Dingle, 2014). Moreover, seasonal migration allows waterfowl to exploit ephemeral food sources. Many dabbling ducks (e.g., mallards) rely on wetland productivity, which fluctuates seasonally (Elmberg et al., 2005). Geese (*Anser* spp.) depend on agricultural fields during winter, where leftover grains provide high-energy food (Fox & Abraham, 2017). Other than providing opportunity for food exploration, migratory behavior reduces predation pressure on breeding grounds.

These birds serve as indicators of the environmental conditions so a stable or increasing population in the wetland reserves along the flyways would indicate about the habitat quality. For instance, shifting temperatures alter migration timing and routes. Some species, such as the barnacle goose (*Branta leucopsis*), now winter farther north due to milder winters (Jonker et al., 2013). Such changes may disrupt synchrony with food availability and, therefore, attention toward conservation of these migratory birds is quite significant. Besides, regulated hunting practices and education programs would help in balancing their conservation and combat traditional practices that contributed to a decline in their abundance (Madsen et al., 2015). Migration in Anatidae family is therefore an essential aspect of ecology, which ensures the survival and reproductive success of these birds. However, anthropogenic threats necessitate urgent conservation measures. Protecting migratory corridors, mitigating climate change effects, and fostering international collaboration are essential for sustaining these iconic waterfowl.

The migratory Anatidae species undertake some of the most extraordinary flights in the avian world that connect breeding sites in the Arctic regions and temperate regions with overwintering areas across the globe, forming ecological bridges between diverse ecosystems (Newton, 2023). However, these remarkable migrations face unparalleled threats from habitat loss, climate change, and human activities (Runge et al., 2014). Such threats demand urgent action for safeguarding their migratory journey, key steps for which include expanding satellite tracking studies (Kölzsch et al., 2016) and strengthening transboundary conservation (Kirby et al., 2008). Current research, although vast and diverse, reveals significant gaps in our understanding of migratory connectivity, stopover ecology, and species-specific responses to environmental change (Kleyheeg et al., 2017). Hence, there is a need toward advocating for enhanced research efforts and conservation initiatives to protect these vital species throughout their annual cycles. Migratory waterfowl play indispensable roles in maintaining healthy ecosystems across continents (Green & Elmberg, 2014). Since Anatidae species cross multiple countries, treaties such as the AEWA are vital for coordinated protection. Establishing and maintaining wetland reserves along flyways, such as the East Asian–Australasian flyway, ensure safe stopovers for these water birds (Bamford et al., 2008).

Despite their ecological importance, our understanding of their complete migratory patterns remains incomplete. While advances in tracking technology have revealed major flyways, critical knowledge gaps persist regarding unpredictability at stopover locations (Lok et al., 2011), energetic demands during long flights (Butler, 2016), and varying responses to environmental changes among populations (Pulido, 2007). These information loopholes hinder our ability to implement effective conservation measures for species that rely on networks of habitats across international borders (Iwamura et al., 2013). To effectively protect migratory Anatidae, several key research areas, such as advanced tracking technologies, including miniaturized GPS-GSM transmitters and isotopic analysis techniques, must be prioritized (Kranstauber et al., 2012). These techniques can provide unprecedented detail about movement patterns and habitat use (Bridge et al., 2011). Such data is essential for identifying critical stopover sites that require protection (Iwamura et al., 2013). Habitat conservation strategies should focus on both preserving existing wetlands and restoring degraded areas through ecological engineering approaches (Zedler & Kercher, 2005). International agreements such as the Ramsar Convention and the AEWA provide frameworks for wetland conservation (Boere & Taylor, 2004). Site-specific protections, such as China's Poyang Lake reserves, are vital for species such as the Siberian crane (*Leucogeranus leucogeranus*; family: Gruidae) (Wang et al., 2018).

Further, despite bans on lead ammunition in some regions, compliance remains inconsistent. Stricter enforcement of hunting regulations and bans on lead ammunition therefore become critical. The AEWA has improved protections, but illegal hunting persists in many regions. Wetland restoration projects, such as those under the Ramsar Convention, can mitigate habitat loss (Gardner et al., 2025). Creating buffer zones around key breeding sites reduces human disturbance. Engaging hunters in conservation through sustainable harvest models, for example, adaptive harvest management in the United States of America, has proven effective (Nichols et al., 2007). Education on nontoxic ammunition alternatives is also essential for their conservation. Translocation and release of wild birds and nest protection boosted hatching success in the Laysan duck (*Anas laysanensis*) (Reynolds & Klavitter, 2006), which is an endangered species. Selective culling of invasive predators aids island-nesting colonies. Restoring wetland mosaics ensures refuge during droughts, as reported by Isola et al. (2000). Protecting stopover sites along migratory routes is critical for long-distance migrants. Monitoring cholera outbreaks (Samuel et al., 2005) and removing carcasses would possibly reduce transmission. Vaccination programs may help manage cholera in concentrated wintering areas. Predators and natural obstacles are intrinsic to Anatidae ecology, but human-induced changes have magnified their impacts. Targeted conservation, including predator control, habitat restoration, and climate adaptation, is essential to mitigate these threats.

International cooperation in this regard is particularly crucial for the successful implementation of several agreements and policies. Successful models such as the East Asian–Australasian flyway partnership demonstrate how collaborative efforts can protect key sites and coordinate conservation actions across political boundaries (Bamford et al., 2008). Climate adaptation plans should specifically incorporate the needs of migratory waterfowl, recognizing their role as indicators of wetland health. Agricultural practices that damage wetland ecosystems must be reformed, with incentives provided for waterfowl-friendly land management (Donald et al., 2001). Hunting and anthropogenic pressures pose severe threats to the Anatidae family, requiring urgent conservation action. Strengthening legal protections, restoring habitats, and reducing pollution are key steps toward ensuring the survival of these ecologically vital birds. At the community level, engaging local people through citizen science programs, hunter education initiatives, and sustainable ecotourism can build support for conservation efforts while providing valuable monitoring data (Sekercioglu, 2002). Increased funding for long-term population monitoring and climate impact modelling will provide the scientific foundation for effective management (Lindenmayer & Likens, 2010). Traditional ecological knowledge from local communities provides significant environmental insights into historical patterns and local conservation solutions. Perhaps most importantly, we must develop climate-resilient habitat networks that can sustain waterfowl populations as environmental conditions continue to change (Heller & Zavaleta, 2009).

While significant progress has been made in understanding and protecting these species, emerging threats demand renewed dedication to research and conservation (Kirby et al., 2008). By implementing strategies such as enhancing research efforts, strengthening international cooperation, and engaging local communities, we can work to ensure that these remarkable migrations endure for generations to come (Piersma, 2007). A recent review by Arya et al. (2025) has highlighted the necessity for an integrated ecosystem strategy addressing multiple concurrent threats, including enhanced pollution mitigation, improved wetland stewardship, and cross-border cooperation for migratory bird conservation. They suggested that comprehensive monitoring initiatives combining traditional and innovative methodologies are essential for assessing waterbird populations and contaminant levels. To safeguard these birds and the ecosystems they support, conservation efforts must also prioritize habitat protection, international cooperation, and adaptive management strategies (Ramsar Convention, 2018). Preserving migratory Anatidae is not just about protecting individual species but also about maintaining the ecological balance of the wetlands they inhabit, which benefits countless other organisms, including humans (Boere & Taylor, 2004). Through sustained, collaborative efforts, we can safeguard not only the Anatidae family but also the interconnected web of life that depends on healthy migratory waterfowl populations.

References

Aagaard, K., Crimmins, S. M., Thogmartin, W. E. et al. (2015). Evaluating predictors of local dabbling duck abundance during migration: Managing the spectrum of conditions faced by migrants. *Wildfowl*, 65, 100–120.

ADB Launches Regional Flyway Initiative to Preserve Priority Wetlands. (2021). *News Release*. Kunming, People's Republic of China. https://www.adb.org/news/adb-launches-regional-flyway-initiative-preserve-priority-wetlands

Aharon-Rotman, Y., McEvoy, J., Zhaoju, Z. et al. (2017). Water level affects availability of optimal feeding habitats for threatened migratory waterbirds. *Ecology and Evolution*, 7(23), 10440–10450. https://doi.org/10.1002/ece3.3566

Alerstam, T., Hedenström, A., & Åkesson, S. (2003). Long-distance migration: Evolution and determinants. *Oikos*, 103(2), 247–260. https://doi.org/10.1034/j.1600-0706.2003.12559.x

Anatidae. (2023, January 9). *New World Encyclopedia*. Retrieved 21:00, July 10, 2025 from https://www.newworldencyclopedia.org/p/index.php?title=Anatidae&oldid=1094461

Arya, A. K., Joshi, K. K., Bachheti, A. et al. (2025). Impact of water pollution on waterbirds: A review. *Environment Conservation Journal*, 26(2), 661–666. https://doi.org/10.36953/ECJ.30333174

Ballard, T., Seager, R., Smerdon, J. E. et al. (2013). Hydroclimate variability and change in the Prairie pothole region, the "Duck Factory" of North America. *Earth Interactions*, 18, 1–28. https://doi.org/10.1175/EI-D-14-0004.1

Ballerini, M., Cabibbo, N., Candelier, R. et al. (2008). Interaction ruling animal collective behavior depends on topological rather than metric distance: Evidence from a field study. *Proceedings of the National Academy of Sciences*, 105(4), 1232–1237. https://doi.org/10.1073/pnas.0711437105

Bamford, M., Watkins, D., Bancroft, W. et al. (2008). *Migratory Shorebirds of the East Asian-Australasian Flyway: Population Estimates and Internationally Important Sites* (pp. 237). Canberra: Wetlands International, Oceania. ISBN: 9789058820082

Bandivadekar, R., Pandit, R., Ponkshe, A. et al. (2012). Veer dam as important winter migratory ground for bar-headed geese *Anser indicus* (Latham, 1790) Family: Anatidae, with special reference to observations of tagged bar-headed geese. *ZOO'S PRINT*, 27(1), 22–23.

Barnea, A., & Nottebohm, F. (1994). Seasonal recruitment of hippocampal neurons in adult free-ranging black-capped chickadees. *Proceedings of the National Academy of Sciences of the United States of America*, 91(23), 11217–11221. https://doi.org/10.1073/pnas.91.23.11217

Bauer, C. M., & Watts, H. E. (2021). Corticosterone's roles in avian migration: Assessment of three hypotheses. *Hormones and Behavior*, 135, 105033. https://doi.org/10.1016/j.yhbeh.2021.105033

Bazely, D. R., & Jefferies, R. L. (1989). Leaf and shoot demography of an arctic stoloniferous grass, *Puccinellia phryganodes*, in response to grazing. *The Journal of Ecology*, 77(3), 811–822. https://doi.org/10.2307/2260987

Beauchamp, G. (2022). Flocking in birds is associated with diet, foraging substrate, timing of activity, and life history. *Behavioral Ecology and Sociobiology*, 76, 74. https://doi.org/10.1007/s00265-022-03183-9

Becker, S. F., Lienau, C., Mouritsen, H., & Radmacher, M. (2009). On the photophysics of migratory bird cryptochrome. *Bachelorarbeit zur Erlangung des Bachelorgrades*. https://uol.de/f/5/inst/physik/ag/uno/download/Bachelorarbeiten/On_the_photophysics_of_migratory_bird_cryptochrome_-_Simon_Becker.pdf

Bellrose, F. C. (1968). Waterfowl migration corridors east of the Rocky Mountains in the United States. *Biological Notes*, no. 061. https://hdl.handle.net/2142/17162

Berthold, P. (2001). *Bird Migration: A General Survey*. Oxford University Press.

BirdLife International. (2025). *Flyway Factsheets*. Downloaded from https://datazone.birdlife.org on 21/07/2025

Biswas, S., & Ch, R. (2025). Birds under threat: First report of titanium and heavy metals in wild birds from Visakhapatnam, India. *Environmental Claims Journal*, 1–36. https://doi.org/10.1080/10406026.2025.2506659

Blake-Bradshaw, A. G., Masto, N. M., Highway, C. J. et al. (2024). Caught out in the cold: *Anas platyrhynchos* (mallard) survival decreased during an extreme climatic event. *Ornithological Applications*, 126(4), duae025. https://doi.org/10.1093/ornithapp/duae025

Blem, C. R. (1990). Avian energy storage. *In:* D. M. Power (Ed.), *Current Ornithology* (pp. 59–113). Plenum Press, New York.

Boere, G. C. (2003). Global activities on the conservation, management and sustainable use of migratory waterbirds: An integrated flyway/ecosystem approach. *Wader Study Group Bulletin*, 100(1). https://digitalcommons.usf.edu/wsg_bulletin/vol100/iss1/20

Boere, G. C., & Taylor, D. (2004). Global and regional governmental policy and treaties as tools towards the mitigation of the effect of climate change on waterbirds. *Ibis*, 146, 111–119. https://doi.org/10.1111/j.1474-919X.2004.00335.x

Boere, G. C., & Stroud, D. A. (2006). The flyway concept: What it is and what it isn't. *In:* G.C. Boere, C.A. Galbraith & D.A. Stroud (Eds.), *Waterbirds Around the World: A Global Overview of the Conservation, Management and Research of the World's Waterbird Flyways* (pp. 40–47). The Stationery Office, Edinburgh, UK.

Boros, E. (2021). Classification method for quantification of waterbird nutrient cycling guilds. *MethodsX*, 8, 101597. https://doi.org/10.1016/j.mex.2021.101597

Boswell, T., & Dunn, I. C. (2015). Regulation of the avian central melanocortin system and the role of leptin. *General and Comparative Endocrinology*, 221, 278–283. https://doi.org/10.1016/j.ygcen.2014.12.009

Brady, C., & Weaver, M. (2025). *California Waterfowl Breeding Population Survey.* California Department of Fish and Wildlife, Sacramento, USA.

Bridge, E. S., Thorup, K., Bowlin, M. S. et al. (2011). Technology on the move: Recent and forthcoming innovations for tracking migratory birds. *BioScience*, 61(9), 689–698. http://www.bioone.org/doi/full/10.1525/bio.2011.61.9.7

Buckton, S. T., & Safford, R. J. (2004). The avifauna of the Vietnamese Mekong Delta. *Bird Conservation International*, 14(4), 279–322. https://doi.org/10.1017/S0959270904000346

Buckton, S. T., Nguyen, C., Ha Q. Q. et al. (1999). Conservation of key wetland sites in Mekong Delta. *BirdLife International Vietnam Programme*. Hanoi, Vietnam.

Buriro, S. A., & Lund, H. G. (2025). Wetlands degradation and its impact on migratory bird populations in Sindh. *International Journal of Social Sciences Bulletin*, 3(1), 597–604. https://doi.org/10.5281/zenodo.14737449

Butler, P. J. (2016). The physiological basis of bird flight. *Philosophical Transactions of the Royal Society B: Biological Sciences*, 371(1704), 20150384. https://doi.org/10.1098/rstb.2015.0384

Carboneras, C. (1992). Family Anatidae (ducks, geese and swans). *In:* J. del Hoyo, A. Elliott & J. Sargatal (Eds.), *Handbook of the Birds of the World* (Vol. 1, pp. 536–628), Lynx Edicions.

Chandra, K., Bharti, D., Kumar, S. et al. (2021). *Faunal Diversity in Ramsar Wetlands of India*. Jointly Published by the Director, Zoological Survey of India and Wetland Division, Ministry of Environment, Forest and Climate Change, Government of India. ISBN: 978-81-8171-567-8

Chastel, O., Fort, J., Ackerman, J. T. et al. (2022). Mercury contamination and potential health risks to Arctic seabirds and shorebirds. *Science of the Total Environment*, 844, 156944. https://doi.org/10.1016/j.scitotenv.2022.156944

Clausen, K. K., & Clausen, P. (2013). Earlier Arctic springs cause phenological mismatch in long-distance migrants. *Oecologia*, 173, 1101–1112. https://doi.org/10.1007/s00442-013-2681-0

Clausen, K. K., Stjernholm, M., & Clausen, P. (2013). Grazing management can counteract the impacts of climate change-induced sea level rise on salt marsh-dependent waterbirds. *Journal of Applied Ecology*, 50(2), 528–537. https://doi.org/10.1111/1365-2664.12043

Convention on Migratory Species (CMS). (2017). Promoting conservation of critical intertidal and other coastal habitats for migratory species (Resolution 12.25). UNEP/CMS Secretariat. https://www.cms.int/sites/default/files/document/cms_cop12_res.12.25_conservation-intertidal-coastal-habitats_e.pdf

Convention on Migratory Species (CMS). (2025). *Convention on the Conservation of Migratory Species of Wild Animals*. https://www.cms.int/en/legalinstrument/aewa

Creaseman, L., Craig, N. J., & Swan M. (1992). The forested wetlands of the Mississippi River: An ecosystem in crisis. *The Nature Conservancy of Louisiana*, Baton Rouge, LA, 4–5.

Cui, L., Wei, Z., Zhou, L. et al. (2024). Effects of constant high water levels in winter on waterbird diversity in Caizi Lakes: A functional perspective. *Global Ecology and Conservation*, 52, e02934. https://doi.org/10.1016/j.gecco.2024.e02934

Dakin, C. L., Small, C. J., Batterham, R. L. et al. (2004). Peripheral oxyntomodulin reduces food intake and body weight gain in rats. *Endocrinology*, 145(6), 2687–2695. https://doi.org/10.1210/en.2003-1338

Davidson, N. C. (2014). How much wetland has the world lost? Long-term and recent trends in global wetland area. *Marine and Freshwater Research*, 65(10), 934–941. https://doi.org/10.1071/MF14173

de Arruda Almeida, B., Sebastián-González, E., dos Anjos, L. et al. (2019). A functional perspective for breeding and wintering waterbird communities: Temporal trends in species and trait diversity. *Oikos*, 128(8), 1103–1115. https://doi.org/10.1111/oik.05903

Denbow D. M. (1989). Centrally and peripherally administered bombesin decreases food intake in turkeys. *Peptides*, 10(2), 275–279. https://doi.org/10.1016/0196-9781(89)90030-2

Deng, X., Zhao, Q., Zhang, J. et al. (2021). Contrasting habitat use and conservation status of Chinese-wintering and other Eurasian greater white-fronted goose (*Anser albifrons*) populations. *Avian Research*, 12(1), 71. https://doi.org/10.1186/s40657-021-00306-0

Dickey, M. H., Gauthier, G., & Cadieux, M. C. (2008). Climatic effects on the breeding phenology and reproductive success of an arctic-nesting goose species. *Global Change Biology*, 14(9), 1973–1985. https://doi.org/10.1111/j.1365-2486.2008.01622.x

Dingle, H. (2014). *Migration: The Biology of Life on the Move*. Oxford University Press.

Docken, N. R. (2011). Effects of block predator management on duck and pheasant nest success in Eastern South Dakota. Electronic Theses and Dissertations, 5135. https://openprairie.sdstate.edu/etd/5135

Doiron, M., Gauthier, G., & Lévesque, E. (2015). Trophic mismatch and its effects on the growth of young in an Arctic herbivore. *Global Change Biology*, 21(12), 4364–4376. https://doi.org/10.1111/gcb.13057

Donald, P. F., Green, R. E., & Heath, M. F. (2001). Agricultural intensification and the collapse of Europe's farmland bird populations. *Proceedings Biological Sciences*, 268(1462), 25–29. https://doi.org/10.1098/rspb.2000.1325

Doxa, A., Robert, A., Crivelli, A. et al. (2012). Shifts in breeding phenology as a response to population size and climatic change: A comparison between short- and long-distance migrant species. *The Auk*, 129(4), 753–762. https://doi.org/10.1525/auk.2012.11213

Drewitt, A. L., & Langston, R. H. (2008). Collision effects of wind-power generators and other obstacles on birds. *Annals of the New York Academy of Sciences*, 1134(1), 233–266. https://doi.org/10.1196/annals.1439.015

Eikenaar, C. (2017). Endocrine regulation of fueling by hyperphagia in migratory birds. *Journal of Comparative Physiology A*, 203, 439–445. https://doi.org/10.1007/s00359-017-1152-1

Elmberg, J., Gunnarsson, G., Pöysä, H. et al. (2005). Within-season sequential density dependence regulates breeding success in mallards *Anas platyrhynchos*. *Oikos*, 108(3), 582–590. https://doi.org/10.1111/j.0030-1299.2005.13618.x

Ely, C. R., & Takekawa, J. Y. (1996). Geographic variation in migratory behavior of Greater white-fronted geese (*Anser albifrons*). *The Auk*, 113(4), 889–901. https://doi.org/10.2307/4088866

Engels, S., Schneider, N. L., Lefeldt, N. et al. (2014). Anthropogenic electromagnetic noise disrupts magnetic compass orientation in a migratory bird. *Nature*, 509, 353–356. https://doi.org/10.1038/nature13290

Eren, S., Beaulieu, A., Piersma, T. et al. (2024). Flyways beyond migratory pathways: The case of waterbird conservation. *Conservation and Society*, 22(2), 74–85. https://doi.org/10.4103/cs.cs_109_22

Falkenberg, G., Fleissner, G., Schuchardt, K. et al. (2010). Avian magnetoreception: Elaborate iron mineral containing dendrites in the upper beak seem to be a common feature of birds. *PLoS One*, 5(2), e9231. https://doi.org/10.1371/journal.pone.0009231

Figuerola, J., & Green, A. J. (2002). Dispersal of aquatic organisms by waterbirds: A review of past research and priorities for future studies. *Freshwater Biology*, 47(3), 483–494. https://doi.org/10.1046/j.1365-2427.2002.00829.x

Fijn, R. C., Hiemstra, D., Phillips, R. A. et al. (2013). Arctic terns *Sterna paradisaea* from the Netherlands migrate record distances across three oceans to Wilkes Land, East Antarctica. *Ardea*, 101(1), 3–12. https://doi.org/10.5253/078.101.0102

Finlayson, C. M., Davidson, N., Pritchard, D. et al. (2011). The Ramsar Convention and ecosystem-based approaches to the wise use and sustainable development of wetlands.

Journal of International Wildlife Law & Policy, 14(3–4), 176–198. https://doi.org/10.1080/13880292.2011.626704

Fluet-Chouinard, E., Stocker, B. D., Zhang, Z. et al. (2023). Extensive global wetland loss over the past three centuries. *Nature*, 614, 281–286. https://doi.org/10.1038/s41586-022-05572-6

Fox, A. D., & Abraham, K. F. (2017). Why geese benefit from the transition from natural vegetation to agriculture. *Ambio*, 46(Suppl 2), 188–197. https://doi.org/10.1007/s13280-016-0879-1

Fox, A. D., & Madsen, J. (2017). Threatened species to super-abundance: The unexpected international implications of successful goose conservation. *Ambio*, 46, 179–187. https://doi.org/10.1007/s13280-016-0878-2

Fox, A. D., Ebbinge, B. S., Mitchell, C. et al. (2010). Current estimates of goose population sizes in Western Europe, a gap analysis and an assessment of trends. *Ornis Svecica*, 20(3–4), 115–127. https://doi.org/10.34080/os.v20.19922

Friedman, J. M., & Halaas, J. L. (1998). Leptin and the regulation of body weight in mammals. *Nature*, 395, 763–770. https://doi.org/10.1038/27376

Fu, M., Wang, J., Hou, C. et al. (2025). Waterbird diversity and its influencing factors in various types of coastal wetlands in the Bohai Rim region. *Global Ecology and Conservation*, 57, e03421. https://doi.org/10.1016/j.gecco.2025.e03421

Gaget, E., Le Viol, I., Pavón-Jordán, D. et al. (2020). Assessing the effectiveness of the Ramsar Convention in preserving wintering waterbirds in the Mediterranean. *Biological Conservation*, 243(5), 108485. https://doi.org/10.1016/j.biocon.2020.108485

Gajjar, J. V., & Bhagat, S. S. (2023). Ramsar convention: Global efforts for conservation and sustainable use of wetlands. *International Journal of Research and Technology Innovations*, 8(4), 1487–1490. https://ijrti.org/papers/IJRTI2304244.pdf

Galbraith, H., Jones, R., Park, R. et al. (2002). Global climate change and sea level rise: Potential losses of intertidal habitat for shorebirds. *Waterbirds*, 25(2), 173–183. https://doi.org/10.1675/1524-4695(2002)025[0173:GCCASL]2.0.CO;2

Gallo-Cajiao, E., Morrison, T. H., & Fuller, R. A. (2024). Agreements for conserving migratory shorebirds in the Asia–Pacific are better fit for addressing habitat loss than hunting. *Ambio*, 53, 1336–1354. https://doi.org/10.1007/s13280-024-02018-3

Gardner, R. C., Caddell, R., & Okuno, E. (Eds.). (2025). *Wetlands and International Environmental Law: The Evolution and Impact of the Ramsar Convention.* Edward Elgar Publishing.

Gardner, R.C., Barchiesi, S., Beltrame, C. et al. (2015). State of the world's wetlands and their services to people: A compilation of recent analyses. Ramsar Briefing Note no. 7. Gland, Switzerland: Ramsar Convention Secretariat. http://dx.doi.org/10.2139/ssrn.2589447

Gauthier, G. (1990). Philopatry, nest-site fidelity, and reproductive performance in Buffleheads. *The Auk*, 107(1), 126–132. http://www.jstor.org/stable/4087810

Gilbert, M., Xiao, X., Domenech, J. et al. (2006). Anatidae migration in the western Palearctic and spread of highly pathogenic avian influenza H5NI virus. *Emerging Infectious Diseases*, 12(11), 1650–1656. https://doi.org/10.3201/eid1211.060223

Gillies, N., Martín López, L. M., den Ouden, O. F. C. et al. (2023). Albatross movement suggests sensitivity to infrasound cues at sea. *Proceedings of the National Academy of Sciences of the United States of America*, 120(42), e2218679120. https://doi.org/10.1073/pnas.2218679120

Giunchi, D., Baldaccini, N. E., Lenzoni, A. et al. (2019). Spring migratory routes and stopover duration of satellite-tracked Eurasian teals *Anas crecca* wintering in Italy. *Ibis*, 161(1), 117–130. https://doi.org/10.1111/ibi.12602

Green, A. J. (1996). Analyses of globally threatened Anatidae in relation to threats, distribution, migration patterns, and habitat use. *Conservation Biology*, 10(5), 1435–1445. https://doi.org/10.1046/j.1523-1739.1996.10051435.x

Green, A. J., & Elmberg, J. (2014). Ecosystem services provided by waterbirds. *Biological Reviews*, 89(1), 105–122. https://doi.org/10.1111/brv.12045

Gwinner, E. (1996). Circadian and circannual programmes in avian migration. *Journal of Experimental Biology*, 199(1), 39–48. https://doi.org/10.1242/jeb.199.1.39

Habib, M. B., & Ruff, C. B. (2008). The effects of locomotion on the structural characteristics of avian limb bones. *Zoological Journal of the Linnean Society*, 153(3), 601–624. https://doi.org/10.1111/j.1096-3642.2008.00402.x

Hagan, A. A., & Heath, J. E. (1980). Regulation of heat loss in the duck by vasomotion in the bill. *Journal of Thermal Biology*, 5(2), 95–101. https://doi.org/10.1016/0306-4565(80)90006-6

Hahn, S., Bauer, S., & Klaassen, M. (2007). Estimating the contribution of carnivorous waterbirds to nutrient loading in freshwater habitats. *Freshwater Biology*, 52(12), 2421–2433. https://doi.org/10.1111/j.1365-2427.2007.01838.x

Harada, Y., Taniguchi, M., Namatame, H. et al. (2001). Magnetic materials in otoliths of bird and fish lagena and their function. *Acta Oto-Laryngolica*, 121(5), 590–595. https://doi.org/10.1080/00016480118968

Harvey, J. T., & Connors, S. (2002). Birds and mammals. *In:* J. M. Caffrey, M. Brown, W. B. Tyler, and M. Silberstein (Eds.), *Changes in a California Estuary: A Profile of Elkhorn Slough* (pp. 187–214). Elkhorn Slough Foundation, Moss Landing, CA.

Hawkes, L. A., Balachandran, S., Batbayar, N. et al. (2013). The paradox of extreme high-altitude migration in bar-headed geese *Anser indicus*. *Proceedings Biological Sciences*, 280(1750), 20122114. https://doi.org/10.1098/rspb.2012.2114

Heim, W., Holtmannspötter, C., Heim, R. J. et al. (2025). High microplastic pollution in birds of urban waterbodies revealed by non-invasively collected faecal samples. *Science of The Total Environment*, 975, 179306. https://doi.org/10.1016/j.scitotenv.2025.179306

Heller, N. E., & Zavaleta, E. S. (2009). Biodiversity management in the face of climate change: A review of 22 years of recommendations. *Biological Conservation*, 142(1), 14–32. https://doi.org/10.1016/j.biocon.2008.10.006

Heyers, D., Manns, M., Luksch, H. et al. (2007). A visual pathway links brain structures active during magnetic compass orientation in migratory birds. *PloS One*, 2(9), e937. https://doi.org/10.1371/journal.pone.0000937

Howard, L. (2003). "Anatidae" (On-line), Animal Diversity Web. Accessed February 11, 2026 at https://animaldiversity.org/accounts/Anatidae/ Last updated: 2003-05-31 / Generated: 2025-11-24 02:42

Isola, C. R., Colwell, M. A., Taft, O. W. et al. (2000). Interspecific differences in habitat use of shorebirds and waterfowl foraging in managed wetlands of California's San Joaquin Valley. *Waterbirds: The International Journal of Waterbird Biology*, 23(2), 196–203. https://www.jstor.org/stable/4641140

Iwamura, T., Possingham, H. P., Chadès, I. et al. (2013). Migratory connectivity magnifies the consequences of habitat loss from sea-level rise for shorebird populations. *Proceedings Biological Sciences*, 280(1761), 20130325. https://doi.org/10.1098/rspb.2013.0325

Jenni-Eiermann, S., & Jenni, L. (1994). Plasma metabolite levels predict individual body-mass changes in a small long-distance migrant, the Garden warbler. *The Auk*, 111(4), 888–899. https://doi.org/10.2307/4088821

Johnson, W. C., Millett, B. V., Gilmanov, T. et al. (2005). Vulnerability of Northern Prairie wetlands to climate change. *BioScience*, 55(10), 863–872. https://doi.org/10.1641/0006-3568(2005)055[0863:VONPWT]2.0.CO;2

Johnson, W. C., Werner, B., Guntenspergen, G. R. et al. (2010). Prairie wetland complexes as landscape functional units in a changing climate. *BioScience*, 60(2), 128–140. https://doi.org/10.1525/bio.2010.60.2.7

Jonker, R. M., Kraus, R. H. S., Zhang, Q. et al. (2013). Genetic consequences of breaking migratory traditions in barnacle geese *Branta leucopsis*. *Molecular Ecology*, 22(23), 5835–5847. https://doi.org/10.1111/mec.12548

Kaiya, H., Kangawa, K., & Miyazato, M. (2013). What is the general action of ghrelin for vertebrates?—Comparisons of ghrelin's effects across vertebrates. *General and Comparative Endocrinology*, 181, 187–191. https://doi.org/10.1016/j.ygcen.2012.10.015

Kamp, J., Urazaliev, R., Donald, P. F. et al. (2011). Post-Soviet agricultural change predicts future declines after recent recovery in Eurasian steppe bird populations. *Biological Conservation*, 144(11), 2607–2614. https://doi.org/10.1016/j.biocon.2011.07.010

Kang, T. H., Yoo, S. H., Yu, J. P. et al. (2011). A study on the community of wintering waterbirds in Saemangeum. *Korean Journal of Environment and Ecology*, 25(1), 81–90.

Kanstrup, N., Fox, A. D., & Balsby, T. J. S. (2020). Toxic lead gunshot persists accessible to waterbirds after a 33-year ban on their use. *Science of The Total Environment*, 714, 136876. https://doi.org/10.1016/j.scitotenv.2020.136876

Kirby, J. S., Stattersfield, A. J., Butchart, S. H. M. et al. (2008). Key conservation issues for migratory land and waterbird species on the world's major flyways. *Bird Conservation International*, 18(S1), S49–S73. https://doi.org/10.1017/S0959270908000439

Kleyheeg, E., van Dijk, J. G. B., Tsopoglou-Gkina, D. et al. (2017). Movement patterns of a keystone waterbird species are highly predictable from landscape configuration. *Movement Ecology*, 5, 2. https://doi.org/10.1186/s40462-016-0092-7

Knudsen, E., Lindén, A., Both, C. et al. (2011). Challenging claims in the study of migratory birds and climate change. *Biological Reviews*, 86(4), 928–946. https://doi.org/10.1111/j.1469-185X.2011.00179.x

Kölzsch, A., Müskens, G. J. D. M., Kruckenberg, H. et al. (2016). Towards a new understanding of migration timing: Slower spring than autumn migration in geese reflects different decision rules for stopover use and departure. *Oikos*, 125(10), 1496–1507. https://doi.org/10.1111/oik.03121

Kranstauber, B., Kays, R., LaPoint, S. D. et al. (2012). A dynamic Brownian bridge movement model to estimate utilization distributions for heterogeneous animal movement. *Journal of Animal Ecology*, 81(4), 738–746. https://doi.org/10.1111/j.1365-2656.2012.01955.x

Krijgsveld, K. L., Akershoek, K., Schenk, F. et al. (2009). Collision risk of birds with modern large wind turbines. *Ardea*, 97(3), 357–366. https://doi.org/10.5253/078.097.0311

Kshatriya, M., & Blake, R. W. (1992). Theoretical model of the optimum flock size of birds flying in formation. *Journal of Theoretical Biology*, 157(2), 135–174. https://doi.org/10.1016/S0022-5193(05)80618-6

Kumar, S., Singh, S., Sharma, L. K. et al. (2024). Assessment of avian diversity and conservation status in Bhindawas Wildlife Sanctuary: A special reference to migratory bird species. *International Journal of Applied Environmental Sciences*, 19(1), 1–18. https://dx.doi.org/10.37622/IJAES/19.1.2024.1-8

Kumawat, S., & Chadoo, S. A. (2025). Echoes across continents: Safeguarding migratory birds in India's Keoladeo wetlands. *Inventum Biologicum*: *An International Journal of Biological Research*, 5(3), 11–19. https://journals.worldbiologica.com/ib/article/view/186

Laguë S. L. (2017). High-altitude champions: Birds that live and migrate at altitude. *Journal of Applied Physiology*, 123(4), 942–950. https://doi.org/10.1152/japplphysiol.00110.2017

Larivière, S., & Messier, F. (1998). Effect of density and nearest neighbours on simulated waterfowl nests: Can predators recognize high-density nesting patches? *Oikos*, 83(1), 12–20. https://doi.org/10.2307/3546541

Lehikoinen, A., Jaatinen, K., Vähätalo, A. V. et al. (2013). Rapid climate driven shifts in wintering distributions of three common waterbird species. *Global Change Biology*, 19(7), 2071–2081. https://doi.org/10.1111/gcb.12200

Lei, J., Jia, Y., Zuo, A. et al. (2019). Bird satellite tracking revealed critical protection gaps in East Asian–Australasian Flyway. *International Journal of Environmental Research and Public Health*, 16(7), 1147. https://doi.org/10.3390/ijerph16071147

Lewis, M. (2016). AEWA at twenty: An appraisal of the African–Eurasian Waterbird Agreement and its unique place in International Environmental Law. *Journal of International Wildlife Law & Policy*, 19(1), 22–61. https://doi.org/10.1080/13880292.2016.1131510

Lindenmayer, D. B., & Likens, G. E. (2010). The science and application of ecological monitoring. *Biological Conservation*, 143(6), 1317–1328. https://doi.org/10.1016/j.biocon.2010.02.013

Lissaman, P. B. S., & Shollenberger, C. A. (1970). Formation flight of birds. *Science*, 168(3934), 1003–1005. https://doi.org/10.1126/science.168.3934.1003

Liu, Y., Holt, P. I., Lei, J. Y. et al. (2006). Distribution, numbers and age structure of Relict gull *Larus relictus* in Bohai Bay, China. *Waterbirds*, 29(3), 375–380. https://doi.org/10.1675/1524-4695(2006)29[375:DNAASO]2.0.CO;2

Livezey, B. C. (1997). A phylogenetic analysis of modern sheldgeese and shelducks (*Anatidae, Tadornini*). *Ibis*, 139(1), 51–66. https://doi.org/10.1111/j.1474-919X.1997.tb04504.x

Lok, T., Overdijk, O., Tinbergen, J. M. et al. (2011). The paradox of spoonbill migration: Most birds travel to where survival rates are lowest. *Animal Behaviour*, 82(4), 837–844. https://doi.org/10.1016/j.anbehav.2011.07.019

Louchart, A. (2008). Emergence of long distance bird migrations: A new model integrating global climate changes. *Naturwissenschaften*, 95(12), 1109–1119. https://doi.org/10.1007/s00114-008-0435-3

Ma, Y., Choi, C. Y., Shang, L. et al. (2024). Mercury contamination is an invisible threat to declining migratory shorebirds along the East Asian-Australasian flyway. *Communications Biology*, 7(1), 585. https://doi.org/10.1038/s42003-024-06254-x

Madsen, J., Guillemain, M., Nagy, S. et al. (2015). Towards sustainable management of huntable migratory waterbirds in Europe: A report by the Waterbird Harvest Specialist Group of Wetlands International. Wetlands International, the Netherlands, pp. 1–43.

Maina, J. N., McCracken, K. G., Chua, B. et al. (2017). Morphological and morphometric specializations of the lung of the Andean goose, *Chloephaga melanoptera*: A lifelong high-altitude resident. *PloS One*, 12(3), e0174395. https://doi.org/10.1371/journal.pone.0174395

Masto, N. M., Robinson, O. J., Brasher, M. G. et al. (2022). Citizen science reveals waterfowl responses to extreme winter weather. *Global Change Biology*, 28(18), 5469–5479. https://doi.org/10.1111/gcb.16288

Mateo, R. (2009). Lead poisoning in wild birds in Europe and the regulations adopted by different countries. *In:* R.T. Watson, M. Fuller, M. Pokras, and W.G. Hunt (Eds.), *Ingestion of Lead from Spent Ammunition: Implications for Wildlife and Humans* (pp. 71–98). The Peregrine Fund, Boise, Idaho, USA, https://doi.org/10.4080/ilsa.2009.0107

McWilliams, S. R., Guglielmo, C., Pierce, B. et al. (2004). Flying, fasting, and feeding in birds during migration: A nutritional and physiological ecology perspective. *Journal of Avian Biology*, 35(5), 377–393. https://doi.org/10.1111/j.0908-8857.2004.03378.x

Meehan, C., Stronach, B. W. H., Huxley, C. et al. (2009). A review of changes in duck populations on Lough Carra, County Mayo 1967–2006. *Irish Birds*, 8, 507–520.

Mengdi, F. U., Jun, W. A. N. G., Chunfei, H. O. U., Junsheng, L. I., & Jiade, B. A. I. (2025). Waterbird diversity and its influencing factors in various types of coastal wetlands in the Bohai Rim region. *Global Ecology and Conservation*, 57, e03421. https://doi.org/10.1016/j.gecco.2025.e03421

Midtgård, U. (1984). Blood vessels and the occurrence of arteriovenous anastomoses in cephalic heat loss areas of Mallards, *Anas platyrhynchos* (Aves). *Zoomorphology*, 104(5), 323–335. https://doi.org/10.1007/BF00312014

Milá, B., Smith, T. B., & Wayne, R. K. (2006). Postglacial population expansion drives the evolution of long-distance migration in a songbird. *Evolution: International Journal of Organic Evolution*, 60(11), 2403–2409.

Mirzaeinia, A., Heppner, F., & Hassanalian, M. (2020). An analytical study on leader and follower switching in V-shaped Canada goose flocks for energy management purposes. *Swarm Intelligence*, 14(2), 117–141. https://doi.org/10.1007/s11721-020-00179-x

Mitsch, W. J., & Hernandez, M. E. (2013). Landscape and climate change threats to wetlands of North and Central America. *Aquatic Sciences*, 75(1), 133–149. https://doi.org/10.1007/s00027-012-0262-7

Miyabayashi, Y., & Mundkur, T. (1999). *Atlas of Key Sites for Anatidae in the East Asian Flyway*. Tokyo, Japan: Wetlands International.

Moayedi, A., Long, J. A., Kölzsch, A. et al. (2025). Multi-modal, interrelated navigation in migratory birds: A data mining study. *Ecological Informatics*, 90, 103218. https://doi.org/10.1016/j.ecoinf.2025.103218

Moon, J. A., Lehnen, S. E., Metzger, K. L. et al. (2021). Projected impact of sea-level rise and urbanization on mottled duck (*Anas fulvigula*) habitat along the Gulf Coast of Louisiana and Texas through 2100. *Ecological Indicators*, 132, 108276. https://doi.org/10.1016/j.ecolind.2021.108276

Moores, N., Kim, A. & Kim, R. (2014). Status of Birds, 2014. *Birds Korea Report on Bird Population Trends and Conservation Status in the Republic of Korea*. Birds Korea.

Mouritsen, H., Heyers, D., & Güntürkün, O. (2016). The neural basis of long-distance navigation in birds. *Annual Review of Physiology*, 78, 133–154. https://doi.org/10.1146/annurev-physiol-021115-105054

Muheim, R., Moore, F. R., & Phillips, J. B. (2006). Calibration of magnetic and celestial compass cues in migratory birds—A review of cue-conflict experiments. *The Journal of Experimental Biology*, 209(1), 2–17. https://doi.org/10.1242/jeb.01960

Murphy-Klassen, H. M., Underwood, T. J., Sealy, S. G. et al. (2005). Long-term trends in spring arrival dates of migrant birds at Delta Marsh, Manitoba, in relation to climate change. *The Auk*, 122(4), 1130–1148. https://doi.org/10.1093/auk/122.4.1130

Murray, N. J., Clemens, R. S., Phinn, S. R. et al. (2014). Tracking the rapid loss of tidal wetlands in the Yellow Sea. *Frontiers in Ecology and the Environment*, 12(5), 267–272. https://doi.org/10.1890/130260

Myers, J. P., McLain, P. D., Morrison, R. I. et al. (1987). The Western Hemisphere shorebird reserve network. *Wader Study Group Bulletin*, 49(2), 19. https://digitalcommons.usf.edu/wsg_bulletin/vol49/iss2/19

Najafabadi, M. S., Darvishzadeh, R., Skidmore, A. K. et al. (2015). Satellite-versus temperature-derived green wave indices for predicting the timing of spring migration of avian herbivores. *Ecological Indicators*, 58, 322–331. https://dx.doi.org/10.1016/j.ecolind.2015.06.005

Newth, J. L., Rees, E. C., Cromie, R. L. et al. (2016). Widespread exposure to lead affects the body condition of free-living whooper swans *Cygnus cygnus* wintering in Britain. *Environmental Pollution*, 209, 60–67. https://doi.org/10.1016/j.envpol.2015.11.007

Newton, I. (2007). Site-fidelity and dispersal. *In: The Migration Ecology of Birds* (Ch 17, pp. 479–522). https://doi.org/10.1016/b978-012517367-4.50017-6

Newton, I. (2023). *The Migration Ecology of Birds* (2nd Ed.) Elsevier, Academic Press.

Nichols, J. D., Runge, M. C., Johnson, F. A. et al. (2007). Adaptive harvest management of North American waterfowl populations: A brief history and future prospects. *Journal of Ornithology*, 148, 343–349. https://doi.org/10.1007/s10336-007-0256-8

Nolet, B. A., Schreven, K. H. T., Boom, M. P. et al. (2020). Contrasting effects of the onset of spring on reproductive success of Arctic-nesting geese. *The Auk*, 137(1), ukz063. https://doi.org/10.1093/auk/ukz063

Norberg, U. M. (1990). Morphological adaptations for flight. *In: Vertebrate Flight: Mechanics, Physiology, Morphology, Ecology and Evolution, Zoophysiology* (pp. 180–236). Berlin, Heidelberg: Springer Berlin Heidelberg. https://doi.org/10.1007/978-3-642-83848-4_11

Nourani, E., Kaboli, M., & Collen, B. (2015). An assessment of threats to Anatidae in Iran. *Bird Conservation International*, 25(2), 242–257. https://doi.org/10.1017/S0959270914000264

Nyeland, J., Fox, A. D., Kahlert, J., & Therkildsen, O. R. (2003). Field methods to assess pectoral muscle mass in moulting geese. *Wildlife Biology*, 9(2), 155–159. https://doi.org/10.2981/wlb.2003.029

Olson, R., & Rohwer, F. C. (1998). Effects of human disturbance on success of artificial duck nests. *The Journal of Wildlife Management*, 62(3), 1142–1146. https://doi.org/10.2307/3802569

Osiak-Wicha, C., Tomaszewska, E., Muszyński, S. et al. (2024). Comparative analysis of morphometric, densitometric, and mechanical properties of skeletal locomotor elements in three duck species (Anatidae: Anatinae). *Animals*, 14(15), 2191. https://doi.org/10.3390/ani14152191

Overton, C. T., Lorenz, A. A., James, E. P. et al. (2021). Megafires and thick smoke portend big problems for migratory birds. *Ecology*, 103(1), e03552. https://doi.org/10.1002/ecy.3552

Owen, M., & Black, J. M. (1990). *Waterfowl Ecology*. Chapman and Hall, New York.

Owusu-Achiaw, R., & Osei-Owusu, Y. (2023). Community-based approach to wetland restoration: Case study of the Songor Wetland, Ghana. *In:* Nishi. M. (Ed.), *Governing*

Sustainability in the Global South (pp. 157–177). Routledge: London, UK. https://doi.org/10.1007/978-981-99-1292-6_9

Parr, N. (2020). *Flights Across the Roof of the World.* University of Exeter (United Kingdom). Thesis. http://hdl.handle.net/10871/120670

Pearson, A. W., Greenwood, N. M., Butler, E. J. et al. (1983). Biochemical changes in layer and broiler chickens when fed on a high-glucosinolate rapeseed meal. *British Poultry Science*, 24(3), 417–427. https://doi.org/10.1080/00071668308416756

Perrins, C. M., Cousquer, G., & Waine, J. (2003). A survey of blood lead levels in mute swans *Cygnus olor*. *Avian Pathology*, 32(2), 205–212. https://doi.org/10.1080/0307946021000071597

Peterson, S. H., Ackerman, J. T., Hartman, C. A. et al. (2024). Duckling survival increased with availability of flooded wetland habitat and decreased with salinity concentrations in a brackish marsh. *Ornithological Applications*, 126(3), duae017. https://doi.org/10.1093/ornithapp/duae017

Piersma, T. (2007). Using the power of comparison to explain habitat use and migration strategies of shorebirds worldwide. *Journal of Ornithology*, 148(Suppl 1), 45–59. https://doi.org/10.1007/s10336-007-0240-3

Piersma, T., & Lindström, Å. (2004). Migrating shorebirds as integrative sentinels of global environmental change. *Ibis*, 146, 61–69. https://doi.org/10.1111/j.1474-919X.2004.00329.x

Portugal, S. J., Hubel, T. Y., Fritz, J. et al. (2014). Upwash exploitation and downwash avoidance by flap phasing in ibis formation flight. *Nature*, 505, 399–402. https://doi.org/10.1038/nature12939

Post, D. M., Taylor, J. P., Kitchell, J. F. et al. (1998). The role of migratory waterfowl as nutrient vectors in a managed wetland. *Conservation Biology*, 12(4), 910–920. https://doi.org/10.1111/j.1523-1739.1998.97112.x

Prop, J., Oudman, T., van Spanje, T. M. et al. (2013). Patterns of predation of Pink-footed goose nests by polar bear. *Ornis Norvegica*, 36, 38–46. https://doi.org/10.15845/on.v36i0.439

Pulido, F. (2007). The genetics and evolution of avian migration. *Bioscience*, 57(2), 165–174. https://doi.org/10.1641/B570211

Quinn, T. W. (1992). The genetic legacy of mother goose–phylogeographic patterns of Lesser snow goose *Chen caerulescens caerulescens* maternal lineages. *Molecular Ecology*, 1(2), 105–117. https://doi.org/10.1111/j.1365-294X.1992.tb00162.x

Ramsar Convention on Wetlands. (2018). *Global Wetland Outlook: State of the World's Wetlands and their Services to People*. Gland, Switzerland: Ramsar Convention Secretariat.

Ramsar Sites Information Service. (2017). *Åsnen Ramsar Site*. https://rsis.ramsar.org/ris/429

Rayner, J. M. V. (1988). Form and function in avian flight. *In:* R.F. Johnston (Ed.), *Current Ornithology* (vol. 5, pp. 1–66) Springer, Boston, MA. https://doi.org/10.1007/978-1-4615-6787-5_1

Reynolds, C., & Ryan, P. G. (2018). Micro-plastic ingestion by waterbirds from contaminated wetlands in South Africa. *Marine Pollution Bulletin*, 126, 330–333. https://doi.org/10.1016/j.marpolbul.2017.11.021

Reynolds, M., & Klavitter, J. (2006). Translocation of wild Laysan duck *Anas laysanensis* to establish a population at Midway Atoll National Wildlife Refuge, United States and US Pacific Possession. *Conservation Evidence*, 3, 6–8.

Richardson, R. D., Boswell, T., Weatherford, S. C. et al. (1993). Cholecystokinin octapeptide decreases food intake in white-crowned sparrows. *The American Journal of Physiology*, 264(5), R852–R856. https://doi.org/10.1152/ajpregu.1993.264.5.R852

Richardson, W. J. (1978). Timing and amount of bird migration in relation to weather: A review. *Oikos*, 224–272. https://doi.org/10.2307/3543482

Richardson, W. J. 1990. Timing of migration in relation to weather: Updated review. *In:* E. Gwinner (Ed.), *Bird Migration: Physiology and Ecophysiology* (pp. 78–101). Springer-Verlag, Berlin. https://doi.org/10.1007/978-3-642-74542-3_6

Robertson, G. J., & Cooke, F. (1999). Winter philopatry in migratory waterfowl. *The Auk*, 116(1), 20–34.

Rocke, T. E., & Bollinger, T. K. (2007). Avian botulism. *In:* N.J. Thomas, D.B. Hunter, and C.T. Atkinson (Eds.), *Infectious Diseases of Wild Birds* (Chapter 21, pp. 377–416). Blackwell Publishing. https://doi.org/10.1002/9780470344668.ch21

Rockwell, R. F., Gormezano, L. J., & Koons, D. N. (2011). Trophic matches and mismatches: Can polar bears reduce the abundance of nesting snow geese in Western Hudson Bay? *Oikos*, 120(5), 696–709. https://doi.org/10.1111/j.1600-0706.2010.18837.x

Runge, C. A., Gallo-Cajiao, E., Carey, M. J. et al. (2017). Coordinating domestic legislation and international agreements to conserve migratory species: A case study from Australia. *Conservation Letters*, 10(6), 765–772. https://doi.org/10.1111/conl.12345

Runge, C. A., Martin, T. G., Possingham, H. P. et al. (2014). Conserving mobile species. *Frontiers in Ecology and the Environment*, 12(7), 395–402. https://doi.org/10.1890/130237

Russell, I. A., Randall, R. M., Zimmerman, D. et al. (2019). Outbreak of avian botulism and its effect on waterbirds in the wilderness lakes, South Africa. *Koedoe*, 61(1), a1553. https://doi.org/10.4102/koedoe.v61i1.1553

Samelius, G., Alisauskas, R. T., & Larivière, S. (2011). Seasonal pulses of migratory prey and annual variation in small mammal abundance affect abundance and reproduction by arctic foxes. *Polar Biology*, 34, 1475–1484. https://doi.org/10.1007/s00300-011-1005-2

Samuel F., & Hussain S. K. (2014). Impacts of unsustainable hunting and poaching practices on the health of river Kabul and its faunal diversity, district Charsadda, Khyber Pakhtunkhwa. *The Pakistan Journal of Forestry*, 64(1), 27–36. https://researcherslinks.com/current-issues/impacts-of-unsustainable-hunting-and-poaching-practices-on-the-health-of-river-kabul-and-its-faunal-/41/5/8482

Samuel, M. D., Shadduck, D. J., Goldberg, D. R. et al. (2005). Avian cholera in waterfowl: The role of Lesser snow and Ross's geese as disease carriers in the Playa Lakes Region. *The Journal of Wildlife Diseases*, 41(1), 48–57. https://doi.org/10.7589/0090-3558-41.1.48

Sanderson, F. J., Kucharz, M., Jobda, M. et al. (2013). Impacts of agricultural intensification and abandonment on farmland birds in Poland following EU accession. *Agriculture, Ecosystems & Environment*, 168, 16–24. https://doi.org/10.1016/j.agee.2013.01.015

Savory C. J. (1987). An alternative explanation for apparent satiating properties of peripherally administered bombesin and cholecystokinin in domestic fowls. *Physiology and Behavior*, 39(2), 191–202. https://doi.org/10.1016/0031-9384(87)90009-6

Sekercioglu, C. H. (2002). Impacts of birdwatching on human and avian communities. *Environmental Conservation*, 29(3), 282–289. https://doi.org/10.1017/S0376892902000206

Semel, B., & Sherman, P. W. (1995). Alternative placement strategies for wood duck nest boxes. *Wildlife Society Bulletin*, 23(3), 463–471. https://www.jstor.org/stable/3782956

Sheta, B., Hassan, A., Sallam, A. E. D., Habbak, L. et al. (2025). Phylogenetic and lipid metabolic differences between migratory and Egyptian-domesticated mallard ducks (*Anas platyrhynchos*). *Comparative Biochemistry and Physiology. Part A, Molecular & Integrative Physiology*, 302, 111814. https://doi.org/10.1016/j.cbpa.2025.111814

Shi, J., Meng, L., Xia, S. et al. (2024). Habitat suitability and determinants for Anatidae in multi-watershed composite wetlands in Anhui, China. *Animals*, 14(7), 1010. https://doi.org/10.3390/ani14071010

Shoaib, M., & Singh, J. K. (2023). Diversity and conservation status of avifauna in Uttar Pradesh with special reference to Patna Bird Sanctuary. *Journal of Emerging Technologies and Innovative Research*, 10(2), d126–d131. https://www.jetir.org/papers/JETIR2302313.pdf

Siegel, S., Gillenwater, D., Toms, C. et al. (2011). *Suisun Marsh Conservation Assessment.* Volume I: Characterization Report. Wetlands and Water Resources, Inc. (pp. 353).

Soons, M. B., Brochet, A. L., Kleyheeg, E. et al. (2016). Seed dispersal by dabbling ducks: An overlooked dispersal pathway for a broad spectrum of plant species. *Journal of Ecology*, 104(2), 443–455. https://doi.org/10.1111/1365-2745.12531

Soulé, M. E., Estes, J. A., Miller, B. et al. (2005). Strongly interacting species: Conservation policy, management, and ethics. *BioScience*, 55(2), 168–176. https://doi.org/10.1641/0006-3568(2005)055[0168:SISCPM]2.0.CO;2

Sovada, M. A., Anthony, R. M., & Batt, B. D. J. (2001). Predation on waterfowl in Arctic tundra and Prairie breeding areas: A review. *Wildlife Society Bulletin*, 29(1), 6–15. https://pubs.usgs.gov/publication/70185400

Spivey, R. J., & Bishop, C. M. (2014). An implantable instrument for studying the long-term flight biology of migratory birds. *Review of Scientific Instruments*, 85(1), 014301. https://doi.org/10.1063/1.4854635

Studds, C. E., Kendall, B. E., Murray, N. J. et al. (2017). Rapid population decline in migratory shorebirds relying on Yellow Sea tidal mudflats as stopover sites. *Nature Communications*, 8, 14895. https://doi.org/10.1038/ncomms14895

Sun, X., Shen, J., Xiao, Y. et al. (2023). Habitat suitability and potential biological corridors for waterbirds in Yancheng coastal wetland of China. *Ecological Indicators*, 148, 110090. https://doi.org/10.1016/j.ecolind.2023.110090

Takekawa, J. Y., Heath, S. R., Douglas, D. C. et al. (2009). Geographic variation in Bar-headed geese *Anser indicus*: Connectivity of wintering areas and breeding grounds across a broad front: *Wildfowl*, 59, 100–123. https://pubs.usgs.gov/publication/70003917

Tape, K. D., Flint, P. L., Meixell, B. W. et al. (2013). Inundation, sedimentation, and subsidence creates goose habitat along the Arctic coast of Alaska. *Environmental Research Letters*, 8(4), 045031. https://doi.org/10.1088/1748-9326/8/4/045031

Tomlik, M. D. (2019). Altered environmental stressors and long-term effects on populations of Common eider (*Somateria mollissima*) on Nova Scotia's Eastern shore. Doctoral dissertation, Acadia University. https://library-archives.canada.ca/eng/services/services-libraries/theses/Pages/item.aspx?idNumber=1224180676

Tozer, D., & Beck, G. (2018). How do recent changes in Lake Erie affect birds? Part one: Invasive *Phragmites*. *Ontario Birds*, 36(3), 161–169. https://digitalcommons.usf.edu/cgi/viewcontent.cgi?article=1627&context=ontario_birds

U.S. Fish and Wildlife Service. 2013. Recovery plan for tidal marsh ecosystems of Northern and Central California. US Fish and Wildlife Service, Sacramento, CA, 77.

Uher-Koch, B. D., Buchheit, R. M., Eldermire, C. R. et al. (2021). Shifts in the wintering distribution and abundance of Emperor geese in Alaska. *Global Ecology and Conservation*, 25, e01397. https://doi.org/10.1016/j.gecco.2020.e01397

UNEP-AEWA. (2023). *Agreement on the Conservation of African–Eurasian Migratory Waterbirds*. https://www.unep-aewa.org/sites/default/files/uploads/aewa_agreement_text_2023-2025_corrected%20version%20as%20of%2010%20August%202023_EN_rev.pdf

Väänänen, V. M. (2000). Predation risk associated with nesting in gull colonies by two *Aythya* species: Observations and an experimental test. *Journal of Avian Biology*, 31(1), 31–35. https://www.jstor.org/stable/3676927

van Bolhuis, G. H., Rijks, J. M., Dorrestein, G. M. et al. (2004). Obliterative endophlebitis in Mute swans (*Cygnus olor*) caused by *Trichobilharzia* sp. (Digenea: Schistosomatidae) infection. *Veterinary Pathology*, 41(6), 658–665. https://doi.org/10.1354/vp.41-6-658

van Toor, M. L., Hedenström, A., Waldenström, J. et al. (2013). Flexibility of continental navigation and migration in European mallards. *PloS One*, 8(8), e72629. https://doi.org/10.1371/journal.pone.0072629

Visser, M. E., Both, C., & Lambrechts, M. M. (2012). Global climate change leads to mistimed avian reproduction. *Advances in Ecological Research*, 35, 89–110. https://doi.org/10.1016/S0065-2504(04)35005-1

Wang, G., Wu, H., Dai, J. et al. (2023). Priorities identification of habitat restoration for migratory birds under the increased water level during the middle of dry season: A case study of Poyang Lake and Dongting Lake wetlands, China. *Ecological Indicators*, 151, 110322. https://doi.org/10.1016/j.ecolind.2023.110322

Wang, X., Kuang, F., Tan, K. et al. (2018). Population trends, threats, and conservation recommendations for waterbirds in China. *Avian Research*, 9, 14. https://doi.org/10.1186/s40657-018-0106-9

Webster, M. S., Marra, P. P., Haig, S. M. et al. (2002). Links between worlds: Unraveling migratory connectivity. *Trends in Ecology & Evolution*, 17(2), 76–83. https://doi.org/10.1016/S0169-5347(01)02380-1

Weller, F. G., Beatty, W. S., Webb, E. B. et al. (2022). Environmental drivers of autumn migration departure decisions in midcontinental mallards. *Movement Ecology*, 10, 1. https://doi.org/10.1186/s40462-021-00299-x

Wilcove, D. S., & Wikelski, M. (2008). Going, going, gone: Is animal migration disappearing. *PLoS Biology*, 6(7), e188. https://doi.org/10.1371/journal.pbio.0060188

Wilcox, K. L., Petrie, S. A., Maynard, L. A. et al. (2003). Historical distribution and abundance of *Phragmites australis* at long point, Lake Erie, Ontario. *Journal of Great Lakes Research*, 29(4), 664–680. https://doi.org/10.1016/S0380-1330(03)70469-9

Wiltschko, R., & Wiltschko, W. (2009). Avian navigation. *The Auk*, 126(4), 717–743. https://doi.org/10.1525/auk.2009.11009

Wiltschko, R., & Wiltschko, W. (2019). Magnetoreception in birds. *Journal of the Royal Society, Interface*, 16(158), 20190295. https://doi.org/10.1098/rsif.2019.0295

Wolff, W. J. (1970). Goal orientation versus one-direction orientation in teal during autumn migration. *Ardea*, 58, 131–141.

Wong, J. B.-Z. (2021). Arctic tern migration: From individual variation to common flyways of trans-equatorial seabirds. Doctoral dissertation, University of British Columbia. https://open.library.ubc.ca/collections/ubctheses/24/items/1.0396330

Work, T. M., Klavitter, J. L., Reynolds, M. H. et al. (2010). Avian botulism: A case study in translocated endangered Laysan ducks (*Anas laysanensis*) on Midway Atoll. *Journal of Wildlife Diseases*, 46(2), 499–506. https://doi.org/10.7589/0090-3558-46.2.499

World Wildlife Fund. (n.d.). *Wetlands for Asian Flyways*. https://wwf.panda.org/discover/our_focus/freshwater_practice/freshwater_inititiaves/asian_flyways/

Xu, Y., Si, Y., Yin, S. et al. (2019). Species-dependent effects of habitat degradation in relation to seasonal distribution of migratory waterfowl in the East Asian–Australasian flyway. *Landscape Ecology*, 34, 243–257. https://doi.org/10.1007/s10980-018-00767-7

Yap, K. N., Tsai, O. H. I., & Williams, T. D. (2019). Haematological traits co-vary with migratory status, altitude and energy expenditure: A phylogenetic, comparative analysis. *Scientific Reports*, 9, 6351. https://doi.org/10.1038/s41598-019-42921-4

Yetter, A. P., Hine, C. S., Lancaster, J. D. et al. (2019). Annual report FY2019, Illinois waterfowl surveys and investigations W-43-R-66. *Technical Report INHS 2019* (21). https://hdl.handle.net/2142/105455

Yong, D. L., Liu, Y., Low, B. W. et al. (2015). Migratory songbirds in the East Asian–Australasian flyway: A review from a conservation perspective. *Bird Conservation International*, 25(1), 1–37. https://doi.org/10.1017/S0959270914000276

Yu, J. P., Han, S. W., Paik, I. H. et al. (2014). Status of wintering populations of the Baikal teal (*Anas formosa*) in Geumgang river, Korea. *Journal of Asia-Pacific Biodiversity*, 7(2), e213–e217. https://doi.org/10.1016/j.japb.2014.04.005

Žalakevičius, M. (1990). The study of mechanisms controlling migratory take-off in spring and autumn. *Acta Ornithologica Lituanica*, 3, 3–52.

Žalakevičius, M. (1993). A study of mechanisms controlling migratory take-off of geese, thrushes and Wood Pigeon in spring and autumn: A radar study. *Acta Ornithologica Lituanica*, 7, 16–26.

Zapka, M., Heyers, D., Hein, C. M. et al. (2009). Visual but not trigeminal mediation of magnetic compass information in a migratory bird. *Nature*, 461, 1274–1277. https://doi.org/10.1038/nature08528

Zarzycki, M. C. (2017). Evidence for cross-seasonal effects: Insights from long-term data on northern pintail. Thesis.

Zhao, S., Xu, H., Zhong, S. et al. (2022). Risk evaluation of onshore wind farms in relation to wild duck (Anatidae) movements in the Yangtze River Mouth, China. *IET Renewable Power Generation*, 16, 470–477. https://doi.org/10.1049/rpg2.12347

Zhu, F., Zou, Y., Zhang, P. et al. (2022). Dyke demolition led to a sharp decline in waterbird diversity due to habitat quality reduction: A case study of Dongting Lake, China. *Ecology and Evolution*, 12(4), e8782. https://doi.org/10.1002/ece3.8782

CHAPTER
7

DNA Barcoding and Phylogenetic Relationships in Anatidae

Bishal Kumar Majhi[1,2], Soumya Sephalika Swain[3,4], Krushna Chandra Giri[5] and Bibhu Prasad Panda[5*]

[1] Govind Ballabh Pant National Institute of Himalayan Environment, North East Regional Centre, Chandranagar, Itanagar – 791113, Arunachal Pradesh, India
[2] Department of Zoology, Mizoram University, Aizawl – 796004, Mizoram, India
[3] Odisha Biodiversity Board, Ekamrakanan, Nayapalli, Bhubaneswar – 751015, Odisha, India
[4] Department of Biotechnology, Utkal University, Bhubaneswar – 751004, Odisha, India
[5] Biodiversity, Ecology and Environmental Laboratory, Environmental Sciences, Department of Chemistry, ITER, Siksha 'O' Anusandhan (Deemed to be University), Bhubaneswar, Odisha, India

1. Introduction

The Anatidae, or the family of ducks, geese, and swans, is the most species-rich family in the order Anseriformes. Mitochondrial genome phylogenetic studies identify it as a sister group to Anseranatidae and Anhimidae (Zhang et al., 2023). Its internal taxonomy is clear and consists of subfamilies Dendrocygninae, Oxyurinae, Anserinae, and Anatinae, which are a response to ecological adaptation and life-history variety (Madsen et al., 1988). The family currently consists of 174 species and 53 genera spread around the world except in the Antarctica (Winkler et al., 2020). The International Union for Conservation of Nature (IUCN) Red List (2023) indicates that approximately 14% of these species are endangered to different extents, and the necessity to take urgent conservation measures. The record of fossil samples dating back to the Eocene (such as Presbyornis) and the Miocene (such as anserines) demonstrates the strong evolutionary ancestry of this group and shows how morphological changes recurred, with several

*Corresponding author: bibhuprasadpanda@soa.ac.in

DOI: 10.1201/9781003669784-7

independent flight losses (Dyke, 2001; Watanabe, 2017; Worthy et al., 2022). Anatidae have a broad range of habitats, including inland wetlands, tundra lakes, and also oceans. They exhibit a great deal of ecological flexibility, with some long-distance Holarctic migrants and others no longer than stationary Southern Hemisphere species. Dabbling, diving, and sea ducks represent feeding guilds that have a separate ecological niche frequently consistent with molecularly motivated clades (Johnson & Sorenson, 1999). Phylogeographic and fossil evidence also suggest that strong biogeographic structuring is reduced by the economic mechanism of high dispersal power in the family (Spaulding et al., 2023). Anatidae are ecosystem engineers, influencing the habitat by their grazing and herbivory to affect the wetland vegetation cover and nutrient cycling (Gendron, 2018; Gordon et al., 2019). Genomic studies have revealed a high level of hybridization and introgression, especially in species of geese, and it is believed that reticulate evolution has contributed significantly to diversification and made the traditional phylogenetic paradigm complex (Ottenburghs, 2016; Ottenburghs et al., 2017; Spaulding et al., 2023). The family Anatidae represents the complexity of the interplay between evolutionary innovation and ecological specialization. This increases the importance of molecular methods in the dissection of the complexities, for example, hybridization, convergence, and cryptic diversity, which are often difficult to measure by morphology only.

1.1. Evolutionary Story

The evolutionary history of the Anatidae family is of great importance as a result of a combination of high rates of hybridization, rapid radiations, ecological differentiation, and a relatively rich fossil record. Comparative phylogenetic studies indicate that the Northern Hemisphere has a more recent and concentrated set of Pleistocene radiations in comparison to the older and more diffused radiations of the Southern Hemisphere. Therefore, conservation should be considered for evolutionary history and functional diversity as well as number of species (Phillimore & Price, 2008; Zeng et al., 2019). Fossil groups such as Presbyornis can be used as important calibration groups to estimate early stem lineages, and morphological and ecological data indicate recurrent convergent evolution in morphological dimensions, including bill morphology, which reflects dietary and habitat adaptations (Livezey, 1997; Olsen, 2017; Tambussi et al., 2019). Though avian remains are usually fragmentary, the fossil record of the waterfowl group is fairly complete, implying that the group experienced significant radiations in the last 60 million years (Delacour & Mayr, 1946; Olson & Feduccia, 1980). Initial molecular research, especially DNA hybridization, introduced the initial large-scale information about phylogenetic relationships of waterfowl and established the basis of further molecular systematics (Sibley & Ahlquist, 1983). Subsequent genomic work has since revolutionized the evolutionary history of the Anatidae, concluding that the phenomena of hybridization and introgression are not isolated and infrequent events but parts of a repetitive evolutionary process in dabbling ducks and geese. Furthermore, the interactions between humans and domestic and wild populations and the introduction of species by humans have weakened the genetic boundaries of taxa, making it more difficult to distinguish domestic and wild species and manage conservation issues (Kraus et al., 2012; Lavretsky et al., 2020; Ottenburghs et al., 2016, 2017). Therefore, Anatidae is a perfect model of studies devoted to the evolutionary processes of speciation, the combination of fossil and molecular evidence, and the use of conservation genomics to save both evolutionary and genetic integrity (Donne-Goussé et al., 2002; Omland, 1994; Peters et al., 2005). Their evolutionary history highlights the complexity of the interaction of historical diversification and contemporary conservation issues.

1.2. Molecular Approaches

Molecular approaches have been indispensable in conservation of Anatidae, a family that experiences extensive hybridization, rapid evolutionary radiations, and ecological specialization that defy conventional taxonomic and management models. Previously, the conservation methods largely relied on morphological and geographical observations, which in many cases were not adequate in revealing the hidden diversity or man-made genetic interference. With the introduction of molecular methods, especially using mitochondrial DNA as a barcoded system on markers such as *cytochrome* c oxidase subunit I (COI) and cytochrome b (Cyt b), a specific identification of a specific species and the discovery of hidden genetic diversity became possible (Hebert et al., 2003). The latter integration of nuclear markers, such as single-nucleotide polymorphisms (SNPs) and microsatellites, further enhanced the study of hybridization patterns, gene flow, and population genetic organization (Haig et al., 2011). Genome-wide studies have been conducted more recently and they have offered a closer insight into genetic integrity and introgression, particularly in regard to the exchange of genes between wild mallards (*Anas platyrhynchos*) and domesticated ones (Lavretsky et al., 2014). Molecular phylogenetics has been used together with fossil data to further define evolutionary time scales and identification of evolutionarily significant units (ESUs) that can be prioritized during conservation efforts (Robertson et al., 2014). Through a combination of molecular, ecological, and historical data, scientists are now capable of more accurately determining flyway connectivity and lineage diversification. Therefore, molecular methods have now expanded their role well beyond the role of species classification to become the basis of evidence-based conservation; essential to counteract the interrelated impact of genetic introgression, habitat modification, and climatic change.

2. DNA Barcoding

Cryptic or concealed diversity is common among the Anatidae family, with morphological similarity, recurring hybridization, and overlapping distribution making differentiation between species a challenging task. Such reasons usually make external features inaccurate to classify and identify the species (Schwarzfeld & Sperling, 2014). To surmount these obstacles, DNA barcoding has come in as a standardized molecular technique worldwide in the exact identification of species. The technique entails sequencing a short, conserved DNA fragment usually the mitochondrial COI gene in animals and comparison with reference databases, in this case, GenBank and the Barcode of Life Data Systems (BOLD) (Ratnasingham & Hebert, 2007; Kerr et al., 2007; Hebert et al., 2003). Due to the existence of a specific barcode gap as the difference of genomic makeup of two individuals of the identical species and those of different species, the determination of species is reliable (Yao et al., 2010). In addition to taxonomy and the identification of cryptic species, DNA barcoding is also used in phylogenetic inference and conservation evaluations of Anatidae. It provides a rapid and economical and generalized, or universal, genetic identity resource when it is used together with ecological and morphological data, which is a crucial basis to integrative biodiversity research and molecular-based conservation planning in waterfowl.

2.1. Key Markers in Birds

1. **Cytochrome c oxidase I (COI):** It is a mitochondrial protein-coding gene widely used for DNA barcoding of animals. It features relatively conserved primer sites across taxa, high interspecific divergence, and extensive reference data in databases such as BOLD and GenBank. In avian studies, COI effectively differentiates species; for example, in psittacines, it exhibited >98% interspecific divergence and produced well-supported phylogenetic trees (Marín-Villa et al., 2025).
2. **Cytochrome b (Cyt b):** Another mitochondrial marker, Cyt b evolves faster than most ribosomal genes, offering strong resolution among closely related species and even within populations. It has proven valuable in population genetics, phylogeography, and assessing genetic diversity, with studies linking reduced diversity in threatened birds to higher extinction risk (Canteri et al., 2021).
3. **16S rRNA (16S):** This mitochondrial ribosomal gene contains both conserved and variable regions, making it suitable for amplifying degraded DNA or when COI primers fail. Although 16S generally shows lower interspecific divergence than COI or Cyt b, it performs well at higher taxonomic levels (e.g., genera, families) and as complementary evidence. In psittacine identification, 16S yielded lower divergence but, when combined with COI, still enabled accurate species-level assignments (Marín-Villa et al., 2025).

Together, these mitochondrial markers have laid the foundation for avian molecular taxonomy, uncovering cryptic diversity and population structure that were previously hidden by morphology alone. Their success has spurred the broader integration of molecular approaches in conservation, an advancement particularly crucial for complex groups such as the Anatidae, where hybridization and recent radiations obscure taxonomic boundaries.

2.2. Strengths

1. **High species discrimination (especially COI and Cyt b):** These mitochondrial markers often display distinct "barcode gaps," where interspecific divergence far exceeds intraspecific variation, allowing reliable identification even among cryptic taxa. For instance, in North American birds, COI successfully distinguished all sampled species with clear interspecific differences (Hebert et al., 2004).
2. **Availability of comprehensive reference databases:** Many avian taxa already have COI and Cyt b sequences in repositories such as GenBank and BOLD, facilitating comparisons, matching of unknowns, and validation of morphological identifications. The increasing completeness of these libraries enhances barcoding efficiency (Hebert et al., 2004).
3. **Applicability to degraded or partial samples:** Mitochondrial genes, particularly 16S, are more readily amplified from degraded materials (e.g., feathers, museum skins, or carcasses) due to shorter amplicons and conserved primer regions. When COI amplification fails because of primer mismatch or DNA degradation, 16S can serve as an effective alternative (Koroiva & Santana, 2022; Marín-Villa et al., 2025).
4. **Utility in conservation, forensic, and biodiversity monitoring:** DNA barcoding aids in species identification in wildlife trade investigations, detection of mislabeled specimens, and recognition of cryptic taxa. The Colombian psittacine study demonstrated how combined COI and 16S data can support forensic identification and conservation law enforcement (Marín-Villa et al., 2025).
5. **Relevance to phylogenetic and population genetic studies (especially Cyt b):** Cyt b variation provides valuable information on population structure, genetic diversity, and historical demography, thereby linking molecular diversity with extinction risk and informing conservation strategies (Canteri et al., 2021).

2.3. Limitations

1. **Maternal inheritance and mitochondrial bias:** As COI, Cyt b, and 16S are maternally inherited, they exclude paternal lineages, potentially misrepresenting gene flow, hybridization, or introgression. In taxa with frequent hybridization (e.g., Anatidae), this may yield misleading signals of species boundaries.
2. **Incomplete lineage sorting and low divergence:** Recently diverged species may lack sufficient mitochondrial differentiation for accurate delimitation. In such cases, interspecific divergence can overlap with intraspecific variation, eliminating a clear "barcode gap" (Koroiva & Santana, 2022).

3. **Primer mismatch and amplification failures:** Primers optimized for one taxonomic group may fail in others or with degraded DNA. Although 16S often amplifies more reliably, its variable regions can complicate sequence alignment and phylogenetic interpretation (Edwards et al., 2023).
4. **Gaps and errors in reference databases:** Even robust markers depend on reliable reference data. Missing, low-quality, or misidentified sequences can lead to erroneous assignments, particularly in poorly studied tropical taxa. 16S references remain scarcer than COI for many bird groups (Gonzalez et al., 2009).
5. **Single-locus limitations:** Mitochondrial genes reflect only maternal lineage history. Nuclear markers may reveal different evolutionary patterns, especially under hybridization, male-mediated dispersal, or selection on mitochondrial loci. Hence, multilocus or genomic datasets offer a more comprehensive understanding of genetic structure.
6. **Variation in mutation rates and selective pressures:** Evolutionary rates differ across markers; Cyt b may undergo strong purifying selection, reducing variation in some lineages, while conserved 16S regions often limit resolution (Wang et al., 2023).

Together, these strengths and limitations underscore why mitochondrial barcoding while powerful for species identification must be complemented by nuclear and genomic data for a full picture of avian diversity and evolutionary dynamics.

3. Phylogenetic Relationships

Phylogenetic relationships among the family members of Anatidae have taken an increasingly precise form over time due to the molecular analysis. Mitochondrial genome studies have been especially helpful, finding the well-supported clades related to morphological differentiation and ecological adaptations (Zhang et al., 2023). The identification of separate phylogenetic lineages by studies of a variety of Anatidae has made it possible to obtain deeper insights into the divergence patterns in the family. Moreover, phylogenetic reconstruction based on fourfold degenerate sites in nine *Anas* species and 16 other species of the family Anatidae has also given an improved resolution of both ancient and more recent splits (Zhang et al., 2025). All these findings indicate that though the large subfamilial groups such as the Anatinae, Anserinae, Dendrocygninae, and Oxyurinae tend to be very strong, it remains difficult to identify the relationships between closely related species.

The application of molecular techniques has altered our understanding of the evolutionary relationships in Anatidae. Due to its maternal inheritance and preservation, mitochondrial DNA (mtDNA) has long been an important part of the phylogenetic analysis (Figure 1). To explain, the mitochondrial genomes of the mallard (*Anas platyrhynchos*) and the Indian spot-billed duck (*Anas*

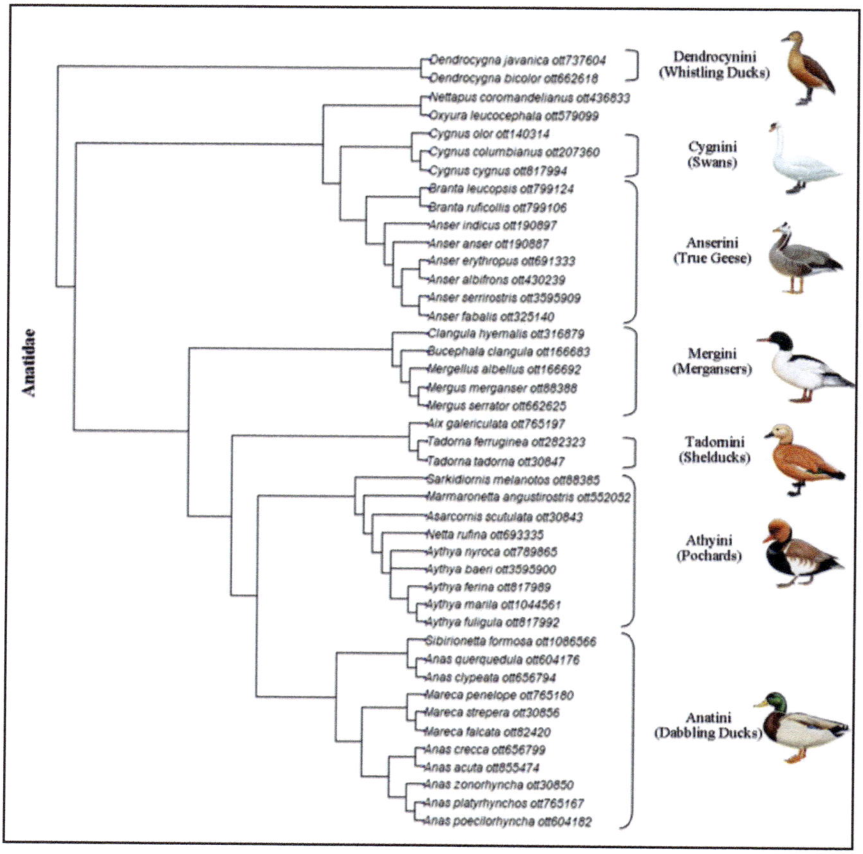

Figure 1. Phylogenetic tree of the Anatidae family showing evolutionary relationships among major tribes based on mitochondrial gene sequences.

poecilorhyncha) have been sequenced and showed specific clades which illuminate the evolutionary separation between them (Nagarajan et al., 2024). The entire genome sequencing has enabled more comprehensive investigations of genetic affiliations beyond the use of the mtDNA. Genome-wide investigations of *Anas* have shown detailed patterns of flow of genes and subtle genetic distinctions that could not be detected by single-locus techniques and give a perspective of the history of domestication and interspecies interaction (Zhang et al., 2025). These developments still make solving Anatidae phylogeny problematic. The obscuration of evolutionary signals by hybridization is very common, and is particularly common in the case of domestic species. An example is that New Zealand makes lineage identification more challenging in cases when native gray ducks (*Anas superciliosa*) are introduced via domestic mallards (Brown et al., 2025). Rapid radiations of species and incomplete lineage sorting can also give rise to conflicting phylogenetic signals. These challenges highlight the importance of integrating morphological, ecological, and historical data with multilocus and genomic data to generate accurate and biologically meaningful phylogenies on the family (Zhang et al., 2025).

4. Field Sampling and Lab Analysis

4.1. Collecting and Handling Samples

In molecular research, both responsible and valid sampling is necessary so that the research process can be credible and in accordance with the conservation laws. This typically involves taking tissues of Anatidae studies, whether feathers, blood, or buccal swabs, and the least amount of stress and disturbance is caused to the birds (Figure 2). Researchers should be guided by ethical principles that are recommended in such organizations as the IUCN, local legislation on the protection of wildlife, and receive the necessary permits (Handel et al., 2006; Licuanan et al., 2017). Proper labeling, strict chain of custody, and ethanol or silica gel preservation are important in order to avoid contamination and maintain traceability. Adherence to these principles ensures that molecular collections are credible, reproducible, and ethically sound which provide a solid foundation to taxonomic and conservation objectives.

4.2. Extracting, Amplifying, and Sequencing DNA

Mitochondrial markers, most commonly COI genes, are often used as avian DNA barcoding. Huang et al. (2016) demonstrated that 694-base pair COI sequence was able to correctly recognize approximately 83.5% of the 79 species of Anatidae. After the procedures were followed, DNA was extracted, and the polymerase chain reaction (PCR) was conducted with the help of COI-specific primers to amplify the target region. Sequencing was done by Sanger technology, and the produced sequences were submitted to a publicly accessible repository, such as the GenBank, which enriched the molecular reference libraries of waterfowl (Tizard et al., 2019). Standardized lab practices not only ensure reproducibility but also facilitate the addition of molecular data to more broadly applicable taxonomic and conservation applications.

4.3. Preparation of Phylogenetic Tree

Phylogenetic reconstruction transforms data on the molecular sequences of organisms into information on the evolutionary relationships. The neighbour-joining technique was used by Huang et al. (2016) to demonstrate that the early-divergent lineages among the Anatidae are represented by such genera as *Dendrocygna* and the clade Nomonyx + Oxyura. The COI barcoding ability to define species was proven by calculating Kimura two-parameter (K2P) distances that showed that intra-species variation was on average 41 times smaller than that between species. Phylogenetic studies also offered a framework to understand

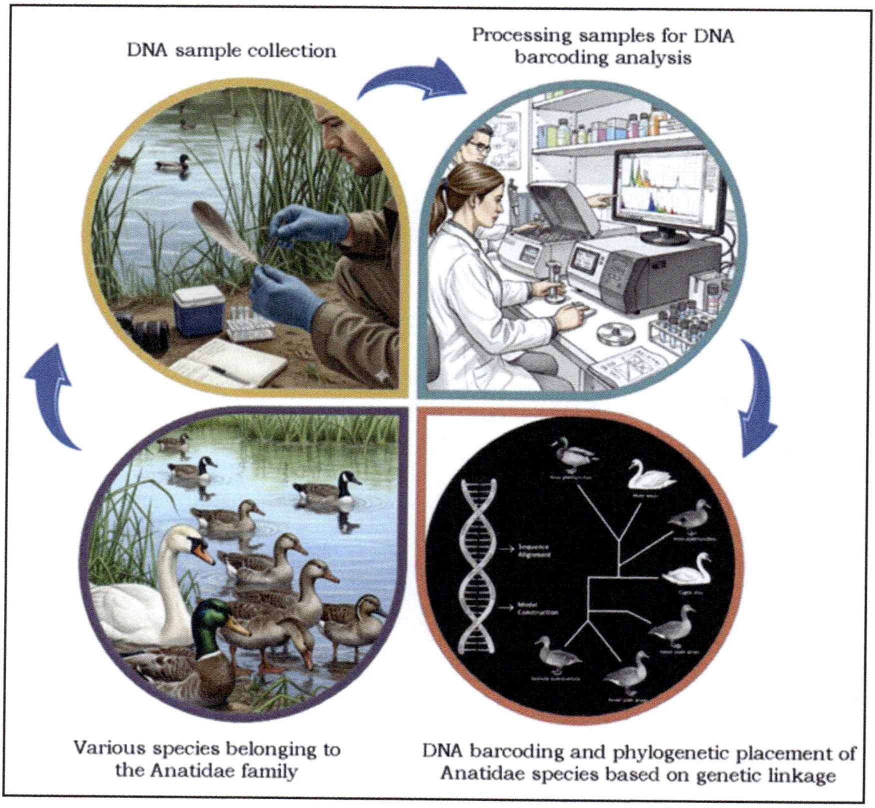

Figure 2. Workflow of DNA barcoding and phylogenetic analysis of Anatidae species from sample collection to genetic lineage determination.

the evolutionary relations within the whole family and further divided the two major subfamilial clades (Figure 2), Anatinae and Anserinae (Tehrani et al., 2025). Overall, all these molecular procedures, ranging between systematic field sampling and tree building, offer the essential methodological basis of credible taxonomic assessment and conservation approaches in the Anatidae.

5. Barcoding in Anatidae

5.1. Identifying Cryptic Species in Anatidae

There is often cryptic diversity with the boundaries of true species being often blurred by overlapping geographic ranges, extensive interbreeding, and striking morphological similarities in the Anatidae. DNA barcoding, especially using mitochondrial DNA markers such as COI and Cyt b, has become an important method of uncovering such hidden lineages by providing standard measures of genetic divergence and making it possible to compare taxa across taxa (Hebert et al., 2003; Kerr et al., 2007). This technique has been helpful in revealing highly similar species within genera such as *Aythya* (diving ducks) and *Anas* (dabbling duck), discovering underexpressed diversity within *Mergus* (mergansers), and solving cases where morphological characteristics have underestimated real species richness. To enhance phylogenetic inferences, it is essential to combine barcoding with multilocus or genomic datasets because such phenomena as introgression and incomplete lineage sorting may dilute mitochondrial DNA signals (Peters et al., 2007; Lavretsky et al., 2014). Despite these challenges, DNA barcoding provides genetic basis, which is essential in conservation planning, and a powerful preliminary model in discovering concealed diversity of Anatidae especially in areas where anthropogenic translocation or hybridization is threatening distinct evolutionary lineages.

5.2. Resolving Taxonomic Confusions with Barcoding

DNA barcoding has also played a significant role in resolving long-standing taxonomic fuzzy areas in Anatidae, besides uncovering obscure species. Unrelated lineages are often highly congruent with convergent evolution, and hybridization also makes boundaries of species difficult to determine, which subsequently complicates morphology-based classification (Hebert et al., 2003). The molecular basis of solving these problems is the strong point of sequencing of conserved mitochondrial markers, especially COI. Barcoding has been used to determine closely related species in the Anatidae and to determine boundaries of a lineage obscured by similar phenotypes and to confirm errors in identification (Tavares & Baker, 2008; Kerr et al., 2009). Phylogeographic patterns are also presented in these studies which indicate historical diversification and dispersion through production of similar sequence information on international collections. Barcoding offers objective molecular evidence, which helps to elucidate boundaries of species, and not to overrule the traditional taxonomy. In conservation, where management should be concerned with genetically distinct lineages as opposed to artificial assemblages, it is essential that taxonomic distinction is accurately defined.

5.3. Regional and Global Initiatives

International and local efforts in habitat conservation, species conservation, and global collaboration have enhanced the conservation of the Anatidae. The Agreement on the Conservation of African–Eurasian Migratory Waterbirds (AEWA) of the Convention on Migratory Species (CMS) establishes the legal basis on the protection, management, and study of cross-continental species, and the Ramsar Convention on Wetlands (Ramsar Convention Secretariat, 2013) ensures the preservation of the important waterbird habitats on the global scale (Boere & Stroud, 2006). Through prioritization of species, publication of action plans, as well as formation of conservation networks, the Threatened Waterfowl

Specialist Group of the IUCN Species Survival Commission further enhances coordination at the global level (IUCN, 2023). Regional initiatives such as the North American Waterfowl Management Plan (NAWMP) can be used to promote habitat rehabilitation and population recovery through cooperative efforts, such as the Black Duck and Pacific Birds Habitat programs (NAWMP, 2018). In the Asia-Pacific region, along the East Asian–Australasian flyway, Migratory Waterbird Conservation Strategy and its Anatidae Working Group promote international cooperation and flyway networks through conservation of wetlands (Wetlands International, 2006). All these multiscale projects demonstrate that molecular tools, international agreements, and region-based actions have to be combined to protect, control, and ensure the long-term sustainability of the Anatidae populations in a diversity of ecosystems and migration pathways.

6. Phylogeography of Anatidae

6.1. Genetic Diversity Across Continents

The spectacular genetic and phylogenetic organization of the family of Anatidae (ducks, geese, and swans) in their world spread has been determined by long-distance migrations and adaptive radiation. Molecular studies in both nuclear and mitochondrial markers have shown deep evolutionary divisions between the major clades of dabbling ducks (*Anas*), diving ducks (*Aythya*), geese (*Anser*), and swans (*Cygnus*). Patterns of diversification in these clades are in close relation to the history of continental biogeography (Gonzalez et al., 2009). Comparative phylogenetic studies suggest a number of events of intercontinental dispersion and hybridization of Eurasian and North American ancestries as a result of which a shared genetic pool and regionally endemic taxa have appeared (Peters et al., 2005; Lavretsky et al., 2014). Those species inhabiting the Southern Hemisphere, such as the shelducks (*Tadorna*) and whistling ducks (*Dendrocygna*), in their turn, show independent evolution directions sustained by regional ecological stress and prolonged isolation on the continent (Donne-Goussé et al., 2002). Recent genome-wide studies have shown that Anatidae phylogenies are currently being influenced by ongoing hybridization and incomplete lineage sorting, which invalidates traditional morphology-based classifications but provides improved insights into their evolutionary history (Ottenburghs et al., 2016). These outcomes, when considered together, demonstrate the importance of molecular tools in reconstructing evolutionary history and informing conservation practices across continents.

6.2. Dispersal and Migration

Ancient dispersal and migration occurred in ancient times and up to the present day; mixtures are still taking place. Early dispersal and migration patterns in the past have greatly determined the evolution and distribution of the Anatidae. According to the fossil records, the group originated during the Late Cretaceous to Early Paleogene where it experienced a massive diversification alongside the expansion of wetland habitats (Ericson, 1997). According to molecular phylogenies, colonization across Eurasia, North America, and the Southern Hemisphere has occurred many times due to the aid of vigorous flight and long-range migration (Donne-Goussé et al., 2002; Gonzalez et al., 2009). Pleistocene climate shifts also affected migration routes whereby subsequent events of secondary contact and population isolation occurred mostly among the Holarctic species such as *Anas* and *Aythya* (Peters et al., 2005). The long-range dispersal events of the past explain the extensive, yet genetically correlated, populations of migratory ducks and geese, and the in-situ diversification through geographic isolation of more sedentary lineages is observed in Australasia, Africa, and South America (Johnson & Sorenson, 1999). The combination of fossil, phylogenetic, and biogeographic evidence identifies the complementary role of the migration and dispersal in the development of the adaptive radiation and the existing diversity of Anatidae.

6.3. Adaptive Radiation in Changing Habitats

Anatidae is a group that has been profoundly influenced by habitat diversity on both geologic and ecological time scales. It is ecologically diverse with its representatives living in a range of environments, such as freshwater lakes and rivers, marine coastal wetlands, tundra, and arid steppe. Phylogenetic analyses demonstrate that morphological and behavioral diversity has always favored habitat changes such as the transition of dabbling feeding in *Anas* ducks to diving in pochards (*Aythya*) (Livezey, 1991; Gonzalez et al., 2009). Climatic swings during the Pleistocene also favored expansions and segregation as well as differentiation, especially among migratory lineages that were moving in response to the seasonal shifts in habitats (Peters et al., 2005). Isolation and distance in shelducks (*Tadorna*), whistling ducks (*Dendrocygna*), and stiff-tailed ducks (*Oxyurinae*) are examples of adaptive radiations in the Southern Hemisphere that promoted the emergence of special feeding behaviors and feather pecking in certain species (Donne-Goussé et al., 2002). Human habitat modification has since imposed more selective pressures in more recent times: domestication of some species to agricultural wetlands, and depression of others due to loss of habitat and reduction of ecological plasticity (Kear, 2005). This developmental history of Anatidae can thus be regarded as an example of how adaptive radiation contributes to existence and diversification in response to natural and man-made environmental changes.

7. Conservation Through DNA Lenses

7.1. Monitoring and Managing of Anatidae

Monitoring and conservation of the Anatidae have been greatly enhanced by the invention of molecular tools. Such approaches are better to identify species and emphasizing genetic variation within and between populations. The field morphological observations have never been able to tell about hybridization events or distinguish cryptic taxa. The DNA barcoding has enabled the establishment of the limits of the species and an understanding of the evolutionary connections, particularly through the use of the mitochondrial markers such as COI. The obvious divergence between the two species was proven through mitochondrial genome sequencing of the relatively closely related mallard (*Anas platyrhynchos*) and the Indian spot-billed duck (*Anas poecilorhyncha*) using two distinct phylogenetic clades (Nagarajan et al., 2024). Microsatellites and SNPs are nuclear markers that can be used to study more extensively the differentiation among populations, genetically interchangeable groups, and genetic variability, beyond the identification of a species. These insights are essential to finding the ESUs and guiding the conservation efforts toward the preservation of the adaptive potential and minimizing the risk of inbreeding (Scribner et al., 2001; Li et al., 2010). Recent advances in high-throughput sequencing now allow obtaining fine-scale information about population connectivity and demographic history. This helps in the evaluation of the effects of human-introduced introductions and habitat fragmentation. Also, noninvasive identification and tracking of waterfowl via water samples has become an effective noninvasive, nondestructive technology, namely, environmental DNA (eDNA). These techniques allow fast assessment of the abundance and relative abundance of numerous species (Honka et al., 2024; Deiner et al., 2017). Ecological information is more accurate and flexible compared to molecular, genomic, and eDNA data when used to make conservation decisions. This is necessary in order to protect Anatidae diversity in the face of increasing environmental and human pressures.

7.2. Detecting Hybridization and Gene Flow

Anatidae are highly hybridized and experience genes exchange, which pose significant conservation issues. Morphological identification usually cannot be used to distinguish hybrids, especially between closely visually similar species such as dabbling duck (*Anas*) and geese (*Anser*). Microsatellites, mitochondrial DNA, and SNP panel-based molecular markers have reported widespread historical and current hybridization regarding the family (Söderquist et al., 2017; Lavretsky, 2014). Genomic studies show that mallards (*Anas platyrhynchos*) interbreed extensively with domestic or feral ones, threatening to lose genetic purity of the wild population in Europe, Asia, or North America (Mank et al., 2004; Ottenburghs et al., 2016). Bird releases due to humans are even more harmful and lead to homogenization of the gene pool, which occurs in hunting and farm releases (Guay et al., 2014). The taxonomic scope of the process has been also indicated in diving ducks (*Aythya*), mergansers (*Mergus*), and shelducks (*Tadorna*). The modern level of high-throughput sequencing, such as RAD-seq (Restriction Site-Associated DNA Sequencing) and whole-genome resequencing, allows the fine-scale evaluation of introgression and distinguishes the ancient evolutionary flows of genes and the contemporary dangers toward genetic uniqueness (Ottenburghs et al., 2016). Hybridization is a key concept that is important to identify ESUs and provide informed conservation management, including translocations, captive breeding, and reintroduction initiatives. Genomic data has already been applied to restoration in the Hawaiian duck (*Anas wyvilliana*), a species on the brink of extinction due to feral mallards genetically swamping its gene pool (Fowler et al., 2009), highlighting the importance of using molecular tools in averting genetic swamping by humans.

7.3. Policy Relevance and Management

The implications of incorporating molecular information into conservation in policy-making have far-reaching implications on the management of the Anatidae at the local, regional, and international levels. DNA barcoding and genomic research underlie reliable taxonomic separation of species to be included in the lists of schemes such as the IUCN Red List, Convention on International Trade in Endangered Species of Wild Fauna and Flora (CITES), and CMS in such a way that conservation objectives reflect the true biological diversity (Haig et al., 2011; Taylor & Larson, 2019). The molecular evidence has been used to redefine the limits among species in genera such as *Anas* and *Aythya*, which are used in management practices and discovery of the so-called cryptic taxa that require conservation (McCarthy, 2006; Kerr et al., 2007). The links among the populations also inform genetic surveillance that can be used in flyway-scale conservation programs such as Ramsar Convention and East Asian–Australian Flyway Partnership (EAAFP) where Anatidae form key migratory animals (Boere & Stroud, 2006). The molecular information is used on national scale to restore the habitat by identifying the breeding and wintering population of high genetic diversity to allocate resources to the most effective conservation (Allendorf et al., 2010). Environmental DNA (eDNA) is also policy relevant by offering a cost-effective, non-destructive methodology to measure species presence in affected wetlands that guides biodiversity measurements according to various plans including the Aichi Biodiversity Targets and post-2020 Global Biodiversity Framework (Deiner et al., 2017; Harper et al., 2019). Conservation genomics will ensure that sound policies are grounded on sound science by linking the genetic data with policy instruments and management approach, which is necessary in the Anthropocene period since destruction of habitat, change in climate, and human-induced species introductions threaten the survival of Anatidae species. Molecular methods are thus providing an invaluable insight into informing adaptive conservation and protecting the evolution of this ecologically important family.

8. Integrative Approaches Beyond Barcoding

Display of the DNA barcoding along with morphological and ecological information has greatly enhanced our understanding of the taxonomy, evolutionary history, and conservation needs of the Anatidae. Even though the sequences of mitochondrial COI remain a highly useful tool of exact species identification, genetic data combined with morphometric factors and environmental aspects, including habitat preferences, migratory traits, and environmental gradients, can be used to give a more definitive picture of boundaries of species and adaptive differentiation. Due to such aspects as convergent patterns in plumage and environmental plasticity, as well as extensive hybridization, particularly among closely related species of *Anas*, morphology alone can be a misleading tool (Livezey, 1996; Hebert et al., 2004). The study of genetic variation combined with ecological evidence and morphological evidence can be used to elucidate population connection along migratory flyways, reveal obscure lineages, and demonstrate that environmental influences impact the movement of genes and divergence (Tavares & Baker, 2008). To strengthen the identification of ESUs and population-level adaptations and implement adaptive management and long-term conservation of waterfowl in the shifting environmental conditions, this integrative approach enhances the role of the latter (Worthy, 2009).

8.1. Whole Genomes and Next-Generation Sequencing

Next-generation sequencing (NGS) and whole-genome sequencing (WGS) have transformed the understanding of Anatidae because they have provided new information on their adaptive evolution, hybridization processes, and population genotype that no one has ever heard of before. Genomic methods allow researchers to study evolution on far smaller scales, by capturing variation across the genome-wide variation as compared with conventional DNA barcoding, depending on short mitochondrial pieces. The high-coverage genomes of the high-altitude adapted species such as mallard (*Anas platyrhynchos*), swan goose (*Anser cygnoides*), and bar-headed goose (*Anser indicus*) have found molecular signatures related to immune functionality, migration, and adaptation to high altitudes (Scott et al., 2011; Wang et al., 2020; Wen et al., 2023). Comparative genomic studies have also depicted by showing reticulate genomic patterns among North American and Eurasian populations how introgressive hybridization has contributed to development of evolutionary relationships (Ottenburghs et al., 2017).

Moreover, genome-wide SNP data generated by reduced-representation sequencing methods such as RAD-seq and GBS (Genotyping-by-Sequencing) has enabled the conducting of fine-scale analyses of population connectivity, demographic history, and the definition of ESUs. The techniques are particularly helpful in establishing the relationships among species with minor morphological dissimilarity or overlap (Lavretsky et al., 2020; Larison et al., 2021). The combination of genomic evidence and ecological as well as behavioral data has now provided a powerful framework to understand the mechanisms of adaptive potential, the identification of genetically vulnerable populations, and the prediction of responses to environmental change. All these changes represent a new beginning in the research of the Anatidae, where extensive genomic expertise is used not only to inform long-term conservation planning but also to inform evolutionary knowledge (Bernatchez et al., 2024).

8.2. eDNA, Metabarcoding, and AI Tools as Future Frontiers

New molecular and computational advances are transforming the research and protection of the Anatidae population as they can be studied and preserved with their noninvasive and remarkably accurate indications of the process of biodiversity and ecosystem. Whether it is through the analysis of eDNA or the application of metabarcoding, these technologies allow species detection and the classification of entire communities using waters or sediment samples without requiring direct capture and visual survey. These methods allow researchers to better and faster evaluate populations in the wetlands that are remote or rapidly altering, where conventional monitoring tools may prove logistically challenging (Deiner et al., 2017; Ruppert et al., 2019). Surveys based on eDNA have shown success in recording waterbird assemblages in freshwater and Ramsar wetlands in India, and show that such techniques can be useful to monitor the ecology of large areas (Bista et al., 2017).

At the same time, the recent advances in the field of artificial intelligence (AI) and machine learning have provided additional opportunities to combine genomic, acoustic, and remote-sensing data to forecast migration patterns, habitat preferences, and demographic trends in a changing climate (Kahl et al., 2021; Sethi et al., 2020). Deep learning models that have been trained on bird sounds and images can now automatically recognize species and track their behavior and offer a continuous surveillance system over a large scale. Combining eDNA, genomics, and AI, it is possible to apply a predictive approach to conservation, which is real-time and minimally invasive, and connects molecular data to ecological processes. These integrative systems represent a transition to adaptive evidence-based management of Anatidae populations such that conservation measures are responsive to the environmental change as well as evolving ecological requirements of the species.

9. Challenges and Opportunities

9.1. Gaps in Barcode Databases

Although DNA barcoding and genomics have been developed, reference databases on Anatidae are still largely missing in places that have been under-sampled, which include South and Southeast Asia, Africa, and portions of South America. Although genetic information on waterfowls has been stored in repositories such as BOLD and GenBank, a significant number of species of waterfowl have not been provided with voucher-verified sequences, which compromises precise taxonomic identification, phylogenetic reconstruction, and biodiversity estimation (Ratnasingham & Hebert, 2007; Porter & Hajibabaei,

2018). Imperfect sampling biases affect major migratory flyways in particular where breeding and wintering sites are linked to each other, and a reference library is not complete enough to result in misidentifying or falsely demarcating species (Kerr et al., 2007; Hebert et al., 2016). Genetic variation within species, due to extensive geographic and hybridization, for example, in *Anas platyrhynchos* and *Anas poecilorhyncha*, is under-reported and this further complicates the use of molecular diagnostics. Barcode gap analyses such as those obtained with BOLD are useful to identify discrepancies but only when using extensive and high-quality datasets. To overcome such gaps, there is a need to survey the global and regional areas with coordination and focus on data-poor flyway corridors, tropical breeding areas, and remote wetlands. Building more connections between morphological taxonomy, molecular data, and voucher specimen with the help of national programs, such as the Zoological Survey of India and the National Biodiversity Authority, will increase the integrity of reference libraries and enhance the quality of molecular-based conservation planning.

9.2. Limits in Phylogenetic Resolution

Despite the potential usefulness of DNA barcoding in rapid and accurate identification of species, it can often be limited in investigating more phylogenetic relationship, particularly in groups with recent radiations, hybrid swarms, and incomplete lineage sorting, including Anatidae. Mitochondrial genes, such as COI and Cyt b which are commonly used, are nevertheless useful in species delimitation, but do not necessarily give a complete evolutionary view due to factors such as introgression, retention of old patterns, and their purely maternal pattern of inheritance (Funk & Omland, 2003; Rubinoff, 2006). These aspects may result in differences between mitochondrial and nuclear phylogenies, that is, cytonuclear incongruence, which makes it more difficult to determine true evolutionary processes over time especially in sympatric and hybridizing species such as *Anas platyrhynchos* and *Anas rubripes* (Mank et al., 2004). To overcome such constraints, modern methods use multilocus and genome-wide datasets, namely, through methods such as RAD-seq, ultra-conserved element (UCE) capture, and whole-genome sequencing, which combine nuclear variation and coalescent modelling. Such techniques improve the precision in identifying the presence of the hybrid ancestry, population split, and reticulate evolution (Ottenburghs et al., 2017; Lavretsky et al., 2020). Nevertheless, their use is still minimal because it requires expenses, large scale of sampling, and computer skills. Integration of genomic data and morphological and ecological data currently provides a more profound path toward revealing the underlying fine-scale processes of speciation, adaptation, and diversification in this ecologically dynamic bird family.

9.3. Directions for Global Collaboration

Close cross-border collaboration in genetic, taxonomic, and ecological studies is the key to the future of the Anatidae taxonomy and conservation genomics. The migration of waterfowl across multiple countries and continents necessitates the need to engage in international partnerships to fully realize the genetic diversity and evolutionary heritage of the migratory waterfowl. Global initiatives such as the Global Genome Biodiversity Network (GGBN) and the International Barcode of Life (iBOL) have introduced considerable requirements toward standardizing the storage of genomic information, metadata standards, and sample collection, and enhanced the interoperability of national and regional research initiatives (Droege et al., 2016). Initiatives such as the EAAFP and conventions such as the CMS encourage the integration of molecular evidence in large-scale waterbird species surveillance and management (Boere & Stroud, 2006). By incorporating molecular archives such as IndiaBOL and other country or nation genome repositories into such global databases, geographic and taxonomic gaps can be bridged (particularly between biodiversity-rich, but underrepresented, areas across Asia, Africa, and South America). Global biodiversity genomics can be fully contributed by researchers in different areas, due to the encouragement of open-access databases, standardized field and sequencing protocols, and training opportunities (Miraldo et al., 2016). Knowledge on gene flow, adaptation, and ability to withstand changes in climate and habitat will be enhanced through the creation of coordinated genomic surveillance of migratory pathways, hybrid zones, and cross-border species. Finally, by integrating high-resolution molecular methods with international collaboration, conservation would ensure that, besides the survival of the Anatidae species, the evolutionary mechanisms, which maintain their abundance across the hemispheres, would be maintained.

10. Conclusion

The best example of how human impact, geography, and evolutionary history combine to determine biodiversity on the planet is Anatidae. Molecular research has altered our understanding of the taxonomy, as well as evolutionary relationships, of these birds. It finds cryptic species which cannot be separated through morphological characteristics only and exposes dramatic divergences within clades. The combination of nuclear and genome-scale data and markers such as COI, Cyt b, and 16S rRNA allowed mapping the gene flow, hybridization patterns, and clearing up difficult taxonomic questions in this varied group. The DNA barcoding plays a very important role in offering quick and effective recognition of species, which is utilized in conserving and ecological monitoring. Molecular tools in combination with traditional taxonomy can help define units of evolutionary significance and direct population management and monitoring throughout an era of shifting anthropogenic and climatic conditions at a rapid rate. Also, it ensured that management decisions are informed by a genetics and evolution knowledge that reinforces the links between ecological studies and conservation policy. Introgression, incomplete lineage sorting, and inadequate barcode coverage across international databases remain the limitations to taxonomic resolution. These issues require more extensive sampling and incorporation of morphological, ecological, and genomic perspectives. The processes of adaptive variation, changing demographics, and the hybridization can now be studied at scales previously unknown due to the continually growing availability of whole-genome sequencing, high-throughput SNP genotyping, environmental DNA surveys, and monitoring through AI. Protecting species is not the only way of conservation but with these technological development advances and the combination of international research and policy effort, protection of evolutionary processes and genetic complexity can be achieved. In order to maintain the genetic diversity and adaptation of Anatidae populations as the climate change and habitat loss continue to transform wetlands ecosystems, a combination of molecular knowledge and ecological and management models will be required. Collectively, this multidisciplinary research makes Anatidae a paradigm to explore evolution of birds, as well as becomes a center of science-based, progressive conservation efforts in the world.

References

Allendorf, F. W., Hohenlohe, P. A., & Luikart, G. (2010). Genomics and the future of conservation genetics. *Nature Reviews Genetics*, 11(10), 697–709.

Bernatchez, L., Ferchaud, A. L., Berger, C. S., Venney, C. J., & Xuereb, A. (2024). Genomics for monitoring and understanding species responses to global climate change. *Nature Reviews Genetics*, 25(3), 165–183.

Bista, I., Carvalho, G. R., Walsh, K., Seymour, M., Hajibabaei, M., Lallias, D. et al. (2017). Annual time-series analysis of aqueous eDNA reveals ecologically relevant dynamics of lake ecosystem biodiversity. *Nature Communications*, 8(1), 14087.

Boere, G. C., & Stroud, D. A. (2006). The flyway concept: What it is and what it isn't. *Waterbirds Around the World*, 40–47.

Brown, J. I., Sheppard, J. L., Mohl, J., Engilis, I. E., Engilis Jr, A., & Lavretsky, P. (2025). Anthropogenic hybridization contributes to the naturalization of introduced domestic mallards (Anas platyrhynchos) at the expense of native New Zealand Gray Ducks (A. s. superciliosa). *Ecology and Evolution*, 15(6), e71536.

Canteri, E., Fordham, D. A., Li, S., Hosner, P. A., Rahbek, C., & Nogués-Bravo, D. (2021). IUCN Red List protects avian genetic diversity. *Ecography*, 44(12), 1808–1811.

Deiner, K., Bik, H. M., Mächler, E., Seymour, M., Lacoursière-Roussel, A., Altermatt, F. et al. (2017). Environmental DNA metabarcoding: Transforming how we survey animal and plant communities. *Molecular Ecology*, 26(21), 5872–5895.

Delacour, J., & Mayr, E. (1946). Supplementary notes on the family Anatidae. *The Wilson Bulletin*, 58(2), pp. 104–110.

Donne-Goussé, C., Laudet, V., & Hänni, C. (2002). A molecular phylogeny of anseriformes based on mitochondrial DNA analysis. *Molecular Phylogenetics and Evolution*, 23(3), 339–356.

Droege, G., Barker, K., Seberg, O., Coddington, J., Benson, E., Berendsohn, W. G. et al. (2016). The global genome biodiversity network (GGBN) data standard specification. *Database*, 2016, baw125.

Dyke, G. J. (2001). The fossil waterfowl (Aves: Anseriformes) from the Eocene of England. *American Museum Novitates*, 2001(3354), 1–15.

Edwards, J., Hoffbeck, C., West, A. G., Pas, A., & Taylor, M. W. (2023). 16S rRNA gene-based microbiota profiles from diverse avian faeces are largely independent of DNA preservation and extraction method. *Frontiers in Microbiology*, 14, 1239167.

Ericson, P. G. (1997). Systematic relationships of the palaeogene family Presbyornithidae (Aves: Anseriformes). *Zoological Journal of the Linnean Society*, 121(4), 429–483.

Fowler, A. C., Eadie, J. M., & Engilis Jr, A. (2009). Identification of endangered Hawaiian ducks (Anas wyvilliana), introduced North American mallards (A. platyrhynchos) and their hybrids using multilocus genotypes. *Conservation Genetics*, 10(6), 1747.

Funk, D. J., & Omland, K. E. (2003). Species-level paraphyly and polyphyly: Frequency, causes, and consequences, with insights from animal mitochondrial DNA. *Annual Review of Ecology, Evolution, and Systematics*, 34(1), 397–423.

Gendron, E. T. E. (2018). Evolutionary Ecology of Host–Parasite Relationships: Role of Host Ecology, Phylogeny, and Demographics in Shaping Parasite Evolution. Doctoral dissertation, The University of New Mexico.

Gonzalez, J., Düttmann, H., & Wink, M. (2009). Phylogenetic relationships based on two mitochondrial genes and hybridization patterns in Anatidae. *Journal of Zoology*, 279(3), 310–318.

Gordon, I. J., Prins, H. H., Mallon, J., Puk, L. D., Miranda, E. B., Starling-Manne, C. et al. (2019). The ecology of browsing and grazing in other vertebrate taxa. *In: The Ecology of Browsing and Grazing II* (pp. 339–404). Cham: Springer International Publishing.

Guay, P. J., Taysom, A., Robinson, R., & Tracey, J. P. (2014). Hybridization between the mallard and native dabbling ducks: Causes, consequences and management. *Pacific Conservation Biology*, 20(1), 41–47.

Haig, S. M., Bronaugh, W. M., Crowhurst, R. S., D'Elia, J., Eagles-Smith, C. A., Epps, C. W. et al. (2011). Genetic applications in avian conservation. *The Auk*, 128(2), 205–229.

Handel, C. M., Pajot, L. M., Talbot, S. L., & Sage, G. K. (2006). Use of buccal swabs for sampling DNA from nestling and adult birds. *Wildlife Society Bulletin*, 34(4), 1094–1100.

Harper, L. R., Buxton, A. S., Rees, H. C., Bruce, K., Brys, R., Halfmaerten, D. et al. (2019). Prospects and challenges of environmental DNA (eDNA) monitoring in freshwater ponds. *Hydrobiologia*, 826(1), 25–41.

Hebert, P. D. N., Stoeckle, M. Y., Zemlak, T. S., & Francis, C. M. (2004). Identification of birds through DNA barcodes. *PLoS Biology*, 2(10), e312.

Hebert, P. D., Cywinska, A., Ball, S. L., & DeWaard, J. R. (2003). Biological identifications through DNA barcodes. *Proceedings of the Royal Society of London. Series B: Biological Sciences*, 270(1512), 313–321.

Hebert, P. D., Ratnasingham, S., Zakharov, E. V., Telfer, A. C., Levesque-Beaudin, V., Milton, M. A. et al. (2016). Counting animal species with DNA barcodes: Canadian insects. *Philosophical Transactions of the Royal Society B: Biological Sciences*, 371(1702), 20150333.

Honka, J., Kvist, L., Olli, S., Laaksonen, T., & Aspi, J. (2024). Optimised PCR assays for detecting elusive waterfowl from environmental DNA. *Ecology and Evolution*, 14(4), e11224.

Huang, Z., Yang, C., & Ke, D. (2016). DNA barcoding and phylogenetic relationships in Anatidae. *Mitochondrial DNA Part A*, 27(2), 1042–1044.

IUCN. (2023). IUCN SSC Threatened Waterfowl Specialist Group. Available at: https://www.iucn.org/our-union/commissions/group/iucn-ssc-threatened-waterfowl-specialist-group

Johnson, K. P., & Sorenson, M. D. (1999). Phylogeny and biogeography of dabbling ducks (Genus Anas): A comparison of molecular and morphological evidence. *The Auk*, 116(3), 792–805.

Kahl, S., Wood, C. M., Eibl, M., & Klinck, H. (2021). BirdNET: A deep learning solution for avian diversity monitoring. *Ecological Informatics*, 61, 101236.

Kear, J. (Ed.). (2005). *Ducks, Geese, and Swans: Species Accounts* (*cairina to mergus*) (Vol. 2). Oxford University Press.

Kerr, K. C. R., Stoeckle, M. Y., Dove, C. J., Weigt, L. A., Francis, C. M., & Hebert, P. D. N. (2009). Comprehensive DNA barcode coverage of North American birds. *Molecular Ecology Resources*, 9(suppl s1), 51–83.

Kerr, K. C., Stoeckle, M. Y., Dove, C. J., Weigt, L. A., Francis, C. M., & Hebert, P. D. (2007). Comprehensive DNA barcode coverage of North American birds. *Molecular Ecology Notes*, 7(4), 535–543.

Koroiva, R., & Santana, D. J. (2022). Evaluation of partial 12S rRNA, 16S rRNA, COI and Cytb gene sequence datasets for potential single DNA barcode for hylids (Anura: Hylidae). *Anais da Academia Brasileira de Ciências*, 94, e20200825.

Kraus, R. H., Kerstens, H. H., van Hooft, P., Megens, H. J., Elmberg, J., Tsvey, A. et al. (2012). Widespread horizontal genomic exchange does not erode species barriers among sympatric ducks. *BMC Evolutionary Biology*, 12(1), 45.

Larison, B., Lindsay, A. R., Bossu, C., Sorenson, M. D., Kaplan, J. D., Evers, D. C. et al. (2021). Leveraging genomics to understand threats to migratory birds. *Evolutionary Applications*, 14(6), 1646–1658.

Lavretsky, P. (2014). Phylogenetics, Population Genetics, and Evolution of the Mallard Complex. Doctoral dissertation. Wright State University.

Lavretsky, P., McCracken, K. G., & Peters, J. L. (2014). Phylogenetics of a recent radiation in the mallards and allies (Aves: Anas): Inferences from a genomic transect and the multispecies coalescent. *Molecular Phylogenetics and Evolution*, 70, 402–411.

Lavretsky, P., McInerney, N. R., Mohl, J. E., Brown, J. I., James, H. F., McCracken, K. G. et al. (2020). Assessing changes in genomic divergence following a century of human-mediated secondary contact among wild and captive-bred ducks. *Molecular Ecology*, 29(3), 578–595.

Li, H. F., Song, W. T., Shu, J. T., Chen, K. W., Zhu, W. Q., Han, W. et al. (2010). Genetic diversity and population structure of 10 Chinese indigenous egg-type duck breeds assessed by microsatellite polymorphism. *Journal of Genetics*, 89(1), 65–72.

Licuanan, A. M., Duya, M. R., Ong, P. S., & Fontanilla, I. K. C. (2017). DNA barcoding, population genetics, and phylogenetics of the illegally hunted Philippine Duck Anas luzonica (Aves: Anseriformes: Anatidae). *Journal of Threatened Taxa*, 9(5), 10141–10150.

Livezey, B. C. (1991). A phylogenetic analysis and classification of recent dabbling ducks (Tribe Anatini) based on comparative morphology. *The Auk*, 108(3), 471–507.

Livezey, B. C., (1996). A phylogenetic analysis of geese and swans (Anseriformes: Anserinae), including selected fossil species. *Systematic Biology*, 45(4), 415–450.

Livezey, B. C. (1997). A phylogenetic analysis of basal Anseriformes, the fossil Presbyornis, and the interordinal relationships of waterfowl. *Zoological Journal of the Linnean Society*, 121(4), 361–428. doi: 10.1111/J.1096-3642.1997.TB01285.X

Madsen, C. S., McHugh, K. P., & de Kloet, S. R. (1988). A partial classification of waterfowl (Anatidae) based on single-copy DNA. *The Auk*, 105(3), 452–459.

Mank, J. E., Carlson, J. E., & Brittingham, M. C. (2004). A century of hybridization: Decreasing genetic distance between American black ducks and mallards. *Conservation Genetics*, 5(3), 395–403.

Marín-Villa, J., López-Herrera, A., Gómez-Ruiz, D. A., Restrepo-Rodas, D. C., Sánchez-Rodríguez, G., & Úsuga-Monroy, C. (2025). Mitochondrial markers (cytochrome c oxidase subunit I and 16S ribosomal RNA) as supporting biomarkers for wild bird identification. *Veterinary World*, 18(5), 1389.

McCarthy, E. M. (2006). *Handbook of Avian Hybrids of the World*. Oxford University Press.

Miraldo, A., Li, S., Borregaard, M. K., Flórez-Rodríguez, A., Gopalakrishnan, S., Rizvanovic, M. et al. (2016). An Anthropocene map of genetic diversity. *Science*, 353(6307), 1532–1535.

Nagarajan, M., Kamalakkannan, R., Prabhu, V. R., Bhavana, K., Gopalakrishnan, A., Thomas, S. et al. (2024). Mitochondrial genome of the Indian spot-billed duck and its phylogenetic and conservation implications. *Scientific Reports*, 14(1), 24572.

North American Waterfowl Management Plan (NAWMP). (2018). *North American Waterfowl Management Plan 2018 Update: Connecting People, Waterfowl and Wetlands*. Canadian Wildlife Service, U.S. Fish and Wildlife Service, Secretaría de Medio Ambiente y Recursos Naturales.

Olsen, A. M. (2017). Feeding ecology is the primary driver of beak shape diversification in waterfowl. *Functional Ecology*, 31(10), 1985–1995.

Olson, S. L., & Feduccia, A. (1980). Presbyornis and the origin of the Anseriformes (Aves: Charadriomorphae). *Smithsonian Contributions to Zoology*, 323, 1–24. Smithsonian Institution Press.

Omland, K. E. (1994). Character congruence between a molecular and a morphological phylogeny for dabbling ducks (Anas). *Systematic Biology*, 43(3), 369–386.

Ottenburghs, J. (2016). Crossing Species Boundaries: The Hybrid Histories of the True Geese. Doctoral dissertation, Wageningen University and Research.

Ottenburghs, J., Megens, H. J., Kraus, R. H., Madsen, O., van Hooft, P., van Wieren, S. E. et al. (2016). A tree of geese: A phylogenomic perspective on the evolutionary history of true geese. *Molecular Phylogenetics and Evolution*, 101, 303–313.

Ottenburghs, J., Megens, H. J., Kraus, R. H., Van Hooft, P., van Wieren, S. E., Crooijmans, R. P. et al. (2017). A history of hybrids? Genomic patterns of introgression in the true geese. *BMC Evolutionary Biology*, 17(1), 201.

Peters, J. L., Gretes, W., & Omland, K. E. (2005). Late Pleistocene divergence between eastern and western populations of wood ducks (Aix sponsa) inferred by the 'isolation with migration' coalescent method. *Molecular Ecology*, 14(11), 3407–3418.

Peters, J. L., Zhuravlev, Y., Fefelov, I., Logie, A., & Omland, K. E. (2007). Nuclear loci and coalescent methods support ancient hybridization as cause of mitochondrial paraphyly between gadwall and falcated duck (Anas spp.). *Evolution*, 61(8), 1992–2006.

Phillimore, A. B., & Price, T. D. (2008). Density-dependent cladogenesis in birds. *PLoS Biology*, 6(3), e71.

Porter, T. M., & Hajibabaei, M. (2018). Over 2.5 million COI sequences in GenBank and growing. *PLoS One*, 13(9), e0200177.

Ramsar Convention Secretariat. (2013). *The Ramsar Convention Manual: A Guide to the Convention on Wetlands* (Ramsar, Iran, 1971) (6th ed., p. 112). Ramsar Convention Secretariat.

Ratnasingham, S., & Hebert, P. D. (2007). BOLD: The Barcode of Life Data System (http:// www. barcodinglife. org). *Molecular Ecology Notes*, 7(3), 355–364.

Robertson, J. M., Langin, K. M., Sillett, T. S., Morrison, S. A., Ghalambor, C. K., & Funk, W. C. (2014). Identifying evolutionarily significant units and prioritizing populations for management on islands. *Monographs of the Western North American Naturalist*, 7(1), 397–411.

Rubinoff, D. (2006). Utility of mitochondrial DNA barcodes in species conservation. *Conservation Biology*, 20(4), 1026–1033.

Ruppert, K. M., Kline, R. J., & Rahman, M. S. (2019). Past, present, and future perspectives of environmental DNA (eDNA) metabarcoding: A systematic review in methods, monitoring, and applications of global eDNA. *Global Ecology and Conservation*, 17, e00547.

Schwarzfeld, M. D., & Sperling, F. A. (2014). Species delimitation using morphology, morphometrics, and molecules: Definition of the Ophion scutellaris Thomson species group, with descriptions of six new species (Hymenoptera, Ichneumonidae). *ZooKeys*, 462, 59.

Scott, G. R., Schulte, P. M., Egginton, S., Scott, A. L., Richards, J. G., & Milsom, W. K. (2011). Molecular evolution of cytochrome c oxidase underlies high-altitude adaptation in the bar-headed goose. *Molecular Biology and Evolution*, 28(1), 351–363.

Scribner, K. T., Petersen, M. R., Fields, R. L., Talbot, S. L., Pearce, J. M., & Chesser, R. K. (2001). Sex-biased gene flow in spectacled eiders (Anatidae): Inferences from molecular markers with contrasting modes of inheritance. *Evolution*, 55(10), 2105–2115.

Sethi, S. S., Jones, N. S., Fulcher, B. D., Picinali, L., Clink, D. J., Klinck, H. et al. (2020). Characterizing soundscapes across diverse ecosystems using a universal acoustic feature set. *Proceedings of the National Academy of Sciences*, 117(29), 17049–17055.

Sibley, C. G., & Ahlquist, J. E. (1983). Phylogeny and classification of birds based on the data of DNA–DNA hybridization. *In: Current Ornithology* (pp. 245–292). New York, NY: Springer US.

Söderquist, P., Elmberg, J., Gunnarsson, G., Thulin, C. G., Champagnon, J., Guillemain, M. et al. (2017). Admixture between released and wild game birds: A changing genetic landscape in European mallards (Anas platyrhynchos). *European Journal of Wildlife Research*, 63(6), 98.

Spaulding, F., McLaughlin, J. F., Cheek, R. G., McCracken, K. G., Glenn, T. C., & Winker, K. (2023). Population genomics indicate three different modes of divergence and speciation with gene flow in the green-winged teal duck complex. *Molecular Phylogenetics and Evolution*, 182, 107733.

Tambussi, C. P., Degrange, F. J., De Mendoza, R. S., Sferco, E., & Santillana, S. (2019). A stem anseriform from the early Palaeocene of Antarctica provides new key evidence in the early evolution of waterfowl. *Zoological Journal of the Linnean Society*, 186(3), 673–700.

Tavares, E. S., & Baker, A. J. (2008). Single mitochondrial gene barcodes reliably identify sister-species in diverse clades of birds. *BMC Evolutionary Biology*, 8(1), 81.

Taylor, S. A., & Larson, E. L. (2019). Insights from genomes into the evolutionary importance and prevalence of hybridization in nature. *Nature Ecology & Evolution*, 3(2), 170–177.

Tehrani, S. J., Rezazadeh, E., Kakhki, N. A., Nourani, L., Ebadi, V., Karimi, S. et al. (2025). DNA barcoding of passerine birds in Iran. *ZooKeys*, 1236, 19.

Tizard, J., Patel, S., Waugh, J., Tavares, E., Bergmann, T., Gill, B. et al. (2019). DNA barcoding a unique avifauna: An important tool for evolution, systematics, and conservation. *BMC Evolutionary Biology*, 19(1), 52.

Wang, W., Wang, F., Hao, R., Wang, A., Sharshov, K., Druzyaka, A. et al. (2020). First de novo whole genome sequencing and assembly of the bar-headed goose. *PeerJ*, 8, e8914.

Wang, X., Bensch, S., Huang, X., & Dong, L. (2023). Purifying selection leads to low protein diversity of the mitochondrial Cytb gene in avian malaria parasites. *BMC Ecology and Evolution*, 23(1), 49.

Watanabe, J. (2017). Quantitative discrimination of flightlessness in fossil Anatidae from skeletal proportions. *The Auk: Ornithological Advances*, 134(3), 672–695.

Wen, J., Li, H., Wang, H., Yu, J., Zhu, T., Zhang, J. et al. (2023). Origins, timing and introgression of domestic geese revealed by whole genome data. *Journal of Animal Science and Biotechnology*, 14(1), 26.

Wetlands International. (2006). *Migratory Waterbird Conservation Strategy: 2006–2010.* Wetlands International.

Winkler, D. W., Billerman, S. M., & Lovette, I. J. (2020). Ducks, geese, and waterfowl (Anatidae), version 1.0. *In:* Billerman S. M., Keeney B. K., Rodewald P. G., & Schulenberg T. S. (Eds.), *Birds of the World.* Ithaca, USA: Cornell Lab of Ornithology.

Worthy, T. H. (2009). Descriptions and phylogenetic relationships of two new genera and four new species of Oligo-Miocene waterfowl (Aves: Anatidae) from Australia. *Zoological Journal of the Linnean Society*, 156(2), 411–454.

Worthy, T. H., Scofield, R. P., Hand, S. J., De Pietri, V. L., & Archer, M. (2022). A swan-sized fossil anatid (Aves: Anatidae) from the early Miocene St Bathans Fauna of New Zealand. *Zootaxa*, 5168(1), 39–50.

Yao, H., Song, J., Liu, C., Luo, K., Han, J., Li, Y. et al. (2010). Use of ITS2 region as the universal DNA barcode for plants and animals. *PloS One*, 5(10), e13102.

Zeng, Q., Reid, J., Saintilan, N., Colloff, M. J., Lei, G., & Wen, L. (2019). Contrasting diversity patterns of breeding Anatidae in the Northern and Southern Hemispheres. *Ecology and Evolution*, 9(17), 9990–10003.

Zhang, L., Xia, T., Gao, X., Yang, X., Sun, G., Zhao, C. et al. (2023). Characterization and phylogenetic analysis of the complete mitochondrial genome of Aythya marila. *Genes*, 14(6), 1205.

Zhang, Z., Ni, Z., Li, T., Ning, M., Gao, C., Hu, J. et al. (2025). Nine high-quality Anas genomes provide insights into Anas evolution and domestication. *Cell Reports*, 44(4). Article number 115477.

CHAPTER
8

Status of Anatidae Family

Byju H.* **and Maitreyi H.**
Centre of Advanced Study in Marine Biology, Annamalai University, Parangipettai – 608502, Tamil Nadu, India

1. Introduction

"Status" of a species can be broadly defined as its interaction with the evolving world, both biotic and abiotic factors, which may reflect on its distribution, population size, and trend over time at regional and global levels. Knowledge of the status of a species will lead to prioritizing its conservation action (Primack, 2014; Sutherland et al., 2009). In this chapter, we discuss the global status of the Anatidae family, encompassing three broader aspects. The first dimension is geographic status, which includes the current distribution and range size of a species, including observed expansions or fragmentation. This spatial component of distribution is crucial for understanding a species' or a family's resilience or vulnerability with respect to the changing environmental factors (Gaston & Fuller, 2009). The second aspect is the demographic status, which includes the absolute population estimates and, more importantly, the trends—whether a population is increasing, stable, or declining over a defined timeframe (e.g., the past 10–30 years) (Buckley & Freckleton, 2010). The third aspect is that of conservation status. This is based on past, present, and future population abundance, and its distribution, from the assessment of the Red List of Threatened Species by the International Union for Conservation of Nature (IUCN). These interconnected dimensions of geographical distribution, demographic trends, and conservation status synthesize the current global status of the Anatidae family.

*Corresponding author: byjuhi@gmail.com

DOI: 10.1201/9781003669784-8

The Anatidae family (ducks, swans, and geese), with its nearly 174 species spanning every continent, presents a unique case study. For protecting ducks, geese, and swans, a clear-cut picture of the challenges they face is needed. Their status serves as a potent bioindicator for the health of the world's freshwater and coastal ecosystems (Green & Elmberg, 2014). From the Arctic tundra to tropical mangroves, the fortunes of waterfowl are inextricably linked to the integrity of wetlands, among the most threatened global habitats (Davidson, 2014).

The global status of Anatidae is not uniform; it includes regional successes and crises, largely reflecting the state of interconnected wetland ecosystems and the intensity of anthropogenic pressures. Hence, this chapter outlines the groundwork of a three-layered status of the family Anatidae, creating a shared reference point that could serve as the base for future research and conservation action globally (Figure 1). Both the Nearctic and Palearctic regions have been combined and represented here as Holarctic, and Australasia is discussed under the broader realm of Oceania for analytical convenience. Many Anatidae species have wide migratory ranges or overlapping distributions that extend across multiple realms. In contrast, few species are restricted to a single realm, representing regional endemics.

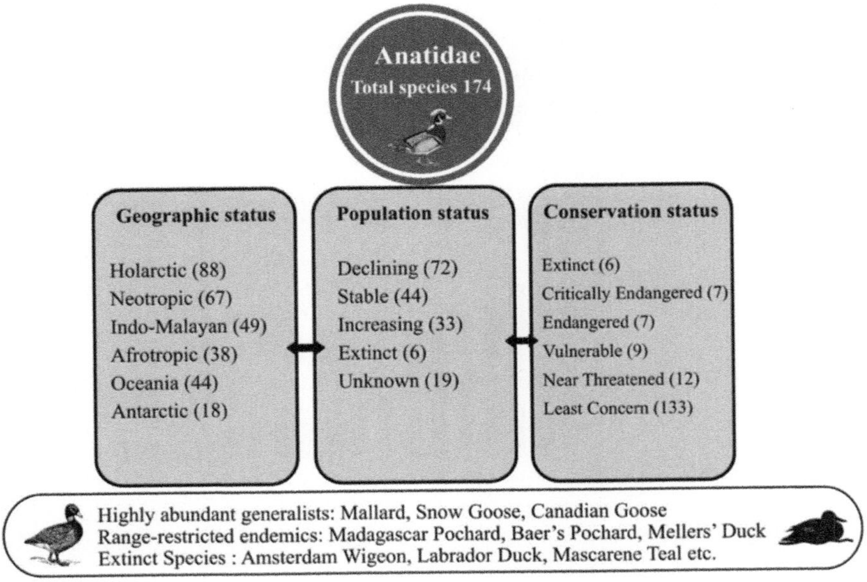

Figure 1. Overview of the status of the Anatidae family with the number of species in parentheses and a few examples for the abundant, endemic, and extinct species. The numbers shown for the geographical realms represent the total species within that realm (both overlapping and endemic), rather than exclusive counts.

2. The Geographic Status

The geographical range of a species is defined as the spatial area where individuals are naturally found, and it forms the fundamental unit of biogeography. It depends on the area of occupancy, extent of occurrence, and its abundance gradient. Understanding the geographical ranges of the species is critical to assess their conservation status, whether these ranges are shrinking or expanding. The habitat specialist or range-restricted species face more risks of extinction in the wild, while some species may expand their range and become invasive as well. The geographical range boundaries are influenced by both abiotic factors (such as climate, water, physiography, etc.) and biotic interactions such as competition, predation, symbiotic relationships, dispersal ability, etc. We have considered six major biogeographical realms worldwide, with a few notable species as examples for the overview of the geographic status of the family Anatidae in them (Figures 2 and 3).

2.1. Holarctic Realm (North America, Europe, and Northern Asia)

The Holarctic region supports the highest number of species (88), but interestingly, 26 anatids are exclusive to this realm and 62 have a wide distribution across multiple realms (IUCN, 2025). In the Holarctic and Indo-Malayan regions, 15 species overlap between them, which is the maximum among realms. This is followed by the Holarctic and Neotropical regions, which have 14 species

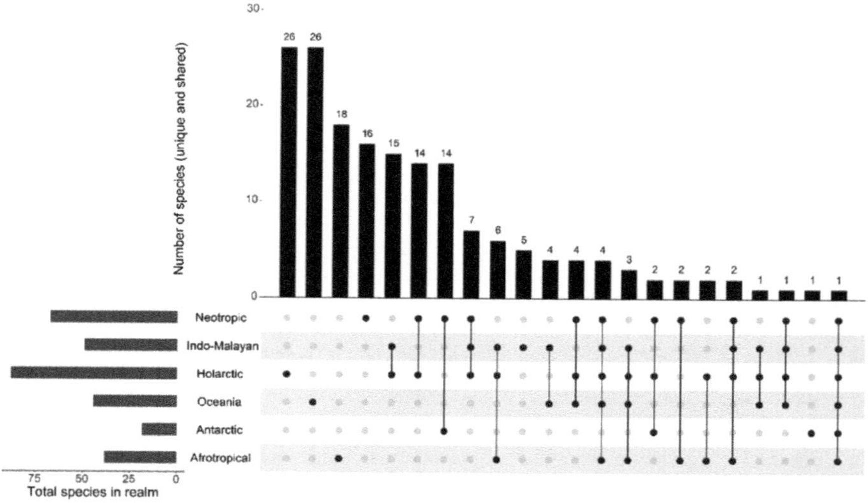

Figure 2. Plot showing the number of unique and shared Anatidae species among various geographical realms with respective intersections.

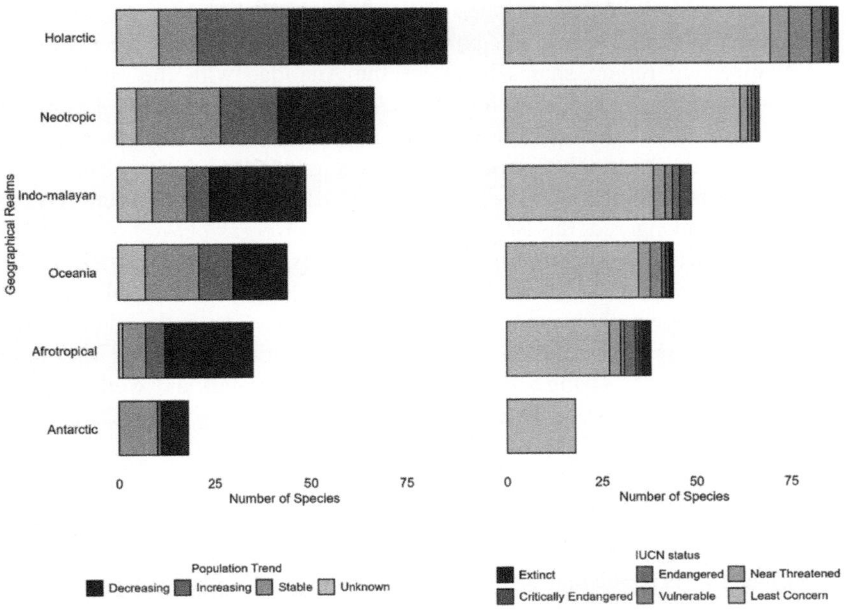

Figure 3. IUCN status and population trends of Anatidae species in each geographical realm.

overlapping. A few species present in this realm also share a few overlapping areas with more than one region. The population abundance of many Anatidae species in this realm shows a generally decreasing trend, while in a few other species, the trends are increasing, and a few others have stable populations when compared to other realms. The major declines are noted for species such as the northern pintail (*Anas acuta*) and certain sea ducks such as long-tailed duck (*Clangula hyemalis*). Among the 14 waterfowl the region supports, some common species include northern pintail, mallard (*Anas platyrhynchos*), canvasback (*Aythya valisineria*), gadwall (*Mareca strepera*), redhead (*Aythya americana*), lesser scaup (*Aythya affinis*), and blue-winged teal (*Anas discors*) (Bryan, 2022). There is an interesting case of a name of a geographical region being dedicated to anatids. The Prairie Pothole Region of North America supports more than half of the continent's breeding populations of ducks, which gives the region its popular name, "duck factory" (Niemuth & Solberg, 2003; Bryan, 2022).

2.2. Neotropical Realm (Central and South America)

The Neotropical region supports 67 anatids, among which 16 are endemics. Among the species restricted to the region, approximately 60% exhibit decreasing population trends, such as the torrent duck (*Merganetta armata*) that shows a decline in population across its range of distribution in the Andes of South America (Johnsgard, 2010a). When only pair-wise inter-realm overlaps are considered, the Neotropical region shares many of the Anatidae with the Holarctic and Antarctic realms (14 species with each) and two species with the Afrotropical realm (white-faced whistling duck (*Dendrocygna viduata*) and southern pochard (*Netta erythrophthalma*)). Even though the status of the majority of the species in the realm is of least concern, one species, which is endemic to South America, is critically endangered, the Brazilian Merganser (*Mergus octosetaceus*) (Gill et al., 2025; IUCN, 2025). Another important species, Mexican duck (*Anas diazi*), is an endemic and endangered dabbling duck of the Mexican highlands. There is not much known about its ecology and population. The endangered status is known to be due to the wetland ecosystem degradation, pollution, and natural habitat loss. Climate-change-induced alterations in rainfall and wetland habitat are the other known causes (Mercado-Reyes et al., 2024).

2.3. Indo-Malayan Realm (South and Southeast Asia)

Globally, the Indo-Malayan region presents a major decline in species of the Anatidae family. Five species endemics to this region are pink-headed duck (*Rhodonessa caryophyllacea*), Indian spot-billed duck (*Anas poecilorhyncha*), Philippine duck (*Anas luzonica*), Andaman teal (*Anas albogularis*), and white-winged duck (*Asarcornis scutulata*). Among this, the critically endangered white-winged duck range in Southeast Asia is severely fragmented and collapsed, the species is probably extinct in many regions, and the population is likely in only a few hundred (BirdLife International, 2024). This realm has a high number of threatened species due to wetland loss owing to agriculture and urban expansion, and hunting pressures. The East Asian–Australasian flyway (EAAF) and Central Asian flyway (CAF) are the world's most threatened migratory routes for waterbirds (Rashiba et al., 2022; Mundkur et al., 2023). The example of an endangered species, swan goose (*Anser cygnoides*), is a classical case in this realm. The species is observed in steppe and forest-steppe zones of Asia and Mongolia to Siberia, with major breeding grounds in China, Mongolia, and Russia. The downgrade of conservation status from near threatened to vulnerable and endangered can be owed to many severe threats throughout its range, including wetland degradation, climate change, hunting, poaching, other anthropogenic disturbances, dogs, pesticides, etc. This species is listed in the Convention on the Conservation of Migratory Species of Wild Animals (CMS) Appendix II, which means the listed migratory species requires international cooperation for its conservation and management. Hence, this species is thereby protected legally in Russia, Mongolia, South Korea, and China (BirdLife International, 2023a).

Although the Anatidae species utilize varied habitats, for example, in the Gulf of Mannar in the southeast coast of India, species such as northern pintail, garganey (*Spatula querquedula*), common teal (*Anas crecca*), etc., inhabit coastal lagoons (Byju et al., 2023a, 2024a, 2025a), islands (Byju et al., 2023b), and irrigation tanks surrounded by agricultural lands (Byju et al., 2024b, 2025b). Some of these habitats are also the breeding grounds for a few species, such as the Indian spot-billed duck, lesser whistling duck (*Dendrocygna javanica*), knob-billed duck (*Sarkidiornis melanotos*) (Byju et al., 2025c). This reflects on their ability to adapt well in human-modified landscapes and overlapping habitat preferences of multiple species in a region. Among the migratory birds, 38 species of the Anatidae family use CAF, and this family has the highest threatened species in CAF (Mundkur et al., 2023).

2.4. Oceanian Realm (Australia, New Guinea, New Zealand, and Eastern Indonesia)

This region comprises mostly countries which are secluded from the mass lands, have species with mixed status, which are more endemic and are range-restricted species. Among the total 44 Anatidae species found in this region, 26 are endemic (approximately 59%) and 18 share their ranges with other realms, including the extinct New Zealand merganser (*Mergus australis*) and one endangered species blue duck (*Hymenolaimus malacorhynchos*) (Tennyson & Martinson, 2006; Gill et al., 2025). The vulnerable Hawaiian duck (*Anas wyvilliana*), near-threatened Hawaiian goose (*Branta sandvicensis*), and critically endangered Laysan duck (*Anas laysanensis*) are restricted to the Hawaiian archipelago, and the latter is now confined to Laysan Island. The Salvadori's teal (*Salvadorina waigiuensis*) is confined to montane forests in New Guinea, representing one of the few riverine teal species in the tropics (Beehler & Pratt, 2016). Species such as the freckled duck (*Stictonetta naevosa*) and blue-billed duck (*Oxyura australis*) are endemic to Southern Australia, primarily in freshwater wetlands and lakes. The few widespread residents in the region include plumed whistling duck (*Dendrocygna eytoni*), grey teal (*Anas gracilis*), hardhead (*Aythya australis*), maned duck (*Chenonetta jubata*), black swan (*Cygnus atratus*), and pink-eared duck (*Malacorhynchus membranaceus*) that occur across mainland Australia, Tasmania, and occasionally New Zealand. Radjah shelduck (*Radjah radjah*) and green pygmy goose (*Nettapus pulchellus*) are distributed across Northern Australia, Southern New Guinea, and adjacent islands, linking the Oceanian and Australasian biogeographic zones.

The intermittently flooded wetlands and the resource-rich floodplains of the Australian continent are known to support diverse breeding populations of Anatidae, such as black swan (*Cygnus atratus*), grey teal, and plumed whistling duck (*Dendrocygna eytoni*) (Reid et al., 2013).

2.5. Afrotropical Realm (Sub-Saharan Africa)

The Afrotropical realm supports a total of 38 Anatidae species, of which 18 are unique to the region. This indicates a relatively high level of endemism. Many of the species in this region show a decreasing population trend. Few generalist species of the region, such as spur-winged goose (*Plectropterus gambensis*), white-backed duck (*Thalassornis leuconotus*), and African black duck (*Anas sparsa*), are resilient and widespread. Four species, Réunion sheldgoose (*Alopochen kervazoi*), Amsterdam wigeon (*Anas marecula*), Mauritius sheldgoose (*Alopochen mauritiana*), and Mascarene Teal (*Anas theodori*) are extinct, and one critically endangered Madagascar pochard (*Aythya innotata*) is known to be nearly extinct and is now restricted to a single lake in the Banemavka area (Gill et al., 2025; IUCN, 2025).

Two other range-restricted endangered species in this region include Maccoa duck (*Oxyura maccoa*) and Meller's duck (*Anas melleri*). Maccoa duck is restricted to the Afrotropical region, inhabiting Ethiopia to South Africa. The species has undergone drastic population declines, and the status has changed constantly from least concern to near threatened, vulnerable, and currently endangered over the span of approximately three decades. The primary reasons for the decline are thought to be pollution, habitat loss, hunting, and poaching (Berruti et al., 2005). The species is known to be more vulnerable to bioaccumulation of the pollutants since it mainly feeds on the benthic invertebrates (BirdLife International, 2021a). Meller's duck is the largest inland freshwater waterfowl in Madagascar. The continued decline is due to hunting, habitat loss, and degradation. Meller's duck is protected through various management actions, such as captive breeding in Europe and North America (Baldassarre, 2014). Forty-six percent of the species' range has been declared as a protected area (BirdLife International, 2021b).

2.6. Antarctic Realm (Antarctica and Island Groups in the Southern Atlantic and Indian Oceans)

This geographic realm supports 18 species of Anatidae, which are shared among other realms, mainly the Neotropical realm. A few species, such as the white-cheeked pintail (*Anas bahamensis*) and cinnamon teal (*Spatula cyanoptera*), extend their ranges northward into the Nearctic realm. Most species occur across Southern Chile, Argentina, Tierra del Fuego, and the Falkland Islands (Gill et al., 2025). All are categorized as least concern under the IUCN Red List. Despite their favorable conservation status, several exhibit declining population trends. Conversely, species such as the rosy-billed pochard (*Netta peposaca*) show increasing trends. One species, Eaton's Pintail (*Anas eatoni*) is restricted only to the Kerguelen Islands and Crozet Islands of the Southern Indian Ocean and hence the only species in the Antarctic realm which is not shared with other geographical regions (IUCN, 2025).

3. The Global Distribution Patterns Across the Realms

The global distribution can still be looked at the hemispheres as migration happens annually twice from north to south and then from south to north. The

Northern Hemisphere, represented largely by the Holarctic realm, exhibits higher speciation and breeding diversity of Anatidae, driven by strong climatic seasonality and the availability of extensive wetland networks (Jetz et al., 2012). High diversity of breeding Anatidae is concentrated across Central and Western North America and Central and Eastern Asia, in regions such as the Yangtze Basin, the major Siberian River basins, and the wetland complexes of Northwestern North America (Zeng et al., 2019). This can be attributed to the habitat heterogeneity of temperate and subarctic wetlands and the strong seasonality of vegetation productivity that supports large breeding and migratory populations.

In contrast, Southern Hemisphere realms, including the Afrotropical, Neotropical, and Oceanian regions, show lower recent diversification rates but retain older, evolutionarily distinct lineages that have long persisted in stable environments and hence show more phylogenetic diversity and functional diversity of Anatidae than taxonomic diversity of the northern realms (Jetz et al., 2012; Zeng et al., 2019). Unlike the north, in the southern realms, diversity is driven largely by interannual variability in resource availability rather than seasonal abundance peaks (Dalby et al., 2014).

Correspondingly, the shared distribution patterns across biogeographic realms in the present analysis reflect these large-scale migratory and historical connections (Figure 2). The Holarctic realm exhibited the greatest degree of overlap with other biogeographic realms, being represented in nearly 80% of all cross-realm species distributions, indicating its dominant and wide-ranging overlap with other realms. The highest number of species is shared between the Indo-Malayan and Holarctic (15), followed by the Holarctic and Neotropic regions (14) and the Antarctic and Neotropic (14). Species occurring in five or all six realms represent a smaller group, but demonstrate remarkable dispersal capability. For instance, four species, northern pintail, Eurasian wigeon (*Mareca penelope*), gadwall, and northern shoveler (*Spatula clypeata*), are distributed across the Indo-Malayan, Holarctic, Neotropical, Afrotropical, and Oceanian realms, and a single species, the highly adaptive mallard, is found in all six geographical regions.

Seven species exhibit a broad distribution pattern across the Indo-Malayan, Holarctic, and Neotropical realms, reflecting extensive migratory routes that link Asia, Eurasia, and the Americas. They are: greylag goose (*Anser anser*), greater white-fronted goose (*Anser albifrons*), brant goose (*Branta bernicla*), common goldeneye (*Bucephala clangula*), tundra swan (*Cygnus columbianus*), common merganser (*Mergus merganser*), and red-breasted merganser (*Mergus serrator*). Similarly, four species are shared between the Holarctic and Neotropical realms along with Oceania (lesser scaup, ring-necked duck (*Aythya collaris*), American wigeon (*Mareca americana*), and blue-winged teal). Another six species occur across the Indo-Malayan, Holarctic, and Afrotropical realms (Eurasian teal (*Anas crecca*), ferruginous duck (*Aythya nyroca*), marbled duck (*Marmaronetta angustirostris*), knob-billed duck, ruddy shelduck (*Tadorna ferruginea*), and common shelduck (*Tadorna tadorna*)). In contrast, several combinations are represented by only a few species, illustrating more localized or specialized distributions. Two species, Egyptian goose (*Alopochen aegyptiaca*) and Cape teal (*Anas capensis*), are shared between the Afrotropical and Holarctic realms. Southern pochard and white-faced whistling duck share between Afrotropical and Neotropical realms, and white-cheeked pintail (*Anas bahamensis*) and cinnamon teal between Antarctic, Neotropical, and Holarctic realms. Three species—common pochard (*Aythya ferina*), tufted duck (*Aythya fuligula*), and garganey—occur in the Indo-Malayan, Holarctic, Afrotropical, and Oceanian regions (Gill et al., 2025).

These patterns collectively reveal that while a few Anatidae species possess wide, intercontinental ranges, a considerable proportion are confined to specific realm combinations, possibly due to habitat specialization, migratory constraints, or historical biogeographic barriers.

4. The Status of Invasive and Problematic Anatidae Species

While we consider the wild populations of the Anatidae family on one side, the invasive species of the family which is affecting the native wild populations cannot be ignored. Human-mediated introductions have created new, resilient populations of certain Anatidae species, whose status has shifted to become highly invasive in their non-native ranges, with significant consequences for ecosystems and native species in many regions globally. A few examples are given in the following text.

The ruddy duck (*Oxyura jamaicensis*), a native of North and Central America (Baldassarre, 2014), has become an invasive species in Europe and is known to threaten the white-headed duck (*Oxyura leucocephala*), regionally. With strict regulations and eradication programs, by 2010, the ruddy duck population was managed, controlled, and subsequently ranged in relatively few countries (Cranswick & Hall, 2010). Another case is that of Canada goose (*Branta canadensis*), which is native to North America but invasive in several European countries (Lukasiewicz et al., 2018). Similarly, the Egyptian goose, which is native to Africa, is expanding as a non-native species in Europe and North America (Henderson & Hart, 2023). High reproductive rates, adaptability, and a lack of natural predators can be attributed to their range expansion and the cause of range restriction of native species, thereby altering the geographical, demographical, and conservation status of such native species.

Conversely, when anatid populations become overabundant, either naturally or due to human-induced changes, their status can shift to that of a pest. In Japan, the spot-billed ducks and mallard were creating a nuisance for the farmers by feeding on rice seed and seedlings and geese species eating on the leaves of wheat (*Triticum aestivum*) and barley (*Hordeum vulgare*) (Lane et al., 1998). Similarly, managing snow goose (*Anser caerulescens*) populations has become a challenge for agriculture and ecology for several decades. The issue was highlighted in 1965 as the population began rapid growth. This unchecked growth led to species being declared as overabundant (Lefebvre et al., 2017). Strict and continuous monitoring led to the conclusion that the population was between 0.6 and 1.0 million and showed declines in the past five years preceding 2017 in Canada and the United States (Lefebvre et al., 2017). By 2020, the snow goose species was widely hunted throughout its range for recreation, and this regulated the rapid population growth (Mowbray et al., 2020). Despite these efforts, the overall population of snow goose reached 16 million individuals, which has led to population management issues (BirdLife International, 2021c). The lesser snow geese constitute more than half of all wintering geese in California. The populations remain consistently high and have resulted in conflicts with farmers and livestock. The severity of these negative interactions has led to liberalizing hunting regulations and financial help for affected agricultural operations (Martinico et al., 2024).

5. Demographic Status—Population Size and Trends

A robust global total count for all Anatidae species is lacking, and only scattered information is available on both the species level and the regional level. Global population trend suggests that overall 72 species of Anatidae show a decreasing trend (41.3%), 33 are increasing (18.9%), and 44 are stable (25.2%); the rest include extinct species, and the trend is unknown (IUCN, 2025) (Figure 1). The complexity of quantifying the global population involves methodological aspects to be addressed. Unlike surveying a single, localized species, generating a global estimate requires integrating a patchwork of disparate data sources, each with its own limitations and inherent biases. The general methodologies followed worldwide are standardized aerial and ground surveys by government agencies and conservation groups, and coordinated counts at major roosting, wintering, and breeding sites. The increasingly valuable, yet unstructured data from citizen science platforms such as eBird and iNaturalist (Sauer et al., 2013; Sullivan et al., 2014) is helping in managing a rough estimate of population. Despite this, the numbers available are approximate counts. For widespread species, these data can refine distribution maps and detect phenological shifts.

One of the major limitations in these population counts is that some specialized species will be missed in these surveys. A few examples are highlighted to understand the shortcomings of only using citizen science data for population assessments (Maitreyi, 2024). Certain species, like the white-winged duck, which inhabits dense forested wetlands, are not easy to obtain citizen science data for. Species that are in remote and inaccessible tropical river systems, such as the Salvadori's teal (*Salvadorina waigiuensis*), and nocturnal feeders such as Baer's pochard (*Aythya baeri*), common pochard, or tufted duck (Wang et al., 2012; Callaghan et al., 2018), are always undercounted and potentially create a false impression of their absence in many habitats and in global estimation of counts. Similarly, a few migratory Anatidae populations have not been sufficiently covered by the Asian Waterbird Census (AWC), such as falcated duck (*Mareca falcata*), greater scaup (*Aythya marila*), harlequin duck (*Histrionicus histrionicus*), scaly-sided merganser (*Mergus squamatus*), and sea ducks (scoters and eiders). This absence of census data and sufficient published literature has led to a broader range of population estimates (Miyabayashi & Mundkur, 1999).

For instance, in North America, the Waterfowl Breeding Population and Habitat Survey (WBPHS), a collaborative effort by the U.S. Fish and Wildlife Service and the Canadian Wildlife Service, has provided foundational data for decades (U.S. Fish and Wildlife Service, 2023). Similarly, Wetlands International coordinates the International Waterbird Census (IWC) across major flyways such as Africa–Eurasia, Asia-Pacific, Caribbean, Neotropics, and Central America, including 143 countries. This survey synthesizes national counts into a cohesive regional picture (Wetlands International, 2023). However, they are financially and logistically intensive, and contain inherent limitations such as visibility bias, where observers miss birds that are present, which is a constant issue, influenced by vegetation, weather, and flock behavior (Scher & Clark, 2023). Even in WBPHS, the sea ducks are not comprehensively recorded by the traditional waterfowl survey methods. These surveys do not cover the entire breeding range of sea ducks, and it can be challenging to identify a few species from the aircraft. Therefore, WBPHS gives a combined population estimate of the sea ducks, such as scoters, goldeneyes, etc., as compared to individual breeding species (Cox et al., 2023). Other poorly surveyed species in Canada include the long-tailed duck, king eider (*Somateria spectabilis*), and the scoters (white-winged scoter (*Melanitta deglandi*), surf scoter (*Melanitta perspicillata*), and black scoter (*Melanitta americana*)), mainly due to their breeding distribution in the Arctic or further north. Some sea duck species are better studied, such as the bufflehead (*Bucephala albeola*) whose population count in 2023 across its range of the Western Boreal Shield region was roughly 33,700. Similarly, in the survey areas of WBPHS, the common goldeneye count was approximately 596 in 2023, reflecting a 4.6% decline compared to the average of 10 years. This species overall shows a stable global population trend, and it spreads across Indo-Malayan, Holarctic, and Neotropical regions.

In 2023, WBPHS estimated about 30.5 million ducks within its traditional survey area (the prairie region of Western Canada and North-Central United States, Western Boreal regions of Canada, some areas of Ontario, and parts of Alaskan tundra), excluding sea ducks. Over the 5 years from 2019 to 2023, populations of green-winged teal (*Anas carolinensis*), American wigeon, gadwall, mallard, and northern shoveler declined, whereas northern pintail, ring-necked duck, blue-winged teal, redhead, canvasback, and ruddy duck exhibited no consistent trends. During 2023, mallards dominated the duck populations, numbering about 6.1 million, while blue-winged teal, greater scaup (*Aythya marila*) and lesser scaup species followed with 5.25 million and 3.5 million, respectively. Northern shovelers and gadwalls were also abundant, with estimated populations of 2.85 million and 2.56 million, respectively. According to the U.S. Fish and Wildlife Service Waterfowl Population Status Report 2024, the total breeding ducks have increased 5% from 2023 (34 million in 2024). Mallard and American black duck

(*Anas rubripes*) top the population abundance in 2024 with 6.6 million and 2.2 million, respectively (U.S. Fish and Wildlife Service, 2024). Analysis of the long-term dataset from 1976 to 2022 indicates population increases in five species in the region—cackling goose (*Branta hutchinsii*), greater white-fronted goose, Mid-Continent lesser snow goose, Western Arctic lesser snow goose (together called *Anser caerulescens caerulescens*), Ross's goose (*Anser rossii*), while the Atlantic brant or the brant goose (*Branta bernicla*) population has remained stable (Cox et al., 2023). The endangered Mexican duck of the Mexican Highlands has an approximate population of 55,500 (Drilling et al., 2020).

Similarly, the Asian survey AWC reported that the Baikal teal (*Sibirionetta formosa*), largely confined to East Asia, was the most abundant duck (316,000–1,065,000) recorded during the AWC (2008 to 2015), although its numbers showed wide annual fluctuations. Among the more widespread dabbling ducks, the northern pintail was commonly observed across East, Southeast, and South Asia (annual count 79,000–521,000). The AWC assessment further noted eleven goose species and four swan species in the Asian region. Among the swans, Bewick's swan (*Cygnus columbianus bewickii*) is the most abundant in East Asia with an annual population of 18,000–25,000. Of eleven goose species, the greater white-fronted goose (198,000–304,000 annual count) is the most widespread species in East Asia and the bar-headed goose (*Anser indicus*) in South Asia (68,000–89,100 as per the original source) (Mundkur & Langendoen, 2016).

The other most pragmatic approach to a global census is to organize data by the world's major migratory flyways. When viewed alongside the flyway level populations which transcend multinational and continental boundaries, it gives a broader global overview of the species status, mirrored by regional analyses. According to the Conservation Status Review of Migratory Waterbirds of the East Asian–Australasian flyway (Mundkur & Langendoen, 2022), reliable population trend data are available for only a small proportion of waterbird populations. Of the assessed 159 biogeographical populations with known trends in the East Asian–Australasian Flyway Partnership (EAAFP), 42% are in decline, 30% are stable or fluctuating, and a mere 27% are increasing. Anatidae is one of the largest families with 63 distinct populations, and 38% of its populations have reliable trend data. In terms of population size estimates, 71% of Anatidae populations are based on census data or expert opinion, a significantly high proportion, unlike other waterbird groups with poor quality data. Notably, within Anatidae, approximately 62% of populations are either stable, fluctuating, or increasing. In the EAAFP assessment, population refers to a biogeographical population, a geographically distinct segment of a species or subspecies, defined primarily by its breeding or non-breeding range for conservation purposes. These units, though not always genetically isolated, provide the basis for population-level conservation assessments. Furthermore, geographical scopes of these reviews, apart from EAAF, include Central Pacific flyway and areas of CAF under Eastern Palearctic, Indo-Malayan, and Australasian/Oceanian realms (Mundkur & Langendoen, 2022). In the African–Eurasian flyway (AEF), according to the agreement on the conservation of African–Eurasian migratory waterbirds 2015 report, Anatidae is the predominant waterbird group (24% of the total) and is the largest group in the African–Eurasian Migratory Waterbird Agreement (AEWA) with the most declining population (43% of 121 Anatidae populations) (Nagy et al., 2015). While swans show a 25% decline, migratory geese and migratory ducks show a population decline of 23% and 44%, respectively. This implies a better population status of swans and geese than ducks in AEF (Davidson & Stroud, 2006).

Overall, for some well-studied species, databases such as BirdLife International provide more accurate global populations. For example, mallard, a global generalist, has a population estimated at 17,260,000–19,300,000 individuals, making it arguably the world's most abundant duck (BirdLife International, 2023b). Similarly, other adaptable ducks have the following numbers, such as the northern pintail (approximately 4,700,000–4,800,000), ruddy shelduck (approximately 134,000–198,000), common shelduck (approximately 415,000–500,000), and several geese including the greylag goose (approximately 710,000–780,000 in Europe alone) and the super-abundant snow goose (approximately 15,000,000 and increasing), and swans such as whooper swan (*Cygnus cygnus*) (approximately 264,600) and mute swan (*Cygnus olor*) (approximately 308,000–348,000) (Wetlands International, 2023; Bird Life Data Zone, 2025). The entire group of swans (Cygnini or *Cygnus* genus), although locally abundant, has a total global population much smaller than that of the most numerous ducks or geese. Most swan species number in the hundreds or thousands (Kear, 2005).

This staggering aggregate number, however, masks profound ecological and conservation disparities. A taxonomic and ecological breakdown reveals that this apparent abundance is heavily skewed, dominated by a relatively small group of highly adaptable, generalist species. While several Anatidae species have stable or increasing populations, many others face severe declines. The Madagascar pochard was rediscovered in 2006 with a wild population of 33–47 mature individuals (Birdlife International, 2023c) and, despite intensive conservation, remains critically endangered (Rasoamihaingo et al., 2024). The Brazilian merganser persists in isolated and extremely fragmented populations totalling fewer than 250 mature individuals (BirdLife International, 2019). The flyway population of Baer's pochard, a critically endangered species in CAF, is 1640 to 1700, whereas the least concern fulvous whistling ducks (*Dendrocygna bicolor*) are 50,000 in number (Wetlands International, 2025). The endangered maccoa ducks (*Oxyura maccoa*) in the Afrotropic region (African–Eurasian) are merely 100–150 (BirdLife International, 2021a), and the estimation of Meller's ducks in the region is 1,300–3,300, with a decreasing trend. The overall population of mature individuals of the endangered swan goose in the Indo-Malayan region was known to be 36,000–43,500. In the Antarctic realm, the Eaton's pintail, a highly restricted species to the Kerguelen Islands, has a population range of 30,000–40,000, with a decreasing trend (BirdLife International, 2025a).

Understanding these extremes and inequality in distribution and population is crucial to evaluating the overall conservation status of the Anatidae family.

6. Species with Improved Status—Some Case Studies

Several Anatidae species show improved population status overall. The following are a few of the case studies of such species:

Laysan duck is a critically endangered species with 500–680 individuals in the Laysan Islands, but the population is believed to be increasing. From a low population during the 20th century, its population has increased gradually on the island, due to the elimination of invasive taxa, especially the rabbits (*Oryctolagus cuniculus*) (Warner 1963; Reynolds et al., 2015).

The wood duck (*Aix sponsa*) population declined during the late 19th century due to over-harvesting and nesting habitat loss. Due to management efforts, the species recovered as many as 4,600,000 mature individuals in breeding pairs across North America today (Bellrose, 1976; BirdLife International, 2021d).

The **Hawaiian goose** is endemic to the Hawaiian Islands. During the 1950s, the wild population of this species declined to around 30 individuals. With extensive conservation measures and management such as enlistment in CITES (Convention on International Trade in Endangered Species of Wild Fauna and Flora) Appendix I and reintroduction efforts, the population started recovering (Marshall et al., 2008). The most recent population estimation available shows that it is around 1700–2200 mature individuals and has been uplisted to near threatened from vulnerable in the Red List classification (BirdLife International, 2021e).

In some instances, the increasing population in one region may not imply an improved overall status within the same flyway itself. One such example is, greater white-fronted geese, whose wintering population showed an increase in Japan and Korea (146,000 to 224,000), while a major decline in China (30,000–55,000) within EAAF (Jia et al., 2016). Later studies showed that the species might imply two distinct biogeographical populations themselves (Cao et al., 2020).

There are several factors which can be attributed to the increased population and improved overall conservation status across their range, such as:

1. **Availability of Alternative Habitats:** The availability of suitable habitat, whether through natural wetland protection, provision of artificial or alternative nesting or foraging habitats (agricultural land for barnacle geese (*Branta leucopsis*)).

   For example, at flyway levels, the interconnected networks of natural and anthropogenic wetlands play a critical role as wintering sites (Byju et al., 2025d), and the availability of food resources determines the distribution and population status of Anatidae (Rajpar et al., 2022). For instance, the rice fields and farmlands have emerged as such wintering sites, supporting more than 100,000 wintering dabbling ducks and several species of geese using the EAAF and East Atlantic Flyway (EAF) (Navedo et al., 2012; Shi et al., 2024).

2. **Regulatory Protection:** Enactment and enforcement of hunting regulations and protective legislation.
3. **Targeted, Science-Based Intervention:** Nest boxes for wood ducks, captive breeding and reintroduction, and habitat management.
4. **Long-Term Commitment and Funding:** These recoveries unfolded over decades, requiring sustained financial and institutional support.

7. Conservation Status

The IUCN Red List of Threatened Species provides the most standardized system for classifying the extinction risk of species and acts as a critical indicator for the global status of biodiversity. The IUCN criteria evaluate species based on quantitative thresholds, including size of the population, rate of decline, extent of geographic range, and extent of population fragmentation.

A comprehensive analysis of the 174 recognized species of Anatidae (Gill et al., 2025) reveals the following distribution across the primary IUCN categories (IUCN, 2025), as done in Figure 4. The list of various categories is mentioned with some important species in the following text.

The modern era has witnessed the definitive extinction (EX) of six anatid species, each loss representing a loss of unique ecological role and evolutionary history. They are the following:

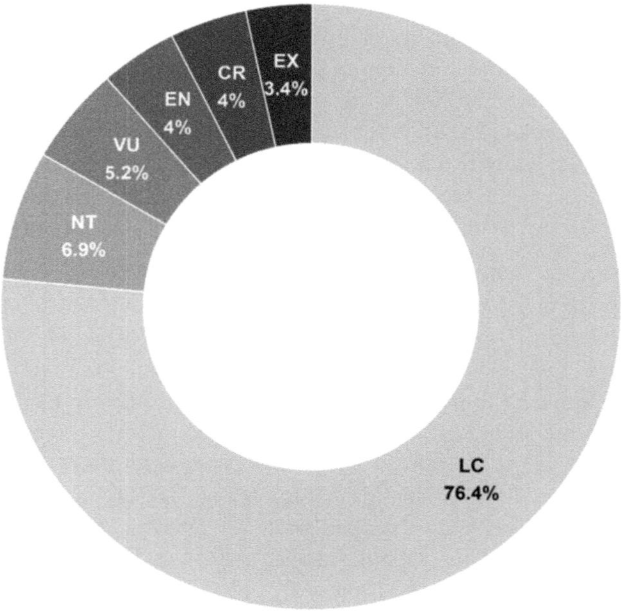

Figure 4. IUCN Red List status and percentage of Anatidae species in each category.

Labrador Duck (*Camptorhynchus labradorius*): This sea duck was known to be wintering in the Atlantic Coast of North America and was thought to breed in Labrador. The last confirmed specimen was collected in 1878. The precise cause of its extinction remains debated, but it was likely a combination of overhunting for food and feathers, and its inherently small, specialized population and specialized diet (believed to be small mollusks) that made it vulnerable (BirdLife International, 2023d).

Mauritius Sheldgoose: This species was endemic to Mauritius and was heavily hunted for food, which is thought to have caused its disappearance. It was last recorded in 1693 (Cheke & Hume, 2008).

Réunion Sheldgoose: This species was endemic to the island of Réunion. The last record of this species was in 1671–1672. Around 1700, overhunting led to the extinction, along with other contributing factors such as introduced predators taking chicks and eggs (Hume, 2017a).

Mascarene Teal: This species, once found in Mauritius, has not been recorded since 1696. Hunting is known to be the major factor behind its extinction (Brooks, 2000).

Amsterdam Wigeon: Once recorded from Amsterdam and St. Paul islands, but last reported in 1793 (Brooks, 2000).

New Zealand Merganser: This species was once endemic to the Auckland Islands, south of New Zealand and was last recorded in 1902. The primary cause of its extinction was believed to be by predators such as pigs and cats introduced to the island by sealers and whalers. The species is a glaring example of the negative impact of invasives on the native-restricted range species (BirdLife International, 2016).

Similarly, more species are facing the threats of extinction, being endangered or critically endangered in the wild.

The critically endangered category has seven species, which include Laysan duck, white-winged duck, Baer's pochard, Madagascar pochard, Brazilian merganser, pink-headed duck, and crested shelduck (*Tadorna cristata*).

The IUCN endangered category of species in the Anatidae family included seven species. They are Bernier's teal (*Anas bernieri*), Mexican duck, Meller's duck, swan goose, blue duck, white-headed duck, and maccoa duck.

The nine vulnerable species are Campbell teal (*Anas nesiotis*), Hawaiian duck, lesser white-fronted goose (*Anser erythropus*), common pochard, red-breasted goose (*Branta ruficollis*), long-tailed duck, velvet scoter (*Melanitta fusca*), Steller's eider (*Polysticta stelleri*), and Chubut steamer duck (*Tachyeres leucocephalus*).

The near-threatened category has twelve species. They include Andaman teal, Auckland teal (*Anas aucklandica*), brown teal (*Anas chlorotis*), ferruginous duck, nene, blue-winged goose (*Cyanochen cyanoptera*), West Indian whistling duck (*Dendrocygna arborea*), marbled duck, black scoter, spectacled eider (*Somateria fischeri*), common eider (*Somateria mollissima*), and bronze-winged duck (*Speculanas specularis*).

Among the least concern category, there are 133 species. A few of them include Andean goose (*Chloephaga melanoptera*), ruddy shelduck, red shoveler (*Spatula platalea*), etc.

8. A Species without a Definite Conservation Status

Pink-headed duck, once recorded from South Asia, including India, Bangladesh, Nepal, and Myanmar, has not been recorded since the beginning of the 20th century, but the data is still considered inadequate to declare this species as extinct. Hence, currently, the status is critically endangered. The causes of decline are known to be excessive hunting and habitat loss (Hume, 2017b).

While it is clear that the majority of Anatidae species (approximately 76%) are classified as least concern, nearly 24% of all Anatidae species are categorized as near threatened, vulnerable, endangered, or critically endangered. The Pacific Americas flyway has 40 species of ducks, geese, and swans among 135 migratory waterbirds, among which 92% are under the least concern category (BirdLife International, 2025b). However, the least concern conservation status does not indicate a positive aspect of the overall Anatidae members. This proportion highlights the peril faced by this wetland-dependent family mainly due to habitat degradation, wetlands loss, pollution, hunting, invasive species, and the impacts of climate change (Nourani et al., 2015).

Genus *Anas* has the highest number of species (32) among all the genera in Anatidae. *Anas* has one extinct (Mascarene teal), one critically endangered (Laysan duck), three endangered (Bernier's teal, Mexican duck, Meller's duck), two vulnerable (Campbell teal, Hawaiian duck), three near threatened (Andaman teal, Auckland teal, brown teal), and 22 least concern species. Fifteen *Anas* species show a decreasing population trend, while four species are increasing and seven are stable, which is again the highest among all other genera. Some genera which have only one species with the threatened status include *Asarcornis* (white-winged duck [CR]), *Cyanochen* (blue-winged goose [NT]), *Hymenolaimus* (blue duck [EN]), *Marmaronetta* (marbled duck [NT]), *Rhodonessa* (pink-headed duck [CR]), *Polysticta* (Steller's eider [VU]), *Speculanas* (bronze-winged duck [NT]), *Clangula* (long-tailed duck [VU]), and genus *Camptorhynchus*, which has one extinct species Labrador duck. The fact that only one threatened or extinct species represents each of these genera presents a matter of critical conservation concern. Genus *Alopochen* has three species, among which two species Réunion sheldgoose and the Mauritius sheldgoose are extinct, and one species Egyptian goose is of least concern, yet with a decreasing population trend (Figure 5).

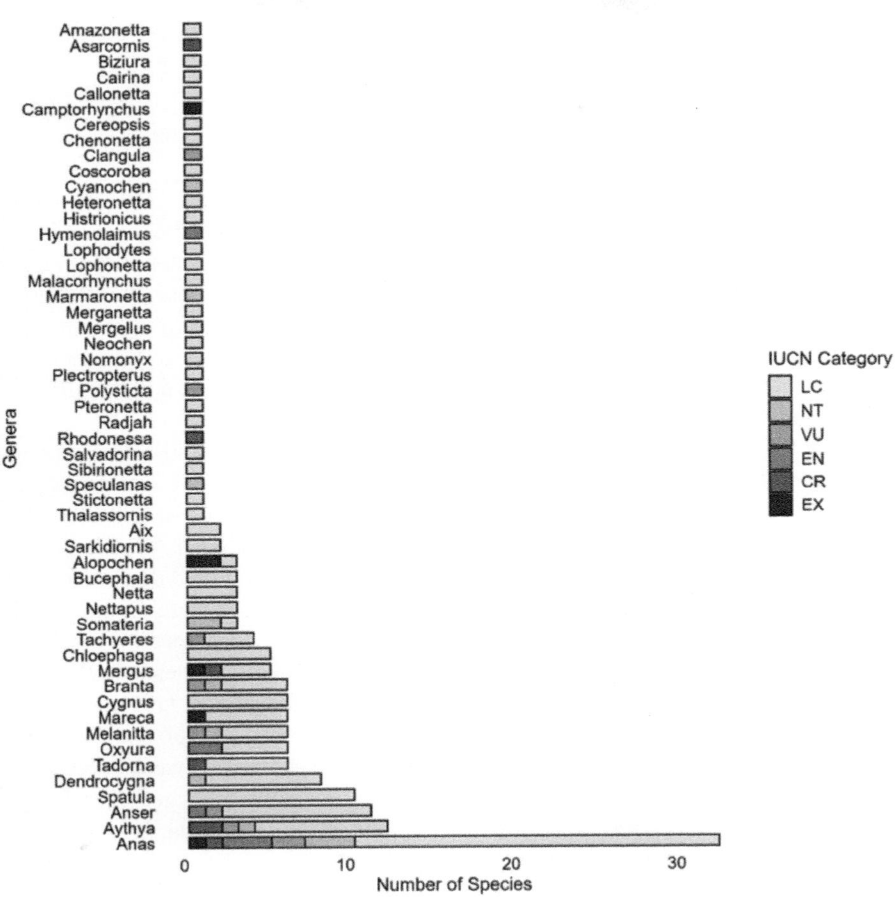

Figure 5. Genera of the Anatidae family with respective number of species in each IUCN Red List status.

9. Status of Anatids in Other Global Conservation Frameworks

The global conservation status, especially for migratory species, is supported by international bodies such as CMS, which also includes instruments such as the AEWA. Among anatids, 61 species are listed in CMS, and 51 species are listed under both CMS and AEWA (CMS, 2025). Four species, swan goose, Baer's pochard, ruddy-headed goose (*Chloephaga rubidiceps*), and Baikal teal, are in the CMS Appendix I list (those which are in danger of extinction), and 10 species are listed in Appendix II (species which are threatened and require international cooperation). Similarly, Arctic Migratory Birds Initiative (AMBI) includes brent goose, lesser white-fronted goose, emperor goose (*Anser canagicus*), common eider, velvet scoter, and long-tailed duck as AMBI priority species for conservation (AMBI, 2025).

10. Socio-Economic Factors Influencing the Anatidae Conservation Status

The perceived value of Anatidae to human societies brings huge economic value, and it's due to priorities in policies. Assessing the ecosystem services of Anatidae species in terms of their economic value can influence the conservation and management actions through prioritizing such species and proper funding allocations. Hunting has been beneficial for population control of certain species, while being harmful in some instances. Of late, ecotourism has added value to its global status.

In North America and Europe, waterfowl hunting is a multi-billion-dollar industry (Johnsgard, 2010b). This economy exists on the stable and managed status of game species such as the mallard, Canada goose, and wood duck. But, often hunting is not selective of the species. Several threatened Anatids in the mixed flocks, for example, vulnerable common pochard and tufted duck, with the declining global population trend, are also hunted in Poland (Marchowski et al., 2025), further affecting the regional and global status of such species. In 2022, an estimated 42,600 sea ducks were harvested in Canada. Buffleheads accounted for the largest share at 30%, followed by common goldeneyes (19%), hooded mergansers (*Lophodytes cucullatus*) (14%), common eiders (12%), and common mergansers (6.5%). During the same year, about 78,800 geese were harvested, with Canada geese making up 63% of the total, followed by snow geese, cackling geese, and Ross's geese. Swan harvesting is still illegal in Canada; however, the swan hunting is permitted in several states within the Central and Atlantic flyways of the United States (Cox et al., 2023).

Certain anatids drive a significant ecotourism economy. Notable examples include – recreational activities associated with the northern pintail in breeding and wintering regions of the USA and Canada were around $26M (Mattsson et al., 2018). The spectacle of millions of snow geese staging in the Skagit Valley (USA), the arrival of whooper swans in wintering sites in Japan, and the iconic status of the mandarin duck (*Aix galericulata*) in China and Japan generate substantial local revenue (Green & Elmberg, 2014) through ecotourism and bird watching. This economic value provides a compelling argument for protecting the habitats and species upon which it depends, affecting its conservation status.

11. Conclusion

The overall status of the Anatidae family is a powerful indicator of the global status of our wetland ecosystems. The widespread declines and extinctions signal the degradation of the vital freshwater and coastal systems. The analysis of the status of Anatidae in various contexts showed that the family is not uniformly in peril, but it faces a deeply polarized and uncertain future, mainly caused by habitat loss and degradation, other than hunting, invasives, and anthropogenic disturbances.

A small number of generalists, adaptable species such as the mallard, barnacle goose, and Canada goose are phenomenally successful, with populations in the millions and expanding ranges. Simultaneously, a disproportionately high number of specialists, range-restricted species such as Madagascar pochard, Baer's pochard, and Brazilian merganser are threatened with extinction; their populations measured in the hundreds or low thousands. This small proportion of threatened species certainly requires more conservation attention, as understanding these trajectories is vital for future species conservation, along with policy interventions required for wetland ecosystems globally.

References

Arctic Migratory Birds Initiative (AMBI). (2025, October 20). *Arctic Migratory Birds Initiative (AMBI)*. Conservation of Arctic Flora and Fauna (CAFF). https://www.caff.is/work/projects/arctic-migratory-birds-initiative/

Baldassarre, G. A. (2014). *Ducks, Geese, and Swans of North America.* Johns Hopkins University Press.

Beehler, B. M., & Pratt, T. K. (2016). *Birds of New Guinea: Distribution, Taxonomy, and Systematics.* Princeton University Press.

Bellrose, F. C. (1976). The comeback of the wood duck. *Wildlife Society Bulletin*, 4(3), 107–110.

Berruti, A., Baker, N., Buijs, D., Colahan, B. D., Davies, C., Dellegn, Y. et al. (2005). *International Maccoa Duck (Oxyura maccoa) Action Plan.* Johannesburg: African Gamebird Research Education and Development Trust.

BirdLife International. (2016). *Mergus australis*. The IUCN Red List of Threatened Species 2016: e.T22680496A92864737. https://dx.doi.org/10.2305/IUCN.UK.2016-3.RLTS.T22680496A92864737.en. Accessed on 02 July 2022.

BirdLife International. (2019). Species factsheet: Brazilian Merganser *Mergus octosetaceus*. Downloaded from https://datazone.birdlife.org/species/factsheet/brazilian-merganser-mergus-octosetaceus on 19/10/2025

BirdLife International. (2021a). Species factsheet: Maccoa Duck *Oxyura maccoa*. Downloaded from https://datazone.birdlife.org/species/factsheet/maccoa-duck-oxyura-maccoa on 19/10/2025

BirdLife International. (2021b). Species factsheet: Meller's Duck *Anas melleri*. Downloaded from https://datazone.birdlife.org/species/factsheet/mellers-duck-anas-melleri on 19/10/2025

BirdLife International. (2021c). Species factsheet: Snow Goose *Anser caerulescens*. Downloaded from https://datazone.birdlife.org/species/factsheet/snow-goose-anser-caerulescens on 29/10/2025

BirdLife International. (2021d). Species factsheet: Wood Duck *Aix sponsa*. Downloaded from https://datazone.birdlife.org/species/factsheet/wood-duck-aix-sponsa on 19/10/2025

BirdLife International. (2021e). Species factsheet: Hawaiian Goose *Branta sandvicensis*. Downloaded from https://datazone.birdlife.org/species/factsheet/hawaiian-goose-branta-sandvicensis on 19/10/2025

BirdLife International. (2023a). Species factsheet: Swan Goose *Anser cygnoid*. Downloaded from https://datazone.birdlife.org/species/factsheet/swan-goose-anser-cygnoid on 19/10/2025

BirdLife International. (2023b). Species factsheet: Mallard Anas *platyrhynchos*. Downloaded from https://datazone.birdlife.org/species/factsheet/mallard-anas-platyrhynchos on 20/10/2025

BirdLife International. (2023c). Species factsheet: Madagascar Pochard *Aythya innotata*. Downloaded from https://datazone.birdlife.org/species/factsheet/madagascar-pochard-aythya-innotata on 19/10/2025

BirdLife International. (2023d). Species factsheet: Labrador Duck *Camptorhynchus labradorius*. Downloaded from https://datazone.birdlife.org/species/factsheet/labrador-duck-camptorhynchus-labradorius on 24/10/2025

BirdLife International (2024). Species factsheet: White-winged Duck *Asarcornis scutulata*. Downloaded from https://datazone.birdlife.org/species/factsheet/white-winged-duck-asarcornis-scutulata on 15/10/2025

BirdLife International. (2025a). Species factsheet: Southern Pintail *Anas eatoni*. Downloaded from https://datazone.birdlife.org/species/factsheet/southern-pintail-anas-eatoni on 31/10/2025

BirdLife International. (2025b). Flyway factsheet: Pacific Americas. Downloaded from https://datazone.birdlife.org/flyway/factsheet/pacific-americas on 21/10/2025

Bird Life Data Zone. (2025). Available at https://datazone.birdlife.org/. Accessed on 22/10/2025

Brooks, T. (2000). Extinct species. *In:* BirdLife International (Ed.), *Threatened Birds of the World* (pp. 701–708). Lynx Edicions and BirdLife International, Barcelona and Cambridge, U.K.

Bryan, T. (2022). Assessment of duckling abundance as a biological indicator of wetland health in the Prairie Pothole Region. Doctoral dissertation, University of Saskatchewan.

Buckley, H. L., & Freckleton, R. P. (2010). Understanding the role of species dynamics in abundance–occupancy relationships. *Journal of Ecology*, 98(3), 645–658. https://doi.org/10.1111/j.1365-2745.2010.01651.x

Byju, H., Raveendran, N., Ravichandran, S., & Kishore, R. (2023a). Importance of conserving a critical wintering ground for shorebirds in the Valinokkam Lagoon—A first study of the avifaunal distribution of the southeastern coast of India. *Journal of Threatened Taxa*, 15(8), 23696–23709. https://doi.org/10.11609/jott.8494.15.8.23696-23709

Byju, H., Raveendran, N., & Ravichandran, S. (2023b). Distribution of avifauna on twenty-one islands of the Gulf of Mannar Biosphere Reserve, India. *Journal of Threatened Taxa*, 15(2), 22574–22585. https://doi.org/10.11609/jott.8112.15.2.22574-22585

Byju, H., Maitreyi, H., Ravichandran, S., & Raveendran, N. (2024a). Avifaunal diversity and conservation significance of coastal ecosystems on Rameswaram Island, Tamil Nadu, India. *Journal of Threatened Taxa*, 16, 26198–26212. https://doi.org/10.11609/jott.9248.16.12.26198-26212

Byju, H., Maitreyi, H., Raveendran, N., & Reshmi, V. (2024b). Avifaunal diversity assessment and conservation significance of Therthangal Bird Sanctuary, Ramanathapuram, Tamil Nadu: Insights about breeding waterbirds. *Journal of Threatened Taxa*, 16(9), 25802–25815. https://doi.org/10.11609/jott.8999.16.9.25802-25815

Byju, H., Maitreyi, H., Ravichandran, S., Raveendran, N., & Vijayan, R. (2025a). Avifaunal diversity and conservation status of waterbirds in Pillaimadam Lagoon, Palk Bay, India. *Journal of Threatened Taxa*, 17, 26789–26802. https://doi.org/10.11609/jott.9425.17.4.26789-26802

Byju, H., Maitreyi, H., Aarif, K. M., & Raveendran, N. (2025b). Disappearing colonies: Temporal decline in abundance and nesting of waterbirds in a key Indian wetland. *Wetlands Ecology and Management*, 33(4), 47. https://doi.org/10.1007/s11273-025-09901-3

Byju, H., Maitreyi, H., Natarajan, R., Vijayan, R., & Venmathi Maran, B. A. (2025c). The avifauna of Ramanathapuram, Tamil Nadu along the southeast coast of India: Waterbird assessments and conservation implications across key sanctuaries and Ramsar sites. *PeerJ*, 13, e18899. https://doi.org/10.7717/peerj.18899

Byju, H., Maitreyi, H., Aarif, K. M., Rubeena, K. A., & Ravichandran, S. (2025d). Diverging roles of natural and artificial lagoons in supporting shorebird communities in the Gulf of Mannar, southeast coast of India. *Journal of Coastal Conservation*, 29(64), 1–15. https://doi.org/10.1007/s11852-025-01163-5

Callaghan, C. T., Nakagawa, S., Mahony, M., & Cornwell, W. K. (2018). Avian monitoring: Comparing structured and unstructured citizen science. *Wildlife Research*, 45(2), 176–184. https://doi.org/10.1071/WR17141

Cao, L., Deng, X., Meng, F., & Fox, A. D. (2020). Defining flyways, discerning population trends and assessing conservation challenges of key Far East Asian Anatidae species: An introduction. *Wildfowl*, 6, 1–12.

Cheke, A. S., & Hume, J. P. (2008). *Lost Land of the Dodo: An Ecological History of Mauritius, Réunion & Rodrigues*. A&C Black.

Convention on Migratory Species (CMS). (2025, October 20). *Convention on Migratory Species*. https://www.cms.int/species

Cox, A., Tremblay, F., & Bianchini, K. (2023). *Population Status of Migratory Game Birds in Canada 2023* (CWS Migratory Birds Regulatory Report No. 58). Canadian Wildlife Service.

Cranswick, P. A., & Hall, C. (2010). *Eradication of the Ruddy Duck (Oxyura jamaicensis) in the Western Palaearctic: A Review of Progress and a Revised Action Plan 2010–2015*. WWT Report to the Bern Convention.

Dalby, L., McGill, B. J., Fox, A. D., & Svenning, J.-C. (2014). Seasonality drives global-scale diversity patterns in Anatidae (Anseriformes) via temporal niche exploitation. *Global Ecology and Biogeography*, 23(5), 550–562. https://doi.org/10.1111/geb.12152

Davidson, N. C. (2014). How much wetland has the world lost? Long-term and recent trends in global wetland area. *Marine and Freshwater Research*, 65(10), 934–941. https://doi.org/10.1071/MF14173

Davidson, N. C., & Stroud, D. A. (2006). African–Western Eurasian flyways: Current knowledge, population status and future challenges. *In:* G. C. Boere, C. A. Galbraith, & D. A. Stroud (Eds.), *Waterbirds Around the World* (pp. 63–73). The Stationery Office.

Drilling, N., Williams, S. O., III, Titman, R. D., & McKinney, F. (2020). Mexican Duck (*Anas diazi*), version 1.0. *In:* P. G. Rodewald & B. K. Keeney (Eds.), *Birds of the World.* Cornell Lab of Ornithology. https://doi.org/10.2173/bow.mexduc.01

Gaston, K. J., & Fuller, R. A. (2009). The sizes of species' geographic ranges. *Journal of Applied Ecology*, 46(1), 1–9. https://doi.org/10.1111/j.1365-2664.2008.01695.x

Gill, F., Donsker, D., & Rasmussen, P. (Eds.). (2025). *IOC World Bird List* (v15.1). https://www.worldbirdnames.org

Green, A. J., & Elmberg, J. (2014). Ecosystem services provided by waterbirds. *Biological Reviews*, 89(1), 105–122. https://doi.org/10.1111/brv.12045

Henderson, I., & Hart, A. (2023). The Egyptian goose (*Alopochen aegyptiaca*) in Britain: A review of its status and ecology. *British Birds*, 116(2), 84–101.

Hume, J. P. (2017a). A high price to pay: New light on the extinction of the pink-headed duck (*Rhodonessa caryophyllacea*). *Forktail*, 33, 56–63.

Hume, J. P. (2017b). *Extinct Birds* (2nd ed.). Bloomsbury Natural History.

International Union for Conservation of Nature (IUCN). (2025). *The IUCN Red List of Threatened Species.* https://www.iucnredlist.org

Jetz, W., Thomas, G. H., Joy, J. B., Hartmann, K., & Mooers, A. O. (2012). The global diversity of birds in space and time. *Nature*, 491(7424), 444–448. https://doi.org/10.1038/nature11631

Jia, Q., Koyama, K., Choi, C. Y., Kim, H. J., Cao, L., Gao, D. et al. (2016). Population estimates and geographical distributions of swans and geese in East Asia based on counts during the non-breeding season. *Bird Conservation International*, 26, 397–417.

Johnsgard, P. A. (2010a). Ducks, geese, and swans of the world: Tribe Merganettini (Torrent Duck). *In: Ducks, Geese, and Swans of the World.* University of Nebraska–Lincoln.

Johnsgard, P. A. (2010b). Waterfowl of North America: Hunting and recreational values of North American waterfowl. *In: Waterfowl of North America* (Rev. ed.). University of Nebraska–Lincoln.

Kear, J. (Ed.). (2005). *Ducks, Geese and Swans* (Vol. 1). Oxford University Press.

Lane, S. J., Azuma, A., & Higuchi, H. (1998). Wildfowl damage to agriculture in Japan. *Agriculture, Ecosystems & Environment*, 70(1), 69–77. https://doi.org/10.1016/S0167-8809(98)00121-5

Lefebvre, J., Gauthier, G., Giroux, J.-F., Reed, A., Reed, E. T., & Bélanger, L. (2017). The greater snow goose (*Anser caerulescens atlanticus*): Managing an overabundant population. *Ambio*, 46(Suppl. 2), 262–274. https://doi.org/10.1007/s13280-016-0883-3

Lukasiewicz, M., Matuszewski, A., Florczuk Kolomyja, P., Kamaszewski, M., & Wardecki, L. (2018). Selected invasive species of the Polish and European avifaunae. *Annals of Warsaw University of Life Sciences – SGGW, Animal Science*, 57(1), 47–56.

Maitreyi H. (2024). State of India's birds report – A brief review. *Zoo's Print*, 39(1), 119–124.

Marchowski, D., Jurszo, R., Stańczak, P., Jasiński, M., & Guentzel, S. (2025). Non-selective waterbird hunting in a Natura 2000 site results in killing of protected species: A case study from western Poland. *Avian Research*, 16, 100276. https://doi.org/10.1016/j.avrs.2025.100276

Martinico, B. L., Busch, R., Maier, G., & Doran, M. (2024). Economic impacts of wintering geese on agricultural operations in the Sacramento Valley of California. *In: Proceedings of the Vertebrate Pest Conference* (Vol. 31, No. 31). https://doi.org/10.5070/V431110510

Marshall, A. P., Banko, P. C., Misajon, K., & Hu, D. (2008). Reintroduction of the Nēnē, a remote island goose, Hawaii, USA. *In:* P. S. Soorae (Ed.), *Global Reintroduction*

Perspectives: Reintroduction Case-Studies from Around the Globe (p. 96). IUCN/SSC Re-introduction Specialist Group.

Mattsson, B. J., Dubovsky, J. A., Thogmartin, W. E., Bagstad, K. J., Goldstein, J. H., Loomis, J. B. et al. (2018). Recreation economics to inform migratory species conservation: Case study of the Northern Pintail. *Journal of Environmental Management*, 206, 971–979. https://doi.org/10.1016/j.jenvman.2017.11.072

Mercado-Reyes, M., Valtierra-Marín, H. E., Delgadillo-Ruiz, L., Valdéz-Romero, E., Chávez-Ruvalcaba, M. I., Chávez-Ruvalcaba, F. et al. (2024). Characteristics of the habitat and population densities of the Mexican duck (*Anas diazi*) in the Plateau of Zacatecas, Mexico. *Birds*, 5(3), 509–521. https://doi.org/10.3390/birds5030038

Miyabayashi, Y., & Mundkur, T. (1999). *Atlas of Key Sites for Anatidae in the East Asian Flyway.* Wetlands International – Japan, Tokyo, and Wetlands International – Asia Pacific, Kuala Lumpur.

Mowbray, T. B., Cooke, F., & Ganter, B. (2020). Snow Goose (*Anser caerulescens*), version 1.0. *In:* P. G. Rodewald (Ed.), *Birds of the World.* Cornell Lab of Ornithology, Ithaca, NY. https://doi.org/10.2173/bow.snogoo.01

Mundkur, T., Ananzeh, A., Chaudhary, A., Evans, M., Jia, Y., Koshkina, A. et al. (2023). *Central Asian Flyway – Situation Analysis: The Status of Migratory Birds and Their Habitats and Recommendations for Their Conservation.* BirdLife International, Cambridge.

Mundkur, T., & Langendoen, T. (Eds.). (2016). *The Asian Waterbird Census 2008–2015: Results of Coordinated Counts in Asia and Australasia* (Draft consultation report). Wetlands International, Ede.

Mundkur, T., & Langendoen, T. (2022). *Report on the Conservation Status of Migratory Waterbirds of the East Asian–Australasian Flyway* (1st ed.). Report to the East Asian–Australasian Flyway Partnership. Wetlands International, Ede, The Netherlands. https://www.wetlands.org/eaaf-conservation-status-review1/

Nagy, S., Flink, S., & Langendoen, T. (2015). *Report on the Conservation Status of Migratory Waterbirds in the Agreement Area* (6th ed.). Agreement on the Conservation of African–Eurasian Migratory Waterbirds (AEWA). Wetlands International.

Navedo, J. G., Masero, J. A., Sánchez-Guzmán, J. M., Abad-Gómez, J. M., Gutiérrez, J. S., Sansón, E. G. et al. (2012). International importance of Extremadura, Spain, for overwintering migratory dabbling ducks: A role for reservoirs. *Bird Conservation International*, 22(3), 316–327. https://doi.org/10.1017/S0959270911000311

Niemuth, N. D., & Solberg, J. W. (2003). Response of waterbirds to number of wetlands in the Prairie Pothole Region of North Dakota, USA. *Waterbirds*, 26(2), 233–238. https://doi.org/10.1675/1524-4695(2003)026[0233:ROWTNO]2.0.CO;2

Nourani, E., Kaboli, M., & Collen, B. (2015). An assessment of threats to Anatidae in Iran. *Bird Conservation International*, 25(2), 242–257. https://doi.org/10.1017/S0959270914000304

Primack, R. B. (2014). *Essentials of Conservation Biology* (6th ed.). Sinauer Associates.

Rajpar, M. N., Ahmad, S., Zakaria, M., Ahmad, A., Guo, X., Nabi, G. et al. (2022). Artificial wetlands as alternative habitat for a wide range of waterbird species. *Ecological Indicators*, 138, 108855. https://doi.org/10.1016/j.ecolind.2022.108855

Rashiba, A. P., Jishnu, K., Byju, H., Shifa, C. T., Anand, J., Vichithra, K. et al. (2022). The paradox of shorebird diversity and abundance on the west and east coasts of India: A comparative analysis. *Diversity*, 14(10), 885. https://doi.org/10.3390/d14100885

Rasoamihaingo, L. A., Razafindrajao, F., De Roland, L. A. R., & Bamford, A. J. (2024). A decade of monitoring the critically endangered Madagascar pochard (*Aythya innotata*): Population trends and duckling survival. *Wildfowl*, 153–164.

Reid, J. R. W., Colloff, M. J., Arthur, A. D., & McGinness, H. M. (2013). Influence of catchment condition and water resource development on waterbird assemblages in the Murray–Darling Basin, Australia. *Biological Conservation*, 165, 25–34. https://doi.org/10.1016/j.biocon.2013.05.009

Reynolds, M. H., Courtot, K. N., Brinck, K. W., Rehkemper, C. L., & Hatfield, J. S. (2015). Long-term monitoring of endangered Laysan Ducks: Index validation and population estimates 1998–2012. *Journal of Fish and Wildlife Management*, 6(1), 92–101. https://doi.org/10.3996/072014-JFWM-050

Sauer, J. R., Link, W. A., Fallon, J. E., Pardieck, K. L., & Ziolkowski Jr, D. J. (2013). The North American breeding bird survey 1966–2011: Summary analysis and species accounts. *North American Fauna*, 79, 1–32. https://doi.org/10.3996/nafa.79.0001

Scher, C. L., & Clark, J. S. (2023). Species traits and observer behaviors that bias data assimilation and how to accommodate them. *Ecological Applications*, 33(3), e2815. https://doi.org/10.1002/eap.2815

Shi, J., Meng, L., Xia, S., Liu, S., & Zhou, L. (2024). Habitat suitability and determinants for anatidae in multi-watershed composite wetlands in Anhui, China. *Animals*, 14(7), 1010. https://doi.org/10.3390/ani14071010

Sullivan, B. L., Aycrigg, J. L., Barry, J. H., Bonney, R. E., Bruns, N., Cooper, C. B., ... & Kelling, S. (2014). The eBird enterprise: An integrated approach to development and application of citizen science. *Biological Conservation*, 169, 31–40. https://doi.org/10.1016/j.biocon.2013.11.003

Sutherland, W. J., Adams, W. M., Aronson, R. B., Aveling, R., Blackburn, T., Broad, S., ... & Watkinson, A. R. (2009). One hundred questions of importance to the conservation of global biological diversity. *Conservation Biology*, 23(3), 557–567. https://doi.org/10.1111/j.1523-1739.2009.01212.x

Sutherland, W. J., Adams, W. M., Aronson, R. B., Aveling, R., Blackburn, T., Broad, S., ... & Watkinson, A. R. (2009). One hundred questions of importance to the conservation of global biological diversity. Conservation biology, 23(3), 557-567.

Tennyson, A. J. D., & Martinson, P. (2006). *Extinct Birds of New Zealand.* Te Papa Press.

U.S. Fish and Wildlife Service. (2023). *Waterfowl Population Status, 2023.* U.S. Department of the Interior, Washington, D.C.

U.S. Fish and Wildlife Service. (2024). *Waterfowl Population Status, 2024.* U.S. Department of the Interior. https://www.fws.gov/media/waterfowl-population-status-2024

Wang, X., Barter, M., Cao, L., Lei, J., & Fox, A. D. (2012). Serious contractions in wintering distribution and decline in abundance of Baer's pochard (*Aythya baeri*). *Bird Conservation International*, 22(2), 121–127. https://doi.org/10.1017/S0959270911000135

Warner, R. E. (1963). Recent history and ecology of the Laysan Duck. *The Condor*, 65(1), 3–23. https://doi.org/10.2307/1365231

Wetlands International. (2023). *Waterbird Population Estimates.* https://wpe.wetlands.org

Wetlands International. (2025). *Waterbird Populations Portal.* https://wpp.wetlands.org

Zeng, Q., Reid, J., Saintilan, N., Colloff, M. J., Lei, G., & Wen, L. (2019). Contrasting diversity patterns of breeding Anatidae in the Northern and Southern Hemispheres. *Ecology and Evolution*, 9(17), 9990–10003. https://doi.org/10.1002/ece3.5401

CHAPTER

9

Human–Anatidae Interactions: Historical, Cultural, and Conservation Perspectives

Harsh Singhal[1,2,3*], Abhishek Mishra[1] and Satya Prakash Mehra[2,3]

[1] Bombay Natural History Society, Mumbai, Maharashtra, India
[2] Global Forum for Sustainable Rural Development, Birbhum, West Bengal, India
[3] Rajputana Society of Natural History, Bharatpur, Rajasthan, India

1. Introduction

Anatidae family are among the most globally distributed avian taxa, occupying wetlands across all continents except Antarctica (Green, 1996). Their migratory movements connect nations, ecosystems, and cultures, making human–Anatidae interactions inherently transboundary and globally relevant (AEWA, 2012). Consequently, pressures affecting Anatidae, such as wetland loss, climate change, agricultural intensification, and disease emergence, must be understood within an international socio-ecological context (Ayanlade et al., 2020; Liang et al., 2021). It is prominent in the avian order Anseriformes (Olson & Feduccia, 1980). As grazers, filter feeders, and bioindicators of wetland health, the roughly 170 species of the Anatidae family are found throughout the world in both freshwater and coastal habitats. They are both ecologically and culturally important because of their reproductive strategies, migratory patterns, and physiological adaptations that enable them to flourish in temperate, tropical, and polar environments (McKinney & Evarts, 1998).

Species in the Anatidae family have long been woven into the fabric of human history, regardless of their biological significance (Forshaw, 1991). The relationship between humans and the Anatidae has changed over centuries across different regions and belief systems, from the ritual offerings of ancient Egypt to

*Corresponding author: hsinghal1200@gmail.com

DOI: 10.1201/9781003669784-9

the domestication of ducks in East Asia and the folkloric symbolism of swans in European mythology (Atkinson, 1993; Andreu, 1997; Aragon, 2006; Coombes, 2007). These birds had undergone domestication over time, from being food, while game hunters pursued them, until the days when they were worshipped as sacred entities. Today, the birds are all respected for photo opportunities and are advertised as symbols of wildlife conservation and ecotourism (Anatidae, 2023). But as landscape change, biodiversity loss, and changing policy frameworks increasingly impact traditional relationships, this Anthropocene era adds yet another layer of complexity to these interactions.

Therefore, studying human–Anatidae interactions offers a globally relevant lens for understanding broader socio-ecological dynamics, particularly in the context of shared migratory systems, international conservation governance, and cross-cultural environmental ethics (Fuentes, 2010; Bhatia et al., 2019). Anatidae are a prominent focus for examining the co-evolution of human cultures and natural ecosystems as the ecological lines separating urban and rural, wild and domesticated, and sacred and utilitarian become increasingly hazy. The need to rethink human–wildlife relationships as overlapping continua shaped by shared histories, mutual dependencies, and emerging challenges are highlighted by recent interdisciplinary scholarship (Fuentes, 2010; Descola, 2013; Bhatia et al., 2019). An excellent group to investigate such questions is the Anatidae family.

Growing environmental pressures make this exploration even more urgent. Critical habitats for Anatidae have been severely reduced due to the estimated 22% global loss of wetlands since 1970, especially for migratory species that depend on interconnected flyways (Davidson, 2014). The timing of migrations, breeding cycles, and species distributions have all been further disturbed by climate change (Ayanlade et al., 2020; Liang et al., 2021). Traditional nesting and feeding grounds are still fragmented or eliminated by factors such as pollution, urban encroachment, dam construction, and intensified agriculture (Anon, 2025). Anatidae species are now found near human-dominated landscapes in many places, which has led to new kinds of coexistence but also sparked disputes over air-traffic hazards, zoonotic disease risks, and crop damage (Mehra, 2019; Otte, 2025).

On the other hand, some species in the family Anatidae have proven highly adaptable, thriving in artificial environments such as city parks, reservoirs, and farmlands. This flexibility has promoted new types of participation, including community conservation initiatives, ecotourism, and citizen science projects (Mehra et al., 2011a, b; Mehra et al., 2012; Mehra & Mehra, 2013; Mehra et al., 2014; Mehra & Mehra, 2014; Upadhyay et al., 2020; Morgan, 2025). In some cases, indigenous knowledge systems—particularly rural and Indigenous ones—still inform sustainable relationships and provide valuable information for contemporary conservation strategies (Mehra et al., 2012; Mills, 2025). However, such heartening stories are often submerged by laws that prioritize human convenience or safety over environmental integrity, leading to management practices such as habitat displacement, culling, or commercial use (Mehra et al., 2012; Mills, 2025).

To explore the conflicts, evolution, and opportunities that occur in human–Anatidae interactions, this chapter situates them within a comprehensive temporal and disciplinary context. It begins by considering how respect and use of Anatidae have developed in various cultures, demonstrating how these birds have been integrated into social, religious, and commercial organizations. The chapter then explores the cultural diversity and symbolic meaning of the Anatidae species using ethnographic, archeological, and artistic records. Understanding how cultural perceptions have influenced and, in some cases, justified human behaviors toward these birds is based on these insights.

The book turns to recent ecological problems and conservation measures in the latter half of the chapter. Along with examining how the trends concern land-use policy, human livelihoods, and biosecurity issues, it integrates recent scientific observations on habitat alteration, migration routes, and population dynamics among the Anatidae family. Interdisciplinary science, community engagement, and new policy are especially emphasized as essential for making more equitable and effective conservation measures.

To illustrate the diversity of human–Anatidae interactions and their distinct contextual dynamics, this study is founded on case studies from diverse geographic contexts, including rice paddies in Southeast Asia, urban wetlands in India, protected migratory corridors in Europe, and traditional hunting grounds in North America. These examples show that no single model applies to coexistence; instead, it arises from the negotiation of competing interests, cultural meanings, and environmental constraints.

Ultimately, this chapter argues that interactions between humans and the Anatidae reveal both the potential and the risks of our contact with nature. These birds have the potential to coexist with other species, but they are also susceptible to the unforeseen effects of human ambition and disregard. Paradigms of conservation, to qualify as such, must recognize the full potential of being both operators in ecosystems and contributors to human society, thus conserving ethically based agents in good conscience.

This chapter contributes to the broader conversation on biodiversity conservation in an increasingly human-centric world by analyzing the evolution of human–Anatidae interactions across time and space. It pushes readers to reconsider their presumptions about wildlife, customs, and the future of shared habitats. These readers include scientists, conservationists, policymakers, and the general public. By doing this, it offers a critical analysis of a particular bird family and a broader appeal to approach conservation as a moral, political, and cultural endeavor.

Accordingly, this chapter adopts a comparative global framework, using regional case studies not as isolated narratives but as interconnected examples of how human societies worldwide negotiate coexistence with Anatidae under changing environmental conditions.

2. Historical Relationships

Historical interactions between humans and Anatidae reveal globally recurring patterns of domestication, ritual use, and symbolic incorporation (Johnsgard, 1975). Despite geographic separation, societies across Eurasia, Africa, and the Americas developed parallel relationships with ducks and geese, reflecting shared ecological opportunities and subsistence needs. The historical relationship between humans and the Anatidae family dates back to ancient times and reflects globally recurring patterns of subsistence, symbolism, and management (Carboneras, 1992; Kear, 2005). This interaction had very important consequences for human subsistence practices and cultural narratives, with some Anatidae species apparently being shaped in their distribution, behavior, and even evolution (Cottam, 1939; Johnsgard, 1975; Jain, 2011). The exploration of this link unravels how Anatidae have historically found their way into human economies and ways of seeing the world, often as concrete-based resources and, at times, as symbols of metaphysics.

2.1. Early Domestication and Subsistence Use

Domestication of waterfowl can be traced back to several millennia; most likely, it occurred independently in different regions (Zhang et al., 2025). The domestic duck (*Anas platyrhynchos domesticus*) is thought to have been first domesticated in Southeast Asia between 2,000 and 4,000 years ago, primarily derived from the wild mallard (*Anas platyrhynchos*) (Kear, 2005). Ducks were integrated into small-scale farming systems, providing meat, eggs, and feathers. In China, ducks became very important for wet rice agriculture inasmuch as they were linked with food systems, as well as weed and pest control in the paddy fields (Hou, 1982).

The greylag goose (*Anser anser*), recognized as the ancestor of most domesticated goose breeds, was domesticated in ancient Europe and the Near East (Buckland & Guy, 2002). Evidence from Egyptian tomb paintings dating back to the Middle Kingdom (approximately 2000 BCE) illustrates the herding and fattening of geese, highlighting their significance in the diets of the elite and as offerings in temples (Houlihan, 1988). Feathers were collected for insulation and bedding, while in some cultures, goose grease was utilized for medicinal purposes (Buckland & Guy, 2002).

Anatidae species also played a crucial role in the subsistence hunting practices of Indigenous communities across North America, Siberia, and Northern Europe (Rowland, 1978; Moguel & Toledo, 1999). In these regions, migratory ducks and geese provided essential protein during periods of resource scarcity (Buckland & Guy, 2002). The native Arctic peoples, such as the Inuit and Chukchi, developed complex hunting methods, using decoys and camouflage to lure waterfowl at their migratory stopovers (Gilchrist et al., 2005). For many of these peoples, the ducks were not only used for sustenance; rather, they were key in sustaining the natural cycles of the ecosystem, with which they practiced their communal hunting.

2.2. Symbolic and Ritualistic Associations

When you look beyond their everyday roles, species from the Anatidae family carry deep symbolic and spiritual significance in many ancient cultures. For instance, in East Asian cosmology, ducks are seen as symbols of marital fidelity and harmony (Eyton, 1838). The mandarin duck (*Aix galericulata*), in particular, has come to embody love and the idea of a lifelong partnership (Beames, 1969). Traditionally, pairs of duck figurines were gifted to couples at Chinese weddings to bless their union with harmony and balance (Ebrey, 1991). The harmonious pairing behavior of ducks served as an archetype for conjugal loyalty in Confucian and Daoist beliefs.

In Ancient Egyptian culture, geese were closely linked to creation myths and fertility (Siegfried, 1962). The goose was seen as a physical representation of the creator god Geb, who was thought to have laid the cosmic egg that gave rise to the Sun god Ra. In funerary rites and temple ceremonies, offerings of geese were quite common, symbolizing the hope for rebirth and plenty (Houlihan, 1988). Additionally, as there were mummified ducks found in tombs, it implies that their flesh was considered good food to sustain someone in the afterlife (Andreu et al., 1997).

In the Greco-Roman world, swans and geese stood at the forefront of popular imagination. They associated their swans with the gods of Apollo and Aphrodite, and from there came the notions of beauty, inspiration, and artistic transcendence (Smith & Trzaskoma, 2022). The tale of Leda and the Swan, where Zeus transforms into a swan, showcases intricate symbolism between species that touches on themes of power and seduction (British Museum, 2021). Romans celebrated the sacred geese of Juno, who were believed to have saved the city during the Gallic siege of 390 BCE by alerting the guards to an impending night attack—an early example of the Anatidae serving as protectors or omens (Toynbee, 1973).

In both Celtic and Germanic traditions, swans often symbolize themes of change and transition, featuring prominently in myths about shape-shifting and journeys to other realms (Smith & Trzaskoma, 2022). A well-known story from Irish folklore, the "*Children of Lir*," tells of royal siblings who are cursed to live as swans, doomed to roam the lakes of Ireland for centuries. These stories tell of the deep connection of swans to sorrow, endurance, and mystical journeys (MacKillop, 1998).

In Indigenous North American traditions, especially among the Anishinaabe and other Algonquian-speaking cultures, the migration of the geese indicates important seasonal cycles and spiritual passage (Mowbray et al., 2003). Their calls are often interpreted as messages from the spirit world, while feathers are utilized in ceremonial practices. Unlike many Western paradigms that view humans as separate from animals, these Indigenous cosmologies regard the Anatidae family as kin and guides (McLandress & McLandress, 1979; Scribner, 2003).

2.3. Literary, Mythological, and Artistic Representations

Anatidae have played significant roles in literature, music, and art throughout various civilizations. Their esthetic appeal and graceful behavior have made them popular subjects in classical poetry, religious iconography, and children's literature (Gangoly, 1963; Sivaramamurti, 1980; Miller, 1983; Raja Dinkar Kelkar Museum, 1988; Jain, 2011). In Hindu iconography, the *Humsa* (often depicted as a swan or goose) represents discrimination and spiritual purity. Saraswati, the goddess of wisdom and art, is often shown riding a *Humsa*, symbolizing transcendental knowledge and the delicate balance between the material and spiritual worlds (Kinsley, 1987).

In Western literary tradition, swans are often seen as symbols of poetic inspiration and noble silence. The phrase "swan song" originates from the ancient belief that swans sing a beautiful melody before their death, a notion embraced by Greek tragedians and later poets such as Tennyson and Yeats. Furthermore, ducks and geese are featured in European fairy tales and fables—most notably in Hans Christian Andersen's The Ugly Duckling, which uses the transformation from a duckling to a swan as a metaphor for self-realization and inner beauty.

Visual art spanning from the medieval to the modern periods frequently portrays Anatidae in both pastoral and allegorical contexts. Ducks also populate water gardens in Mughal miniatures, where their symbolism meanders between sensuality and the more concrete plane of esthetic beauty. In a similar fashion, Japanese and Korean ink paintings always employ pairs of mandarin ducks amid lotus blossoms to represent harmony and prosperity.

The use of Anatidae feathers or effigies in Native American regalia, perhaps as well as shamanic practices in Siberia and animistic traditions in Siberian and Mongolian cultures, underlines the global and cross-cultural presence of these birds in the ranch of ritual life (Vitebsky, 2005). The symbolism here becomes an issue of cosmological mediation and socio-religious expression rather than just mere decoration.

The history of human–Anatidae relations reveals a dynamic interplay that goes far beyond matters of utility and symbolism. These birds have sustained human bodies while also nourishing minds and spirits—shaping cultural metaphors, seasonal calendars, and ethical frameworks. While later sections of this chapter will touch upon the contemporary ecological pressures and conservation challenges that ran counter to such historical bonds, it must first be acknowledged that contemporary interactions with the Anatidae are rooted deeply in those rich and manifold legacies.

3. Cultural Significance Across Regions

Across cultures, Anatidae have served as powerful symbolic agents, representing fertility, fidelity, transition, and moral order (Lyver & Moller, 2010; Descola, 2013). While these meanings vary regionally, their recurrence across continents suggests convergent human responses to wetland-dependent life forms. The Anatidae family has not only fulfilled various ecological roles but has also deeply influenced the cultural frameworks of human societies worldwide. Ducks, geese, and swans have become iconic species within traditional knowledge systems, rituals, folklore, and social symbolism (Kear, 2005). The cultural meanings attributed to these birds differ across geographical and historical contexts, reflecting local ecologies, spiritual beliefs, and the relationships between humans and wildlife. This section presents a comparative overview of the symbolism and representation of Anatidae in select regions—South and East Asia, Europe, and Indigenous cultures of North America—emphasizing how these birds serve as cultural indicators and agents of meaning-making.

3.1. South Asia: Ritual Ecology and Symbolic Integration

In the Indian subcontinent, Anatidae's symbolic meaning can be found in relation to fauna, which is interlaced with Indian art and philosophy (Aragon, 2006). The *Humsa*, which is normally considered as a singing bird or swan, has importance in Hinduism, Buddhism, and Jainism. Unlike its Western counterparts, which are more rigidly defined biologically, the *Humsa* in Indian thought serves as a metaphysical symbol embodying the soul (*ātman*), purity, and discernment (Yeomans & Mazzucato, 2024). The Bhagavata Purana describes the *Humsa* as having the unique ability to separate milk from water—metaphorically representing the spiritual capacity to distinguish the eternal from the transient (Kinsley, 1987).

Saraswati, the goddess of learning, speech, and music, often riding a *Humsa*, is depicted as a spiritual guide, highlighting the role of a bird and symbol of transcendent wisdom (Bhatt, 2010). In rural folklore and seasonal festivals throughout eastern and northern India, geese and ducks are closely associated with the monsoon cycle and fertility, particularly in agricultural communities, where these wetland birds signal ecological renewal (Bhatt, 2010; Mehra et al., 2017).

Furthermore, traditional Indian miniature paintings—especially in the Mughal and Rajput styles—often illustrate pairs of ducks swimming together in lotus-laden water bodies, symbolizing romantic harmony and sensory balance (Losty, 1982). This integration of Anatidae into both high art and rural traditions suggests a distinctive form of "ritual ecology," where the biological traits of these birds are harmonized with spiritual and esthetic values. Comparable symbolic roles of Anatidae are documented in other world regions, highlighting cross-cultural continuity in human–waterfowl relationships.

3.2. East Asia: Marital Symbols and Harmonious Coexistence

East Asian cultures, especially in China and Korea, highly honor the species of the Anatidae family as a symbol of loyalty and marital harmony, especially mandarin ducks (*Aix galericulata*). Mandarin ducks' remarkable pairing behavior and tendency to have a mate to live with have contributed to their cultural significance as an example of ideal relationships (Ebrey, 1991). During traditional Chinese weddings, wooden or ceramic representations of mandarin ducks are often presented to couples to bless their union with loyalty and affection (Hunn & Thornton, 2010). Such symbolic associations continue to thrive in popular culture, wedding décor, and visual art.

East Asian ink paintings frequently portray mandarin ducks amid reeds or lotus blossoms—settings that symbolize peace, natural abundance, and mutual understanding. Within Confucian ethics, which prioritize family stability as essential to social order, the duck transcends its status as a biological species and emerges as a moral archetype (Li, 1999).

In Japanese Shinto traditions, ducks also play a ceremonial role in agricultural rituals. Seasonal observances that involve offerings of duck feathers or effigies are designed to ensure bountiful rice harvests, linking the Anatidae family directly to food security and ecological harmony (Moeran, 1984). These examples highlight the interconnectedness of Anatidae with seasonal cycles, ethical ideals, and community rituals within East Asian cultural systems. Comparable symbolic roles of Anatidae are documented in other world regions, highlighting cross-cultural continuity in human–waterfowl relationships.

3.3. Europe: Symbolism, Myth, and Class Representation

European cultural narratives have long integrated Anatidae, particularly swans and geese, as metaphors within myth, literature, and folklore (Irwin, 1984; Lyver & Moller, 2010). In classical antiquity, swans were associated with the Greek god Apollo and were thought to sing just before death—a motif that inspired the enduring metaphor of the "swan song," symbolizing final creative expression before demise (Toynbee, 1973).

Roman accounts celebrated the geese of Juno as protectors of the Capitoline Hill during the Gallic siege, embedding these birds into Roman civic mythology as sentinels and guardians (Morford et al., 2019). Swans and geese also appeared in medieval European bestiaries, serving as moral allegories. For example, the swan's pure white feathers were likened to the soul's innocence, while its dark skin beneath symbolized the hidden sinfulness of humanity—illustrating how birds were employed in religious instruction (Flores, 1996).

In subsequent centuries, the European aristocracy adopted swans as symbols of nobility and esthetic grace. Private estates and royal parks often featured ornamental ponds stocked with mute swans (*Cygnus olor*), which became markers of elite taste and territorial claims (Thomas, 1983). This esthetic appreciation of Anatidae continues today in urban park design and public art throughout Europe.

Meanwhile, fairy tales such as Hans Christian Andersen's The Ugly Duckling revive the duck–swan dichotomy as a story of personal change, flexibility, and hidden ability. This story is resonant globally and has shaped the notion of swans as aspirational and redemptive figures. Comparable symbolic roles of Anatidae are documented in other world regions, highlighting cross-cultural continuity in human–waterfowl relationships.

3.4. Indigenous North America: Ecological Timekeepers and Sacred Messengers

Below North America, geese and ducks are seen by several Indigenous groups, especially those who speak Algonquian and Athabaskan languages, not simply as game animals but as elements of the ecological and spiritual calendars. Geese's spring and fall migrations are deeply integrated into Indigenous cosmologies as signs of seasonal change and spiritual thresholds (Nelson, 1983; Brown, 1992; Bonta, 2010; Hunn & Thornton, 2010).

For example, the Ojibwe and Cree nations interpret the migratory return of geese as a renewal and a guarantee of nourishment after winter. Ceremonial dinners and shared hunts are initiated to celebrate this return as an expression of thanks and respect for these bird entities. Geese and duck feathers are used in regalia, healing ceremonies, and storytelling traditions and are seen as channels of prayer and communication to the spiritual realm (Beck et al., 1990).

Essentially, these societies tend to view Anatidae not only as biological entities but also as family members with consciousness and agency. This ontology often differs from Western categorizations and suggests a protection morality that favors mutuality, interpretation, and co-stewardship of habitats (Watson, 2015). Comparable symbolic roles of Anatidae are documented in other world regions, highlighting cross-cultural continuity in human–waterfowl relationships.

3.5. Contemporary Cultural Continuities

While modernity has transformed many traditional relationships with Anatidae, cultural associations continue to endure and adapt. In urban India, for instance, ducks in temple ponds are still regarded as symbols of good fortune. Similarly, in various regions of China, duck feather products remain deeply connected to heritage practices. In North America, indigenous youth programs are rapidly gaining scientific recognition and are incorporating traditional knowledge of waterfowl, demonstrating the capacity for biocultural protection (Reo & Whyte, 2012).

Additionally, the global appeal of waterfowl in children's literature, ecotourism, and wildlife photography has made their symbolic importance more comprehensive. Ducks, geese, and swans are often featured in educational and emotional narratives, enabling their cultural relevance to resonate across various age groups, subjects, and geographical areas.

4. Changing Landscapes and Interactions

The relationships between humans and Anatidae have undergone significant changes in recent decades, driven rapidly by ecological, social, and technological changes. These shifts have increased both the proximity and complexity of interactions between various Anatidae species and human societies. While earlier human–Anatidae relationships were often shaped by cultural traditions and seasonal cycles, contemporary interactions are increasingly influenced by land-use changes, habitat fragmentation, decreasing water level, industrial agriculture, climate change, and urbanization (Izadian & Tayefeh, 2024). These factors not only reshape the ecological niche of ducks, geese, and swans but also modify human perceptions, practices, and management reactions.

4.1. Habitat Loss and Wetland Degradation

The primary breeding habitats of several Anatidae species are wetlands, where they obtain vital water resources for feeding, breeding, and migration. However, there has been a drastic decline in the area of the world's wetlands. At a global scale, approximately 22% of wetlands have been lost since 1970, with severe consequences for Anatidae populations across major migratory flyways, where wetland-dependent species have declined by over 80% (Davidson, 2014; Ramsar Convention Secretariat, 2025). This loss is primarily caused by agricultural drainage, urbanization, and infrastructure development (Ramsar Convention Secretariat, 2018; 2025; Mehra, 2019). In Asia, where many migratory waterfowl depend on the East Asian–Australasian flyway, wetland loss is severe. For instance, in the Yangtze River basin, drainage and conversion of floodplains and lakes to agricultural land have negatively affected wintering numbers of species such as the lesser white-fronted goose (*Anser erythropus*) and Baer's pochard (*Aythya baeri*) (Wang et al., 2017).

In India, as urbanization and pollution took their toll on the wetlands, it directly impacted the Anatidae, both migratory and resident, leading to habitat loss. Among the wetlands, Najafgarh Lake in Delhi and Pallikaranai in Chennai were important habitats for waterfowl that no longer exist; they were drastically reduced through landfilling, sewage discharge, and land reclamation (Thirunavukkarasu, 2022; Bhanot & Chatterjee, 2023). Shrinkage of the wetlands has decreased biodiversity and affected traditional interactions among communities and waterbirds, several of which traditionally included subsistence fishing, hunting, or seasonal celebrations (Mehra et al., 2014).

4.2. Agricultural Intensification and Land-Use Change

Historically, the growth of agriculture has also supported and interrupted the freedom of the Anatidae population. There existed, however, a certain classification of agricultural and flood irrigation activities that provided suitable habitats for several species of ducks and geese; however, the aggressive forms of agricultural enhancement, including pesticides and mechanization, have rendered such options unsuitable for wildlife. Changes in industrial agricultural practices from traditional agricultural systems have reduced water quality, food availability, and nesting cover for Anatidae (Vickery et al., 2004).

Furthermore, the application of pesticides and herbicides has been linked to declines in aquatic invertebrates and plant species that are essential to the diets of Anatidae. Cases of secondary poisoning, resulting from the ingestion of pesticide-contaminated plants or prey, have been reported, particularly in areas of intensive agriculture in China and India (Parsons et al., 2010).

Land conversion because of oil palm (*Arecaceae*) plantations in Southeast Asia and soybean (*Glycine max*) production in South America has resulted in the destruction of significant wetland complexes, adversely impacting both resident species and migratory populations. In Europe, modifications to agri-environment schemes and subsidies under the Common Agricultural Policy have led to variable habitat quality in crucial wintering grounds, such as the Camargue in France and Doñana in Spain, thereby affecting populations of greylag geese and Eurasian wigeons (Holopainen et al., 2018).

4.3. Urbanization and Novel Ecosystems

As wetlands get drained and rural landscapes intensify, many Anatidae species transform into urban areas. Urban lakes, parks, golf courses, and man-made wetlands now support secondary habitats for birds such as the mallard (*Anas platyrhynchos*), greylag goose (*Anser anser*), and mute swan (*Cygnus olor*). In most examples, human presence has habituated these species, altering their foraging habits, avoidance of hunters, and reproductive tactics (Marzluff, 2001).

While this flexibility highlights ecological resilience, it also adds new dynamics. Urban populations at high density can lead to human–wildlife conflict, including pollution of water bodies, breeding-season aggression, and destruction of decorative plants. In addition, park-visitor-led artificial feeding can lead to overpopulation and malnutrition among waterfowl and increase the susceptibility to disease transmission (Fuller et al., 2008).

In many cities, Anatidae emerged as politically relevant symbols, winning public affection and civic pride, but also sparking controversies over control. For example, in Vancouver's Stanley Park and Central Park in New York, city geese are subjected to protective activities and killing programs, reflecting the contradictions between nature management and cultural feeling (Jerolmack, 2008).

4.4. Migration Disruption and Climate Change

Anatidae species are most vulnerable to the effects of climate change due to their dependence on migratory corridors and seasonal wetlands. Increased temperatures have brought about earlier springs in northern breeding sites, culminating in phenological mismatches—situations in which birds arrive too early or too late to benefit from the best food supply (Both et al., 2006; Singh et al., 2020; Mehra & Singhal, 2025). This has been noted in Arctic-breeding geese, such as the barnacle goose (*Branta leucopsis*) and the pink-footed goose (*Anser brachyrhynchus*), whose breeding success depends on the timing of plant emergence (Madsen et al., 1999; Madsen, 2016; Madsen & Johnson, 2017).

Modified precipitation patterns, droughts, and shifting river regimes threaten staging and molting habitats. Climate-driven desiccation of lakes such as Lake

Balkhash and Lake Issyk-Kul in Central Asia has affected the migration routes of species such as the ruddy shelduck (*Tadorna ferruginea*) (Seri & Rahman, 2021).

Additionally, rising sea levels present further challenges to coastal wetlands, such as estuaries and salt marshes, utilized by Anatidae during migration or overwintering. For example, in the Bay of Bengal, saline intrusion and erosion are diminishing the extent of Sundarbans mangroves and associated wetlands, adversely affecting the diversity of Anatidae in the region (Mahadevia & Vikas, 2012).

4.5. Conflict, Dependence, and Human–Wildlife Interactions

The evolving landscapes have intensified both reliance on and conflict with Anatidae species in human interactions. In certain agricultural regions, Anatidae are perceived as pests, particularly when large flocks graze on cereal crops or freshly sown fields. Farmers in the Netherlands, the UK, and northern India have reported economic losses due to crop damage caused by geese and ducks (Williams et al., 2013). Compensation schemes and deterrent measures have yielded mixed results and often necessitate careful consideration of ecological, economic, and ethical factors.

On the other hand, specific populations of Anatidae have adapted to survive using anthropogenic sources. For example, landfill and aquaculture ponds have become foraging grounds for the greater white-fronted goose (*Anser albifrons*) and the northern shoveler (*Anas clypeata*) (Kear, 2005). Such dependencies make conservation more complex because management must consider ecological appropriateness and human health or nuisance issues.

Secondly, Anatidae occurrence in peri-urban and rural areas raises concerns about disease transmission, especially avian influenza. Asymptomatic carriers of ducks, in particular, may act as reservoirs of the H5N1 virus and other influenza viruses and pose a risk to poultry farms and, potentially, human populations (Chen et al., 2005). These health threats have led to stepped-up surveillance, biosecurity measures, and, at times, culling, adding complexity to the human–Anatidae relationship.

The landscape transformation due to agriculture, urbanization, and climate change has dramatically altered the ecological and cultural dynamics of human–Anatidae interaction. In this world, some species are showing extreme adaptability, while others are increasingly being exposed. These changes create contradictions in promoting conservation paradigms and interdisciplinary approaches that blend ecological information, societal values, and creative policies. The following section explores these solutions, which place conservation contexts and models of local stewardship under special focus. Because Anatidae migrations routinely cross political boundaries, landscape change and climate-driven disruptions increasingly demand coordinated international monitoring and adaptive management across flyways.

5. Conservation Contexts

Conservation of Anatidae operates primarily within global governance frameworks, as migratory populations require cooperation among nations, institutions, and local stakeholders (AEWA, 2012; Johnson, 2016; Madsen et al., 2017; Ramsar Convention Secretariat, 2018; 2025).

5.1. Protected Areas and Ramsar Sites

Under the Ramsar Convention, a wetland may be designated as a Ramsar Site if it regularly supports 20,000 or more individual waterbirds (Criterion 5) or 1% of the individuals in a population of one species or subspecies of waterbird (Criterion 6), among the nine criteria used for identifying Wetlands of International Importance (Ramsar Convention Secretariat, 2018). Wetlands designated as Ramsar sites are afforded legal protection, ensuring the conservation of their ecosystems and biodiversity. This designation plays a critical role in safeguarding the ecological integrity of these vital habitats. Thus, Ramsar designation reflects not only national conservation priorities but also a global commitment to sustaining international waterbird populations.

For instance, Baer's pochard (*Aythya baeri*), a critically endangered duck, faces severe habitat threats: a 2022 assessment found that most of its confirmed breeding sites are not within protected areas (BirdLife International, 2019). In addition to the formal governance framework of the Anatidae site network under the East Asian–Australian flyway, this initiative facilitates coordinated efforts to protect crucial wetlands at a flyway scale. The network connects migratory routes across various countries, enhancing habitat protection, wetland management, and the designation of essential habitats (Anatidae Working Group, EAA Flyway).

5.2. Role of Community Based Organization (CBOs), Local Stewardship, and Community-Based Conservation

The previous study underscores the significance of community-based conservation measures, particularly in regions where formal protection methods are less effective and underfunded (Jaman et al., 2025). These community-driven efforts are not only highly efficient but also require fewer financial resources. Moreover, such conservation approaches can be readily replicated and implemented in different areas.

For instance, in the Brazilian Amazon, regions employing community-based conservation methods support 80% of migratory waterbirds. The effectiveness of this approach has led to substantial increases in species populations, such as the black skimmer and various tern species, with breeding numbers rising by up to 135%. These findings clearly demonstrate the potential of community-based conservation for both local and migratory birds (Campos-Silva et al., 2021). It's truly inspiring when communities take the initiative in conservation efforts, setting an excellent example globally.

In California's Central Valley, innovative conservation strategies for migratory birds are not only attracting attention but also proving effective. Techniques such as citizen science, big data analysis, and habitat marketplaces enable conservationists and researchers to create temporary wetlands when they are most needed by the birds. This innovative approach not only addresses existing gaps but also operates with reduced funding requirements. These strategies illustrate how collaboration and creativity can lead to more successful conservation outcomes for wildlife (Hickey et al., 2017).

5.3. Ecotourism, Citizen Science, and Awareness Campaigns

Ecotourism serves as a valuable strategy in advancing conservation efforts. Activities such as non-captive bird feeding can significantly contribute to these initiatives by inspiring individuals, particularly young people, to disseminate information about wildlife conservation. Emotional responses, such as a deep concern for nature, are powerful motivators for engaging in conservation practices. However, translating these feelings into meaningful actions often presents a challenge (Massingham et al., 2019).

When ecotourism is systematically and scientifically managed, with attention to infrastructure and effective tourism guidelines, it can be implemented in a manner that does not adversely affect the Anatidae population in a preserved area over the long term. While increased ecotourism may lead to temporary disturbances due to a higher number of tourists, these impacts are generally reversible. High tourist numbers do not contribute to the decline of wildfowl populations if activities are well-regulated (Lucas et al., 2007). This clearly indicates that ecotourism can support conservation efforts and raise awareness for wildlife protection without harming ecosystems when properly managed and planned.

Furthermore, ecotourism has the potential to provide economic opportunities for local communities, allowing them to thrive while preserving nature. By offering financial support, ecotourism decreases communities' reliance on natural resources, which in turn helps conserve the environment. With increased financial backing and autonomy, communities are often more willing to engage in conservation efforts. It is also crucial to ensure that targeted programs and support enable social and cultural benefits to reach their full potential. Targeted interventions can play a significant role in ensuring that communities genuinely reap the economic and social rewards of ecotourism (Das & Chatterjee, 2022).

Citizen science plays a vital role in gathering data on the population trends, distribution, and survival of Anatidae and other waterbirds (Mehra et al., 2019). This community-driven approach bridges gaps in scientific knowledge and provides valuable insights that inform conservation strategies. By engaging the public in citizen science, we not only collect ecologically significant data but also empower individuals from diverse backgrounds to contribute to wildlife conservation and preservation efforts (Callaghan et al., 2021). Recent observations indicate an increasing trend in public participation in citizen science, which not only aids in data collection but also raises awareness and fosters a commitment to the conservation of waterbirds (Wang & Ma, 2025).

5.4. International Union for Conservation of Nature (IUCN) Status for the Species

The Anatidae family is facing many different threats, which have created a serious risk to the species at the global level. Among all the threats, habitat loss stands out to be the most affecting threat, which includes degradation and reduction of wetlands. These wetlands are not only important for sustaining the aquatic life but also for the migration patterns and breeding of the Anatidae species. Previous study shows that loss of habitats due to dam constructions, drainage, and land reclamation has come to be the most severe threat faced by Anatidae, affecting nearly 73% of Anatidae taxa (Su et al., 2024). Hunting also poses a considerable threat to the Anatidae family, affecting nearly half of the species categorized as threatened, particularly in areas where these birds hold economic or cultural significance for local communities (Green, 1996). This confluence of habitat destruction, water management challenges, and hunting pressures not only threatens individual species but also disrupts the delicate balance of ecosystems that rely on these waterfowl. It is crucial that we address these threats through concerted conservation efforts to ensure the sustainability of Anatidae and their habitats for future generations.

Aythya baeri, commonly known as Baer's pochard, is a critically endangered bird species with an estimated global population of 150–700 individuals and continuing to decline (BirdLife International, 2019). The primary factors contributing to this alarming decline are habitat loss and hunting activities (Green, 1996).

In particular regions such as Iran, the situation is dire; approximately 72% of nonbreeding Anatidae populations with sufficient data are classified as threatened, and a similar concern persists with breeding populations, where 67% are also assessed as threatened. However, many breeding populations remain listed as data-deficient, reflecting inadequate monitoring (Collen et al., 2017). In Algeria's Ziban Oasis, similar situation was observed by the presence of two species of Anatidae, the marbled teal (*Marmaronetta angustirostris*) and the ferruginous duck (*Aythya nyroca*); both are listed as vulnerable (Marouani et al., 2019). In India, threatened species of Anatidae include the ferruginous duck (*Aythya nyroca*, classified as near threatened) and the falcated duck (*Mareca falcata*, also classified as near threatened). This study indicates the need for a conservative approach toward the protection of vulnerable bird populations and their habitat, supported by modern taxonomic and molecular tools that improve species identification, population assessment, and conservation prioritization (Sharma & Kumar, 2019; Mehra & Singhal, 2025).

6. Emerging Challenges and Biosecurity

Emerging disease risks associated with Anatidae are increasingly recognized as global biosecurity concerns, linking wildlife conservation with public health and agricultural systems worldwide (Chen et al., 2005; Baele et al., 2015). The Anatidae family faces various emerging challenges, both directly and indirectly related to changes in their habitat and human interference. These wetland birds are not only crucial for maintaining the health of the wetland, but also for managing the spread of epidemic diseases. Understanding the complex interplay among the abovementioned factors is crucial for developing biosecurity measures and protecting species and their habitats.

6.1. Disease Transmission

In the global spread of highly pathogenic avian influenza (HPAIV), Anatidae function as key transboundary reservoirs. The spread of this virus is significantly facilitated by the migratory pattern of these birds along the major flyways, leading to the faster expansion of epidemics, especially noted in the region of the Central Asian flyways. A previous study indicated that wild Anatidae help in the increase of H5N1 outbreaks, often the study links the geographically distant locations where the virus is present through the migratory patterns of these birds (Baele et al., 2015).

Moreover, the migratory birds have been found to carry various pathogens, including Giardia, Cryptosporidium, Cyclospora, and Microsporidia. The shedding of such pathogens in the droppings of these birds can contaminate soil and water, especially in urban and suburban areas. This contamination can be indirectly transferred to humans and animals through contact with water or the soil (Heddergott et al., 2023).

6.2. Zoonotic Risk and Policy Response

The Anatidae family of birds faces a significantly high risk from a pathogenic virus called HPAI, including strains such as H5N6. These birds carry and transmit this virus over longer distances without showing any symptoms, making them a potential vector for interspecies transmission of this disease, particularly during the outbreak period (Usui et al., 2022). Additionally, the outbreak of *Clostridium botulinum* has caused a large mortality rate among the Anatidae family. Moreover, the Anatidae are known to carry multidrug-resistant bacteria, raising concerns about transmission to humans and other animals (Shaker et al., 2020).

To address and control these challenges, a molecular-based approach is important for early detection and suppression of zoonotic diseases affecting the Anatidae family (Heddergott et al., 2023). There are a few biosecurity measures that can be implemented. First, the separation of captive ducks during viral outbreaks can help reduce the spread of the virus. Second, the scientific quarantine for the birds that have been transferred between the facilities is crucial for minimizing the viral outbreak. Third, rigorous hygiene and practice for biosecurity in both wild and captive birds are crucial for environmental contamination (Usui et al., 2022).

7. Case Studies of Coexistence and Conflict

Ecomorphological differences in the Anatidae family, which means differences in body size and lamellar density in the bills, enable species of this family to coexist by resource partitioning, which reduces interspecific competition. Difference in body length allows the species to access the food resources at different depths of the water level, whereas differences in their lamellar density allow different species to filter and feed on different sizes of food particles. Such morphological difference helps in reducing the interspecific competition and allow the coexistence of different species within a particular habitat (Simon et al., 2002). Coexistence and conflict case study from India shows that breeding sites of the waterbirds are found near the villages and coexist with the villagers; these sites are protected and remain safe due to the cultural belief of villagers, but due to the changing tradition and erosion of the tradition, conflicts between waterbirds and humans are increasing (Kannan, 2023).

In recent years, the barnacle goose has started migrating and breeding further north, a change that's largely attributed to climate change. This shift has had a noticeable impact on agricultural lands in Finland, especially in regions such as North and South Karelia (Yrjölä et al., 2017). Farmers have been facing some tough challenges, as these grazing flocks can significantly reduce crop yields by as much as 85% in certain areas. The compensation claimed regarding the issue reached over €2.7 million within a year, with around €1 million coming from North Karelia alone (Anon, 2025). For the resolution of this problem, the researcher tested a coexistence model which included the "accommodation field" and "repelling field"; the former is a safe feeding zone for the geese, and on the other hand, a "repelling field" which is designed for the protection of the crops. The geese's foraging pattern was also monitored with the help of GPS tracking. The findings of this study reveal that geese preferred the accommodation field; it clearly shows that, with smart planning and better spatial management, conservation and crop protection are both possible (Seltmann et al., 2024).

Another example is of the pink-footed goose; this goose breeds in Svalbard, and a remarkable increase in their population has been noted over the years. The count of these geese was about 15,000 during the 1960s, but by 2010, the count for the same increased to 80,000 (Madsen et al., 2017). On one side, it was good to see the growth of these geese, but soon it became a tough challenge for the farmers. Farmers in Norway, Denmark, the Netherlands, and Belgium suffered serious crop damage from the same geese.

To address the problem, an agreement under the Convention on the Conservation of African–Eurasian Migratory Waterbirds (AEWA) was reached. AEWA launched the International Species Management Plans in 2012, which were later solidified in 2016. The planned launch of the first-ever adaptive harvest management (AHM) was implemented for migratory birds in Europe. Under this plan, regulated hunting of pink-footed geese was allowed to control and stabilize the population, aiming to keep it around 60,000 (Jensen et al., 2014). This approach demonstrates a wonderful collaboration between conservation and community involvement, setting an example of how such collaboration can reduce human–wildlife conflict. Taken together, these geographically distinct case studies demonstrate that coexistence strategies such as spatial zoning, adaptive harvest management, and community engagement are transferable across regions when informed by ecological data and cultural context (AEWA, 2012; Jansen et al., 2014; Madsen et al., 2017).

8. Toward a Future of Ethical Coexistence

With ongoing conservation efforts, the Anatidae family of birds is making a strong comeback. We are at a very sensitive and crucial moment. While it is inspiring to see the birds thriving, it also raises a question about how we plan to conserve this wild species while supporting human communities. How do we plan and implement the policies that maintain a balance between both wildlife and human coexistence?

8.1. Principle for Ethical Coexistence

As previously discussed in this chapter, we have examined the examples of the goose from North and South Karelia and the pink-footed goose, from which four core principles emerge that guide us in maintaining coexistence between the Anatidae and the human community.

1. **Respect for All**: We should see the wildlife as our friend and partner. Our existence is due to their existence. As in the example of the goose from Karelia, which was provided with a separate feeding area and was not just seen as a pest, it helps control crop damage without harming wildlife.
2. **Flexibility in Strategies:** Our planning, strategies, and rules should be flexible and should be adjusted based on real-time data available. The land separation for geese feeding in Finland was seasonal. And the hunting quota for the pink-footed goose was based on real-time data collected.
3. **Diverse Involvement:** Involvement of conservationists, communities, farmers, NGOs, and government entities creates a strong framework. This is essential for promoting conservation and ensuring coexistence. These two efforts can work together hand in hand.
4. **Everyone's Responsibility:** Ethical coexistence is not just a responsibility of the government or NGOs, but every citizen's contribution is required for the

coexistence of any wildlife with humans. It's very important that everyone work together to find ways for coexistence to be possible and sustained.

8.2. Way Forward

1. **Replicating the Local Solution on a Large Scale:** Once success is achieved on any local coexistence solution, it should be replicated and implemented on a larger scale. Finland's success in solving the goose problem should be replicated, and farmers should be supported and incentivized.
2. **Building Trust:** For the success of AEWA to be continued, there is a need for data transparency and power sharing between stakeholders. For maintaining the collaboration workshop, joint monitoring programs are important.

8.3. Adapting to Future Challenges

As we face challenges such as climate change and changes in land use, there is a need for flexibility in policy and scientific approaches for wildlife conservation and human interests. A solution to the human–wildlife conflict should be presented in a way that protects wildlife while maintaining coexistence.

References

AEWA. (2012). *International Species Management Plan for the Svalbard Population of the Pink-Footed Goose (Anser brachyrhynchus)* (AEWA Technical Series No. 48). African-Eurasian Migratory Waterbird Agreement. Retrieved from the AEWA website.

Anatidae. (2023, January 9). New World Encyclopedia. Retrieved 19:04, June 4, 2025 from https://www.newworldencyclopedia.org/p/index.php?title=Anatidae&oldid=1094461.

Andreu, G., Rutschowscaya, M., & Ziegler, C. (1997) *Ancient Egypt at the Louvre*. Hachette, Paris.

Anon (2025, June 2). Birds struggle to keep up with rapid climate change and face new threats; here's why. *Times of India*. Retrieved: 4 June 2025.

Anon, (2025). Designated fields could reduce crop damage caused by barnacle geese. Retrieved: 12 February 2026 from https://phys.org/news/2025-01-fields-crop-barnacle geese.html

Aragon, L. V. (2006). Bird omens and metaphors in central Sulawesi ritual songs. *In:* Le Roux, P., & Sellato, B. (Eds.), *Divine Messengers: Bird Symbolism and Aesthetics in Southeast Asia*. Connaissances & Savoirs, Bangkok, Thailand.

Atkinson, R. (1993). Killing the kereru. *Forest and Bird*, 267, 18–23.

Ayanlade, A., Sergi, C. M., Di Carlo, P. et al. (2020). When climate turns nasty, what are recent and future implications? Ecological and human health review of climate change impacts. *Curr Clim Change Rep*. 6, 55–65. https://doi.org/10.1007/s40641-020-00158-8

Baele, G., Gilbert, M., Lemey, P., Suchard, M., & Trovão, N. (2015). Bayesian inference reveals host-specific contributions to the epidemic expansion of influenza A H5N1.

Molecular Biology and Evolution, 32(12), 3264–3275. https://doi.org/10.1093/molbev/msv185.
Beames, I. (1969). Welcome immigrants: Mandarin duck. *Animals*, 12, 315–317.
Beck, P. V., Walters, A. L., & Francisco (Eds.). (1990). *The Sacred: Ways of Knowledge, Sources of Life*. Navajo Community College Press.
Bhanot, C., & Chatterjee, S. (2023). Conservation of Urban Wetland with potential international significance: A case study of Najafgarh Jheel, Delhi, India. *International Journal of Conservation Science*, 14(2), 1057–1070. https://dx.doi.org/10.36868/IJCS.2023.03.18
Bhatia, S., Redpath, S., Suryawanshi, K., & Mishra, C. (2019). Beyond conflict: Exploring the spectrum of human–wildlife interactions and their underlying mechanisms. *Oryx*, 54, 621–628. https://doi.org/10.1017/S003060531800159X.
Bhatt, P. M. (2010). Birds and nature in the step wells of Gujarat, Western India. *Ethno-ornithology: Birds, Indigenous Peoples, Culture and Society*. Earthscan. ISSN 978-1-84407-783-0.
BirdLife International (2019). Species factsheet: Baer's pochard Aythya baeri. Downloaded from https://datazone.birdlife.org/species/factsheet/baers-pochard-aythya-baeri. Retrieved: 18 June 2025.
Bonta, M. (2010). Ethno-ornithology and biological conservation. *In:* Tidemann, S., & Gosler, A. (Eds.), *Birds, Indigenous Peoples, Culture and Society*. Earthscan. ISSN 978-1-84407-783-0.
Both, C., Bouwhuis, S., Lessells, C. M., & Visser, M. E. (2006). Climate change and population declines in a long-distance migratory bird. *Nature*, 441(7089), 81–83. https://doi.org/10.1038/nature04539
British Museum. (2021, May 7). *God and Goddesses of the Greek and Roman Pantheon.* Retrieved: 5 June 2025. https://www.britishmuseum.org/blog/gods-and-goddesses-greek-and-roman-pantheon
Brown, C. H. (1992). British names for American birds. *Journal of Linguistic Anthropology*. 2, 30–50.
Buckland, R., & Guy, G. (Eds.). (2002). *Goose Production System*. FAO Animal Production and Health Paper-154. ISSN 0254-6019
Callaghan, C., Cornwell, W., Fuller, R., Watson, J., & Lyons, M. (2021). Conservation birding: A quantitative conceptual framework for prioritizing citizen science observations. *Biological Conservation*, 253, 108912. https://doi.org/10.1016/j.biocon.2020.108912.
Campos-Silva, J., Abrahams, M., Davenport, L., Peres, C., Andrade, P., & Hawes, J. (2021). Community-based conservation with formal protection provides large collateral benefits to Amazonian migratory waterbirds. *PLoS ONE*, 16. https://doi.org/10.1371/journal.pone.0250022.
Carboneras, C. (1992). Family Anatidae (Ducks, Geese and Swans). *In:* del Hoyo, Josep, Elliott, Andrew & Sargatal, Jordi (Eds.), *Handbook of Birds of World* (Volume 1: *Ostrich to Ducks*) (pp. 536–629). Lynx Edicions, Barcelona. ISBN 84-87334-10-5
Chen, H., Smith, G. J. D., Zhang, S. Y., Qin, K., Wang, J., Li, K. S. et al. (2005). Avian flu: H5N1 virus outbreak in migratory waterfowl. *Nature*, 436(7048), 191–192. https://doi.org/10.1038/nature03974
Collen, B., Farhoodinia, M., Kaboli, M., & Nourani, E. (2017). National assessment of threatened species using sparse data: IUCN Red List classification of Anatidae in Iran. *Animal Conservation*, 20. https://doi.org/10.1111/acv.12282.

Coombes, B. (2007). Postcolonial conservation and kiekie harvests at Morere New Zealand—Abstracting indigenous knowledge from indigenous polities. *Geographical Research*, 45, 186–193. https://doi.org/10.1111/j.1745-5871.2007.00450.x

Cottam, C., (1939). Food habits of North American diving ducks. *USDA Technical Bulletin*, p. 643.

Das, M., & Chatterjee, B. (2022). Ecotourism a solution or deception for conservation: A case of Bhitarkanika Wildlife Sanctuary, Odisha, India. *Journal of Hospitality and Tourism Insights*. https://doi.org/10.1108/jhti-12-2021-0336.

Davidson, N. C. (2014). How much wetland has the world lost? Long-term and recent trends in global wetland area. *Marine and Freshwater Research*, 65(10), 934–941. https://doi.org/10.1071/MF14173

Descola, P. (2013). *Beyond Nature and Culture* (J. Lloyd, Trans.). University of Chicago Press.

Ebrey, P. B. (1991). *Confucianism and Family Rituals in Imperial China: A Social History of Writing About Rites*. Motilal Banarsidass and Princeton University Press. https://doi.org/10.2307/2719457

Eyton, T. C. (1838). *A Monograph on the Anatidae, or Duck Tribe*. London, Longman, Orme, Brown, Green, & Longman and Eddowes, Shreswburry, 1838. https://doi.org/10.5962/bhl.title.51971

Flores, N. C. (Ed.). (1996). *Animals in the Middle Ages*. Routledge Medieval Casebook.

Forshaw, Joseph (Ed.). (1991). *Encyclopaedia of Animals: Birds*. London: Merehurst Press. pp. 81–87. ISBN 1-85391-186-0.

Fuentes, A. (2010). Natural cultural encounters in Bali: Monkeys, temples, tourists, and ethnoprimatology. *Cultural Anthropology*, 25(4), 600–624. https://doi.org/10.1111/j.1548-1360.2010.01071.x

Fuller, R. A., Warren, P. H., Armsworth, P. R., Barbosa, O., & Gaston, K. J. (2008). Garden bird feeding predicts the structure of urban avian assemblages. *Diversity and Distributions*, 14(1), 131–137.

Gangoly, O. C. (1963). *Landscape in Indian literature and Art*. University of Lucknow, Lucknow. http://primo.getty.edu/GRI:GETTY_ALMA21119825410001551

Gilchrist, H. G., Mallory, M. L., & Merkel, F. (2005). Can local ecological knowledge contribute to wildlife management? Case studies of migratory birds. *Ecology and Society*, 10(1), 20. www.ecologyandsociety.org/vol10/iss1/art20/, accessed 6 June 2025

Green, A. (1996). Analyses of globally threatened anatidae in relation to threats, distribution, migration patterns, and habitat use. *Conservation Biology*, 10, 1435–1445. https://doi.org/10.1046/J.1523-1739.1996.10051435.X.

Heddergott, M., Wojtkowiak-Giera, A., & Solarczyk, P. (2023). Migrating anatidae as sources of environmental contamination with zoonotic giardia, cryptosporidium, cyclospora and microsporidia. *Pathogens*, 12. https://doi.org/10.3390/pathogens12030487.

Hickey, C., Sullivan, B., Golet, G., Reiter, M., Reynolds, M., Kelling, S. et al. (2017). Dynamic conservation for migratory species. *Science Advances*, 3. https://doi.org/10.1126/sciadv.1700707.

Holopainen, S., Arzel, C., Elmberg, J., Fox, A. D., Guillemain, M., Gunnarsson, G. et al. (2018). Sustainable management of migratory European ducks: Finding model species. *Wildlife Biology*, 2018(1), 1–11.

Hou, J. (1982). On the origin and history of domestic duck in China. *Acta Zoologica Sinica*, 28(2), 185–190.

Houlihan, P. F. (1988). *The Birds of Ancient Egypt*. Aris & Phillips Ltd.

Hunn, E. S., & Thornton, T. F. (2010). Tlingit birds: An annotated list with a statistical comparative analysis. *Birds, Indigenous Peoples, Culture and Society*. Earthscan. ISSN 978-1-84407-783-0. https://faculty.washington.edu/hunn/vitae/Hunn%20&%20Thornton%202010%20Tlingit%20Birds.PDF

Irwin, J. (1984). *An Introduction to Maori Religion—Its Character before European Contact and Its Survival in Contemporary Maori and New Zealand Culture*. Australian Association for the Study of Religions, Bedford Park. https://openjournals.library.sydney.edu.au/AASR/issue/view/1145

Izadian, M., & Tayefeh, F. (2024). Population decline of anatidae species in the Parishan Wetland (SW Iran) resulting from a decreasing water level. *Ardea*, 112, 229–237. https://doi.org/10.5253/arde.2024.a2.

Jain, P. (2011). Dharma and ecology of Hindu communities: Sustenance and sustainability. *In: New Critical Thinking in Religion, Theology and Biblical Studies Series*. Farnham: Ashgate. xii, pp. 213. https://doi.org/10.4324/9781315576916

Jaman, M., Shome, A., Alam, M., Hossain, M., Mia, T., Rabbe, M., & Ahmed, A. (2025). Community structure, threats and conservation issues of migratory birds in the south-central coastal area of Bangladesh. *Dhaka University Journal of Biological Sciences*. https://doi.org/10.3329/dujbs.v34i1.79868.

Jensen, G. H., Madsen, J., & Wisz, M. S. (2014). Prioritizing refuge sites for migratory geese to balance hunting and agriculture. *Biological Conservation*, 173, 144–152.

Jerolmack, C. (2008). How pigeons became rats: The cultural–spatial logic of problem animals. *Social Problems*, 55(1), 72–94.

Johnsgard, P. A. (1975). *Waterfowl of North America*. Bloomington: Indiana University Press.

Kannan, V. (2023). Coexistence and conflict—Case study on colonial waterbirds in southern India. *In:* Nautiyal, S., Gupta, A. K., Goswami, M., Imran Khan, Y. D. (Eds.), *The Palgrave Handbook of Socio-Ecological Resilience in the Face of Climate Change*. Palgrave Macmillan, Singapore. https://doi.org/10.1007/978-981-99-2206-2_9

Kear, J. (Ed.). (2005). *Ducks, Geese and Swans: Species Accounts* (Vols. 1–2). Oxford University Press.

Kinsley, D. (1987). *Hindu Goddesses: Visions of the Divine Feminine in the Hindu Religious Tradition*. University of California Press.

Li, C. (1999). *The Tao Encounters the West: Explorations in Comparative Philosophy*. SUNY Press.

Liang, J., Peng, Y., Zhu, Z., Li, X., Xing, W., Li, X., & Yuan, Y. (2021). Impacts of changing climate on the distribution of migratory birds in China: Habitat change and population centroid shift. *Ecol. Ind.*, 127, 107729. https://doi.org/10.1016/j.ecolind.2021.107729.

Losty, J. P. (1982). *The Art of the Book in India*. British Library Publishing.

Lucas, C., Guillemain, M., Blanc, R., & Lepley, M. (2007). Ecotourism disturbance to wildfowl in protected areas: Historical, empirical and experimental approaches in the Camargue, Southern France. *Biodiversity and Conservation*, 16, 3633–3651. https://doi.org/10.1007/s10531-007-9160-z.

Lyver, P. O'B., & Moller, H. (2010). An alternative reality: Maori spiritual guardianship of New Zealand's native birds. *Birds, Indigenous Peoples, Culture and Society*. Earthscan. ISSN 978-1-84407-783-0.

MacKillop, J. (1998). *Dictionary of Celtic Mythology*. Oxford University Press.

Madsen, J. (2016). Regulation of the hunting season as a tool for adaptive harvest management—First results for pink-footed geese (Anser brachyrhynchus). *Wildlife Biology*, 22(5), 204–208. https://doi.org/10.2981/wlb.00234

Madsen, J., & Johnson, F. A. (2017). Adaptive harvest management for the Svalbard population of pink-footed geese (Anser brachyrhynchus). EGMP Technical Report No. 5, AEWA European Goose Management Platform.

Madsen, J., Cracknell, G., & Fox, T. (Eds.). (1999). *Goose Populations of the Western Palearctic: A Review of Status and Distribution*. Wetlands International.

Madsen, J., Henty Williams, J., Johnson, F. A., Tombre, I. M., Dereliev, S., & Kuijken, E. (2017). Implementation of the first adaptive management plan for a European migratory waterbird population: The case of the Svalbard pink-footed goose (Anser brachyrhynchus). *Ambio*, 46(S2), 275–289. https://doi.org/10.1007/s13280-016-0888-0

Madsen, J., Tombre, I. M., & Williams, B. K. (2017). Goose management schemes to resolve conflicts with agriculture: Theory, practice and effects. *Ambio*, 46, 231–240. https://doi.org/10.1007/s13280-016-0884-4.

Mahadevia, K., & Vikas, M. (2012). Climate Change—Impact on the Sundarbans. *InternationalScientificJournal*. https://www.researchgate.net/publication/311607858_Climate_Change_-_Impact_on_the_Sundarbans_A_case_study

Marouani, S., Saheb, M., Aras, R., Bensaci, E., Mimeche, F., & Nouidjem, Y. (2019). Check list of waterbirds at Wadi Djedi in Ziban Oasis-Algeria. *Arxius de Miscel·lània Zoològica*. https://doi.org/10.32800/amz.2019.17.0034.

Marzluff, J. M. (2001). Worldwide urbanization and its effects on birds. In: Marzluff, J. M., Bowman, R., & Donnelly, R. (Eds.), *Avian Ecology and Conservation in an Urbanizing World*. Springer, Boston, MA. https://doi.org/10.1007/978-1-4615-1531-9_2

Massingham, E., Fuller, R., & Dean, A. (2019). Pathways between contrasting ecotourism experiences and conservation engagement. *Biodiversity and Conservation*, 28, 827–845. https://doi.org/10.1007/s10531-018-01694-4.

McKinney, F., & Evarts, S. (1998). Sexual coercion in waterfowl and other birds. *Ornithological Monographs*, 49(49), 165–193. doi:10.2307/40166723

McLandress, M. R., & McLandress, I. (1979). Blue-phase Ross' geese and other blue-phase geese in western North America. *Auk*, 96, 544–550.

Mehra, S. P., Mehra, S., Uddin, M., Verma, V., Sharma, H. et al. (2017). Waste as a resource for avifauna: Review and survey of the avifaunal composition in and around waste dumping sites and sewage water collection sites (India). *Int J Waste Resour*., 7, 289.

Mehra, S., Mehra, S. P., & Sharma, K. K. (2011). Aquatic avifauna of Aravalli Hills Rajasthan, India. *In:* Gupta, V. K. & Verma, A. K. (Eds.), *Animal Diversity, Natural History and Conservation* (Vol. I, pp. 145–167). Daya Publishing House, Delhi (ISBN: 978-81-7035-752-0).

Mehra, S., Mehra, S. P., & Sharma, K. K. (2011). Aquatic avifauna: Its importance for wetland conservation in Rajasthan, India. *In:* Mathur, S. M., Shrivastava, V. K., & Purohit, R. C. (Eds.), *Conservation of Lakes and Water Resources Management Strategies* (pp. 179–190). Himanshu Publications, Udaipur (978-81-7906-263-0).

Mehra, S., Mehra, S. P., & Sharma, K. K. (2012). Importance of aquatic avifauna in southern Rajasthan, India. *In:* Rawat., M. & Dookia, S. (Eds.), *Biodiversity of Aquatic Resources* (pp. 159–183). Daya Publishing House, Delhi (978-81-7035-789-6).

Mehra, S. P., & Mehra, S. (2013). *Avian Diversity of Vagad Region with Main Emphasis on Aquatic Species*. Dungarpur Bird Fair 2013, pp. 11–15.

Mehra, S. P., & Mehra, S. (2013). Short study to assess the potential of wetlands of Dholpur, Rajasthan, India. *In:* Sheikh, M. M. (Ed.), *Environmental Consciousness and Human Perceptions* (pp. 306–313). LAP LAMBERT Academic Publishing, Germany (978-3-659-49573-1).

Mehra, S. P., & Mehra, S. (2014). Perspective on water and biodiversity issues: A case study of Keoladeo National Park, Bharatpur, India. *In:* Maheshwari, B. L., Purohit, R. C., Malano, H. M., Singh, V. P., & Amerasinghe (Eds.), *Securing Water, Food, Energy and the Liveability of Cities: Challenges and Opportunities for Peri-Urban Futures* (pp. 419–434). Springer Science+Business Media B.V. Dordrecht, The Netherlands. (https://link.springer.com/chapter/10.1007/978-94-017-8878-6_32)

Mehra, S. P., Mehra, S., & Sharma, K. K. (2014). Importance of urban biodiversity—A case study of Udaipur, India. *In:* Maheshwari, B. L., Purohit, R. C., Malano, H. M., Singh, V. P., & Amerasinghe (Eds.), *Securing Water, Food, Energy and the Liveability of Cities: Challenges and Opportunities for Peri-Urban Futures* (pp. 403–418). Springer Science+Business Media B.V. Dordrecht, The Netherlands. https://link.springer.com/chapter/10.1007/978-94-017-8878-6_31

Mehra, S. P., Mehra, S., Khamesra, P., Joshi, P., Oza, K., & Kashyap, S. (2019). *Perspectives of Citizen Science: Importance of the Urban Avifaunal Diversity*. Udaipur Bird Festival 2019, pp. 3–4.

Mehra, S. P. (2019). Editorial: Ecology vis-à-vis economy: The Ramsar Site—Sambhar (India). *Pollution, Bioremediation & Biodegradation Journal*, 2(1), 180007. https://chembiopublishers.com/PBBJ/index.php

Mehra, S. P., & Singhal, H. (2025). *Changing Climate & Its Impact on Migratory Birds*. Udaipur Bird Festival, 11th Edition, pp. 22–24. https://www.researchgate.net/publication/389272075_Changing_climate_and_its_impact_on_migratory_birds

Miller, B. S. (ed.). (1983). *Exploring India's Sacred Art: Selected Writings of Stella Kramrisch*. University of Pennsylvania Press, Philadelphia, PA.

Mills, E. (2025, June 3). Time to listen up: What animals are telling us about climate change. *Nature and Biodiversity*. Retrieved: 4 June 2025.

Moeran, B. (1984). Individual, group, and Seishin: Japan's internal cultural debate. *Man*, 19(1), 252–266.

Moguel, P., & Toledo, V. M. (1999). Biodiversity conservation in traditional coffee systems of Mexico. *Conservation Biology*, 13, 11–21.

Morford, M. P. O., Lenardon, R. J., & Sham, M. (Eds.). (2019). *Classical Mythology: Greek & Roman Myths*. 11th Edition. Oxford University Press.

Morgan, J. (2025, March 16). 20 Examples of animals adapting to climate-induced habitat changes. *Climate Cosmos: Nature and Wildlife Conservation*. Retrieved: 4 June 2025.

Mowbray, T. B., Ely, C. R., Sedinger, J. S., & Trost, R. E. (2003). *In:* A. Poole & F. Gill, (Eds.), *The Birds of North America*, No. 682. Philadelphia. PA: The Birds of North America, Inc.

Nelson, R. K. (1983). *Make Prayers to the Raven: A Koyukon View of the Northern Forest*. University of Chicago Press, Chicago, IL.

Olson, S. L., & Feduccia, A. (1980). Presbyornis and the origin of the Anseriformes (Aves: Charadriomorphae). *Smithsonian Contributions to Zoology*, 323(323). Smithsonian Institution, pp. 1–24. doi:10.5479/si.00810282.323.

Otte, J. (2025, May 16). *16 Creatures That adapted to Climate Change—Animals Around the Globe*. Retrieved: 4 June 2025.

Parsons, K. C., Mineau, P., and Renfrew, B. (2010). Effects of pesticides use in rice fields on birds. *Waterbirds*, 33(sp1), 193–218. DOI: 10.1675/063.033.s115

Raja Dinkar Kelkar Museum. (1988). *Birds and Animals in Indian Art and Crafts*. Raja Dinkar Kelkar Museum, Pune, India.

Ramsar Convention Secretariat. (2018). *Global Wetland Outlook: State of the World's Wetlands and Their Services to People*. Gland, Switzerland: Ramsar Convention Secretariat.

Ramsar Convention Secretariat. (2025). *Global Wetland Outlook: Special Edition 2025*. Gland, Switzerland: Ramsar Convention on Wetlands.

Reo, N. J., & Whyte, K. P. (2012). Hunting and morality as elements of traditional ecological knowledge. *Human Ecology*, 40(1), 15–27. https://doi.org/10.1007/s10745-011-9448-1

Rowland, B. (1978). *Birds with Human Souls: A Guide to Bird Symbolism*. University of Tennessee Press, Knoxville, TN.

Scribner, E. T., Talbot, S. L., Pierson, J. M., Bollinger, K. S., & Derkson, D. V. (2003). Phylogeography of Canada geese (Branta canadensis) in western North America. *Auk*, 120, 889–907.

Seltmann, M. W., Ylitalo, A. K., Piironen, A., Store, R., Heikkinen, J., Heim, W. et al. (2024). Arctic migrating barnacle geese utilize accommodation fields in a new agricultural staging area. *Journal of Applied Ecology*, 62, 317–328. https://doi.org/10.1111/1365-2664.14838.

Seri, N. A., & Rahman, A. A. (2021). Impact of climate change on migratory birds in Asia. *Pertanika J. Sci. Technol.*, 29, 2937–2965.

Shaker, A., El-Hamid, M., Awad, N., Elsohaby, I., Ahmed, H., & Mohamed, R. (2020). Pet birds as potential reservoirs of virulent and antibiotic resistant zoonotic bacteria. *Comparative Immunology, Microbiology and Infectious Diseases*, 75, 101606. https://doi.org/10.1016/j.cimid.2020.101606.

Sharma, A. & Kumar, P. (2019). Wetland birds assemblages in man-made sacred ponds of Kurukshetra, India. *Proceedings of the Zoological Society*, 72, 61–73. https://doi.org/10.1007/s12595-018-0259-x.

Siegfried, W. R. (1962). Observations on the post-embryonic development of the Egyptian goose Alopochen aegyptiacus (L.) and the redbill teal Anas erythrorhyncha (Gmelin). *Investigational Report*, Dept. of Nature and Conservation (Republic of South Africa) 2, 9–17.

Simon, G., Guillemain, M., Fritz, H., & Guillon, N. (2002). Ecomorphology and coexistence in dabbling ducks: The role of lamellar density and body length in winter. *Oikos*, 98, 547–551. https://doi.org/10.1034/J.1600-0706.2002.980321.X.

Singh, T., Upadhyay, M., & Mehra, S. P. (2020). *Climate Change: A New Paradigm of Threat for the Birds*. Udaipur Bird Festival 2020, pp. 21–22.

Sivaramamurti, C. (1980) *Approach to Nature in Indian Art and Thought*. Kanak Publications, New Delhi.

Smith, R. S., & Trzaskoma, S. M. (Eds.). (2022). *The Oxford Handbook of Greek and Roman Mythography*. Oxford University Press.

Su, Y., Wang, C., Dong, B., Sun, G., Wang, B., Duan, H., Xia, S. et al. (2024). Study on the impact of reclamation and development of Yancheng coastal wetlands on the spatio-temporal evolution of wintering Anatidae habitat. *Ecological Indicators*. https://doi.org/10.1016/j.ecolind.2024.111892.

Thirunavukkarasu, N. (2022). Assessment of the conservation value of wetland birds on the Pallikaranai Marsh, Chennai, Tamil Nadu, India. *International Journal of Zoology Investigation*. 01(02), 117–129. https://doi.org/10.33745/ijzi.2022.v08i02.015

Thomas, K. (1983). *Man and the Natural World: Changing Attitudes in England* (pp. 1500–1800). Oxford University Press.

Toynbee, J. M. C. (1973). *Animals in Roman Life and Art*. Thames and Hudson. Cambridge University Press. ISBN : 0500400245.

Upadhyay, M., Vyas, R., Sharma, V., & Mehra, S. P. (2020). Avifaunal composition of municipal areas of Ajmer District, Rajasthan, India. *Journal of Ecology and Natural Resources*, 4(2), 000192. https://doi.org/10.23880/jenr-16000192

Usui, T., Yamaguchi, T., Matsushita, M., Hidaka, C., Tomioka, Y., & Soda, K. (2022). Susceptibility of common family Anatidae bird species to clade 2.3.4.4e H5N6 high pathogenicity avian influenza virus: An experimental infection study. *BMC Veterinary Research*, 18. https://doi.org/10.1186/s12917-022-03222-7.

Vickery, J. A., Bradbury, R. B., Henderson, I. G., Eaton, M. A., & Grice, P. V. (2004). The role of agri-environment schemes and farm management practices in reversing the decline of farmland birds in England. *Biological Conservation*, 119(1), 19–39. DOI:10.1016/j.biocon.2003.06.004

Vitebsky, P. (2005). *The Reindeer People: Living with Animals and Spirits in Siberia*. Houghton Mifflin.

Wang, W., Fraser, J. D., & Chen, J. (2017). Wintering waterbirds in the middle and lower Yangtze River floodplain: Changes in abundance and distribution. *Bird Conservation International*, 27(2), 167–186.

Wang, R., & Ma, K. (2025). Trends in the application of citizen science in waterbird conservation: A bibliometric analysis. *Animals: An Open Access Journal*, 15. MDPI, https://doi.org/10.3390/ani15030368.

Watson, I. (2015). *Aboriginal Peoples, Colonialism and International Law: Raw Law*. Routledge.

Williams, D. R., Pople, R. G., Showler, D. A., Dicks, L. V., Child, M. F., Ermgasses, E. K. H. J. et al. (2013). Bird conservation: Global evidence for the effects of interventions. *Exter, Synopses of Conservation Evidence*, Pelagic Publishing, ISBN: 978-1-907807-19-0

Yeomans, L., & Mazzucato, C. (2024). Human–bird interactions in the Levant during the Late Pleistocene to Early Holocene: Multi-scalar analysis of avifaunal remains. *Archaeol Anthropol Sci*., 16, 185. https://doi.org/10.1007/s12520-024-02090-6

Yrjölä, R. A., Holopainen, S., Pakarinen, R., Tuoriniemi, S., Luostarinen, M., Mikkola-Roos, M. et al. (2017). The Barnacle Goose (Branta leucopsis) in the archipelago of southern Finland—Population growth and nesting dispersal. *Ornis Fennica*, 94(4), 161–171. https://doi.org/10.51812/of.133921

Zhang, Z., Ni, Z., Li, T., Ning, M., Gao, C., Hu, J. et al. (2025). Nine high-quality Anas genomes provide insights into Anas evolution and domestication. *Cell Reports*, 44(4), 115477. ISSN 2211–1247, https://doi.org/10.1016/j.celrep.2025.115477

CHAPTER

10

Diseases Associated with the Anatidae Family

Sumit Kumar[1*], Ajit Kumar[2], Akshatha Soratur[2] and Ankit Sarkar[2]

1 Department of Industrial Fish and Fisheries, Babasaheb Bhimrao Ambedkar Bihar University, Muzaffarpur – 842001, Bihar, India

2 Department of Ocean Studies and Marine Biology, Pondicherry Central University, Port Blair – 744103, Andaman and Nicobar Islands, India

1. Introduction to the Anatidae Family and their Ecological Significance

1.1. Overview of the Anatidae Family

Ducks, geese, whistling ducks, and swans are members of the family Anatidae, which is a varied group classified within the order Anseriformes. This order encompasses over 170 species distributed across 42 genera, with the most prominent being Anatinae, Anserinae, and Dendrocygninae (Anatidae taxonomy) (G. Liu et al., 2014; Livezey, 1997; Pavia, 2013). Phylogenetic work maintains that the remaining Anseriformes diverged from the rest of the family and began adaptive radiation because of the unique structure of bills with deep specialization, such as filters in dabbling ducks (*Anas platyrhynchos*), serrated edges in geese (*Branta* spp.) for grazing, and osmoregulatory adaptations such as salt glands in marine species (*Somateria mollissima*) to excrete excess salt for colonization of saline ecosystems roughly 30–40 million years ago during the Oligocene epoch (Livezey, 1996; Sun et al., 2017). These birds are distributed almost across the globe except for Antarctica, and thrive in numerous types of freshwater wetlands, coastal estuaries, and even marine environments, from

*Corresponding author: isumitvsingh@gmail.com

DOI: 10.1201/9781003669784-10

rivers and lakes to oceans (Parker et al., 1992). The excess salt excretion through salt glands in marine species, such as *Somateria mollissima*, illustrates their osmoregulatory adaptations enabling them to colonize saline ecosystems. This adaptability to these ecosystems further highlights their ecological importance. Anatidae, in turn, help biodiversity by vascular wetland flora and shrub growth by transporting plant propagules in their body and feathers aiding in Angola's plant biodiversity (CABI, 2019; Goldstein et al., 1990; Van Leeuwen et al., 2012). For example, research on mallards shows that 12–15% of ingested seeds pass through the gut and remain viable, resulting in the proliferation of *Potamogeton* spp. and *Sparganium* spp. The organisms also enhance the nutrient cycle of the ecosystem by adding nutrients through feeding and excretion and increasing organic matter in water systems (Bartel et al., 2018). Furthermore, they are integral to the food web, being hunted by eagles, foxes, crocodiles, and large fish, which maintains balance in the ecosystem. Snow geese (*Chen caerulescens*) eggs and goslings make up to 30% of the summer diet of Arctic foxes (*Vulpes lagopus*), demonstrating the importance of these birds in sustaining predator populations (Ives et al., 2005).

1.2. Socioeconomic and Cultural Importance

Species within the Anatidae family are of great ecocultural and socioeconomic importance around the globe. Some subdivisions of birdwatching ecotourism serve as a prime example for duck- and goose-related ecotourism in the Danube Delta in Romania and Bosque del Apache National Wildlife Refuge in New Mexico, with the former earning $120 million annually and the latter bringing in thousands of visitors (Dima et al., 2020) (Figure 1). *Cygnus olor*, commonly referred to as the mute swan, along with various other prominent species, attracts over 1 million visitors annually to the Wildfowl & Wetlands Trust (WWT) reserve, contributing £60 million to the economy. Commercial poultry farming is highly dependent on ducks and geese for meat, eggs, and feathers, which supports diverse rural and industrial economies. Goose down feathers are also utilized in clothing and bedding insulation, supporting diverse rural and industrial economies (Knapp, 2000; Maurer, 2003). The duck meat production surpasses 4.5 million metric tons all around the globe, with China holding 65% of production with Pekin ducks (*Anas platyrhynchos domestica*) which are known for their rapid growth and efficient feeding. In the United States of America, waterfowl hunting contributes extensively to rural areas economically, with hunters spending roughly $900 million each year, which contributes toward 27,000 jobs (waterfowl hunting economic impact). Since 1934, the Federal Duck Stamp program, which funds wetland conservation, has preserved more than 6 million acres of habitat, thereby establishing a balance between economic investment and ecological conservation (Bennett & Whitten, 2003). North America's wetland conservation efforts further contribute $60 million annually to the socioeconomic valuation of Anatidae (Figure 1). Culturally speaking,

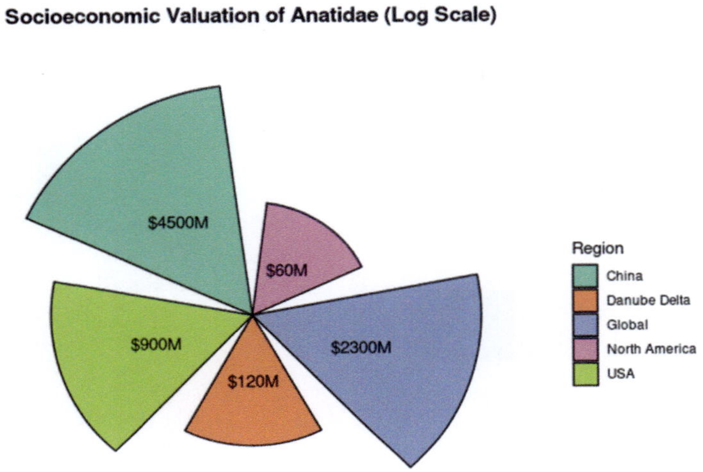

Figure 1. Logarithmic-scale pie chart illustrating the socioeconomic valuation of Anatidae across key regions: In the world, China dominates with $4.5 billion from duck meat production, followed by the global down industry at $2.3 billion. The USA contributes $900 million via waterfowl hunting, while the Danube Delta and North America add $120 million and $60 million through ecotourism and wetland conservation, respectively. Data analyzed from Grado et al. (2011); Santos et al. (2016); and Vrtiska et al. (2013).

Anatidae are significant anthropomorphically; swans are graceful examples of love and fidelity to the European folklore, appearing in tales such as "The Ugly Duckling" as well as Wagner's operas (swan symbolism). In Hindu mythos, the *hamsa*, which in many interpretations is taken as a goose or swan, is said to symbolize spiritual wisdom and is honored as a mount of goddess Saraswati. Culturally, ducks represent being part of a group as well as being adaptable, while geese are associated with strong family ties and their ability to navigate expertly (Gilbert et al., 2006).

1.3. Vulnerability to Diseases

The family Anatidae suffers from a range of diseases because of both biological and human factors. Biologically, the reason would be their high-density breeding colonies—such as of eider ducks (*Somateria mollissima*) and snow geese (*Chen caerulescens*)—that create ideal conditions for pathogen transmission with proximity facilitating rapid spread (Benzarti et al., 2019). For example, outbreaks of avian cholera (*Pasteurella multocida*) in snow-goose colonies have staggering transmission rates of over 70% during the nesting season (Figure 2), worsened by contaminated water sources and fecal–oral pathways. Their migratory behavior that stretches for thousands of kilometers along pathways such as the East Asian–Australasian flyway enables the transport of pathogens

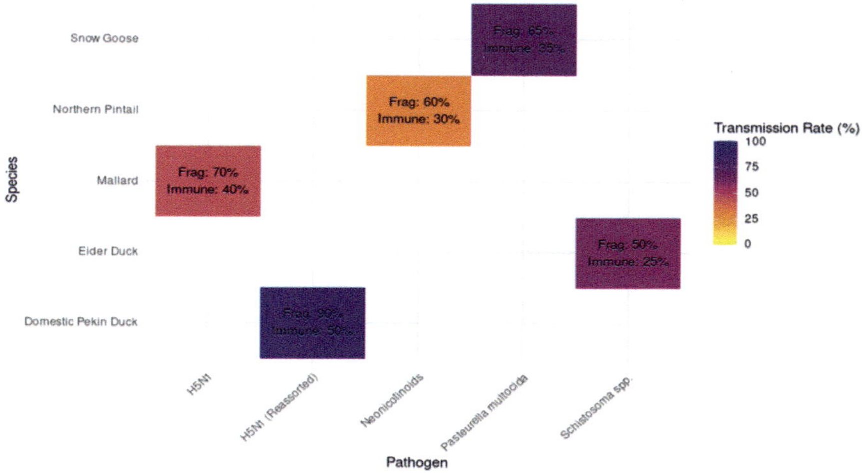

Figure 2. Disease transmission dynamics in Anatidae: The transmission rates of pathogens among Anatidae species. Snow goose, northern pintail, mallard, eider duck, and domestic pekin duck being the species and *Pasteurella multocida*, neonicotinoids, H5N1 avian influenza, *Schistosoma* spp., and H5N1 reassorted as the pathogens on the x-axis. All the squares are sized according to the estimate of their transmission rate (%), which is represented in the color scale ranging from yellow (0%) to dark purple (100%) with breaks at notable highs. The data is analyzed from Blanchong et al. (2006); Breban et al. (2009); and Si et al. (2010).

introducing them to new regions (Samuel et al., 2005). The satellite tracking of northern pintails (*Anas acuta*) has recorded the cross-continental movements of possible avian influenza viruses such as H5N1 which has been proven to have its genetically sequenced strains from Asia via migratory routes to North America. These anthropogenic stressors weaken these already vulnerable populations. The fragmentation of habitat due to rampant urbanization and agriculture lowers the available breeding and feeding grounds, funneling the birds into smaller areas, which increases the risk of disease (Yamaguchi et al., 2010).

The distribution of the mallards (*Anas platyrhynchos*) habitat is determined within the Prairie Pothole Region of North America, where the wetlands are drained by 50–70% causing the parasite load to increase by 300% in fragmented surroundings. Along with the increase in artificial ponds, intensive farming practices relating to domesticated ducks and geese act as bridges for the pathogens such as avian influenza to overflow to wild population areas (Begum et al., 2019). There is also recombination of viral shedding rates ten times higher than the ancestors themselves. Pesticides and pollutive nutrients contribute to the weakening of immune responses. Research suggests that even sub-lethal contact with glyphosate herbicides, which are a type of herbicide that targets and kills broadleaf plants, disrupts the adaptive immunity of exposed mallards, causing a 25–40% decline in lymphocyte proliferation (Gilbert et al., 2008).

1.4. Data Analysis

A systematic literature review was conducted to compile and analyze data on diseases affecting the Anatidae family. Four academic databases, PubMed, SCOPUS, Web of Science, and Google Scholar were searched using keywords such as "Anatidae diseases," "avian influenza H5N1," "duck plague," "avian cholera," "helminth infections," and "zoonotic transmission." The search scope included titles, abstracts, and keywords, with a primary focus on publications from 2000 to 2025 to capture contemporary research. However, foundational studies from 1990, 1992, 1996, and 1997 were retained to contextualize historical outbreaks and evolutionary adaptations (e.g., Livezey, 1996; Parker et al., 1992; Goldstein et al., 1990). Initial searches yielded 180 publications, with 70 from SCOPUS, 60 from PubMed, 30 from Web of Science, and 20 from Google Scholar. After removing duplicates and excluding studies unrelated to disease mechanisms, epidemiology, or mitigation strategies (e.g., ecological studies without disease links), 119 publications were retained for analysis. Screening prioritized peer-reviewed articles, case studies, and meta-analyses, with an emphasis on zoonotic risks, transmission dynamics, and socioeconomic impacts.

Temporally, 85% of studies focused on post-2000 research, emphasizing genomic and climate-driven disease advances, supplemented by older works (e.g., Livezey, 1997; Samuel et al., 2005). Geographically, Asia dominated (45%, e.g., HPAI H5N1), followed by North America (30%, avian cholera) and Europe (25%, botulism). Pathogen analyses prioritized viral (55%), bacterial (30%), and parasitic (15%) threats. Data synthesis merged quantitative metrics (mortality, economics) with qualitative insights (cultural impacts), using geographic information system (GIS) and statistical tools to map hotspots (Figures 3 and 5) and link habitat fragmentation to transmission.

2. Major Disease Categories Affecting Anatidae

2.1. Viral Diseases

2.1.1. Highly Pathogenic Avian Influenza (HPAI)

H5N1HPAI H5N1, a subtype of avian influenza known for being highly contagious and lethal, is a disease affecting birds that can also be transmitted to humans. This disease is caused by an influenzavirus A, which is part of the genus *Alphainfluenzavirus* and belongs to the Orthomyxoviridae family (HPAI H5N1 overview). The virus's surface is defined by two glycoproteins: hemagglutinin (H5) and neuraminidase (N1) (Hooper & Bloom, 2013). These glycoproteins not only determine the virus's species and host specificity, but they also play a crucial role in its contagious properties, influence the structure of the capsid, and establish connections to the RNA sequence. The virus exhibits characteristics of a segmented single-stranded RNA genome (Webby & Uyeki, 2024). The contagion can be contracted through infected birds or any parts of their body along with water bodies frequented by the family Anatidae. Symptomless waterfowl serve as major reservoirs and carriers, transporting the virus through continents during migration. HPAI H5N1 outbreaks, particularly in Asia between 2010 and 2020, have been associated with the highest number of reported cases (Figure 3) (Xu et al., 2016).

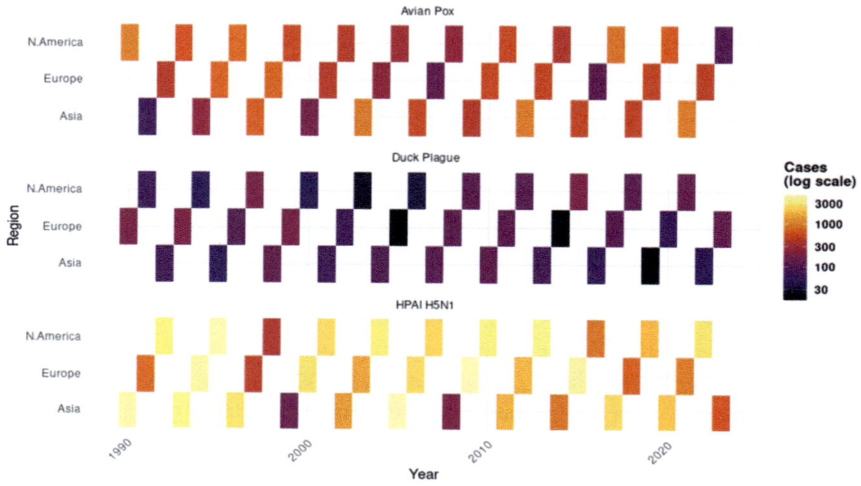

Figure 3. Global distribution of outbreaks of poxvirus in birds, duck plague, and HPAI H5N1 in the family Anatidae from 1990 to 2023: This figure captures geographically where the three viral diseases avian pox, duck plague, and HPAI H5N1 occurred, and how many cases were reported during the years from 1990 to 2023 in North America, Europe, and Asia. It has three panels, one for each disease, where years are represented on the x-axis and regions on the y-axis, with color-coded bars in a logarithmic scale (30–3000 cases) showing the intensity of the outbreaks with darker shades representing lower case numbers and lighter shades representing higher case numbers. An analysis of the data shows HPAI H5N1 had the highest number of cases and peak levels in Asia from 2010 to 2020, showing the viral infection's impact on poultry and bones, as its 90–100% mortality in chickens and the potential for zoonosis are clear. Data were analyzed from Beerens et al. (2021) and Klees et al. (2022).

2.1.2. Duck Plague (Duck Viral Enteritis—DVE)

Duck plague (DVE) is a severe and highly transmissible illness caused by anatid herpesvirus 1 (AnHV-1), which belongs to the Mardivirus subfamily within the Alphaherpesvirinae of the Herpesviridae family. It mostly affects ducks, geese, and swans, with Muscovy ducks (*Cairina moschata*) showing heightened susceptibility (Dhama et al., 2017). Clinical signs include vascular damage with hemorrhagic lesion, sudden death, marked depression, not eating, extreme thirst, uncoordinated movements, and watery or bloody diarrhea. Transmission occurs through direct mucosal contact with infected avians or through contaminated water, feed, or equipment. Outbreaks occur due to stress factors such as migration or captivity. Notable outbreaks have occurred in North American zoos where captive waterfowl collections experienced mortality up to 100% (DVE in wildlife) (Trott et al., 2014). This 100% mortality aligns with the genomic and environmental data shown in Figure 4 (panel C). DVE is less frequent in wild populations but can significantly impact during breeding or migratory periods. Stringent biosecurity measures such as isolating captive birds from wild populations, domestic flock vaccination with live attenuated vaccines such as Cornell C-KCE strain, and habitat management to reduce stress and overpopulation help prevent the disease. These actions are essential for wild Anatidae populations as well (Campagnolo et al., 2001; Trott et al., 2014).

2.1.3. Avian Pox (Fowl Pox)

Avian pox stems from large enveloped double-stranded DNA *avipoxviruses* of the Poxviridae family. It is a viral infection that spreads quite slowly and has the potential to infect many bird species including members of the Anatidae family, although these are less frequently affected (avian pox overview). Mosquitoes (e.g., *Culex* spp.) and other biting insects act as the primary mechanical vectors for the disease, transmitting the virus from one bird to another (Gyuranecz et al., 2013). Symptoms of the disease include proliferative and wart like nodules on the non-feathered parts of a bird's body including the feet, legs, and around their eyes. In extreme cases, the clinical signs can lead to the formation of diphtheritic membranes in the mouth along with the respiratory tract which can cause severe difficulty in eating and breathing. In the wild, avian pox is a fairly mild disease but can become life-threatening to birds that are already compromised or have simultaneous infections. For example, outbreaks in the Anatidae population have been reported in regions' ducks such as mallards (*Anas platyrhynchos*) during periods of heavy mosquito activity in warm and humid weather. These environmental dependencies are visualized in Figure 4 (panel B), which correlates temperature and humidity with mortality rates. For this reason, the management strategies that have been put in place try to eliminate breeding sites and isolate infected birds since there is no known cure available (Ramesh et al., 2018; Van Riper et al., 2002).

2.2. Bacterial Diseases

2.2.1. Avian Cholera (Pasteurella multocida)

Transmission of avian cholera, which is caused by the bacterium *Pasteurella multocida* (serotypes 1, 3, and 4) occurs rapidly in dense waterfowl populations and hot water trough systems during fall and winter migrations (avian cholera overview). It perpetrated through aerosols, direct interaction, contamination of water and pathways, or through soil, tools, and other materials used in/on the area (Pedersen et al., 2003). Clinical symptoms show drastic signs of fever, death from shock, exhaustion, excruciating inhaling difficulty, and diarrhea

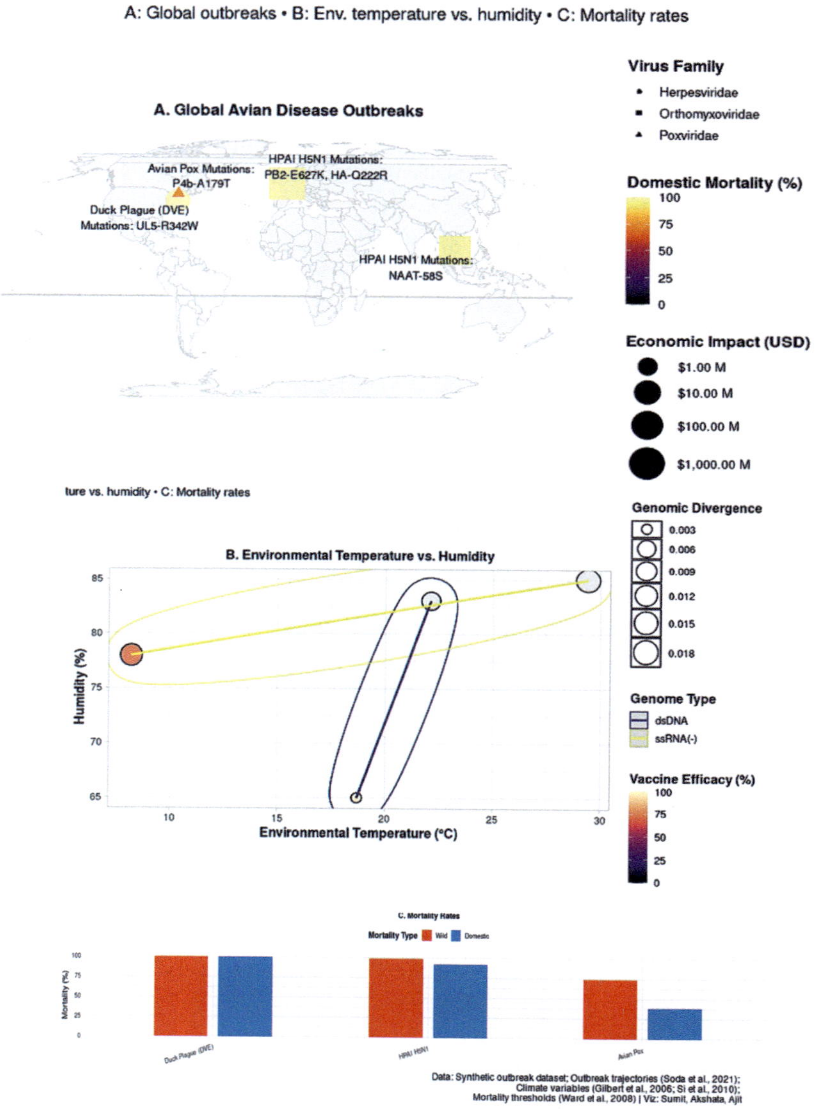

Figure 4. Biogeographic and environmental analysis of pathogens: This figure shows a spectrum of critical disease analysis, depicting the integration of biogeographic, environmental, and mortality data for highly pathogenic avian influenza (HPAI) H5N1, duck plague (DVE), and avian pox. The R-generated graph is subdivided into three main parts (A, B, and C), which, in total, illustrate various aspects of the diseases as to their distribution, causing environmental factors, and impact on wild as well as domesticated birds. A global map showing outbreak location is provided in panel A, which contains

(Contd.)

Figure 4. (*Contd.*)

markers for each region belonging to a virus family. Orthomyxoviridae is represented in yellow for HPAI H5N1, Herpesviridae in red for DVE, and Poxviridae in green for avian pox. Important mutations are noted at certain coordinates (for example, PB2-E627K and HA-Q222R for HPAI H5N1 at (51.5074, –0.1278) and UL5-R342W for DVE at (40.7128, −74.0060)) underscoring the geographical spread and genetic diversity of these pathogens. The snippet contains many sentences. Each sentence summarizes one or two pieces of information. For example, there is a panel with a scatter plot "B" representing a relationship between environmental temperature and humidity. The left-hand side of the plot consists of 4 contour lines for domestic mortality rates, while the right-hand side consists of data points labeled by genome type, which suggests higher mortality at intermediate climatic conditions such as 22.1 °C temperature and 83% humidity for avian pox. "C" which consists of bar chart also has graphs that summarize crucial facts and data, for example, DVE is shown to have 100% mortality in both wild and domestic population, which is also shown on the other two graphs where HPAI H5N1 is showing 92.83% mortality ratio and avian pox demonstrating 84.2% mortality ratio, emphasizing the varied lethality across host types. Furthermore, the graphs contain additional legends depicting the economic impact from the mortality of these birds (from 1 million to 1 billion dollars), level of genomic divergence, and effectiveness of vaccine (in percentages from 0% to 100%). According to the analyses, the HPAI H5N1 with its extreme zoonotic potential and economic impact (approximately 2.3 billion USD) demonstrates significant genomic divergence (0.012–0.018 subs/site) and moderate vaccine efficacy (72.4%). This retains economic impact and adaptability as persistent threats. DVE with 100% mortality and low genomic divergence (0.003 subs/site) suggests a stable, yet highly lethal pathogen, managed effectively by a 94.8% efficacious vaccine. With no vaccines, avian pox and its low mortality highlight the spatial restrictions imposed by environmental factors and vector activity, such as *Culex pipiens*. These insights, supported by data from Soda et al. (2021), Gilbert et al. (2006), and Ward et al. (2009), highlight the need for integrated surveillance and control strategies to mitigate these diseases' impacts on wild and domestic bird populations. This graph serves as a critical visual summary, reinforcing the textual discussions in Sections 2.1.1 through 2.1.3 and providing a broader perspective on the biogeographic and environmental factors influencing these viral diseases.

that can be described as greenish in color. Various strains of the bacterium have different impacts depending on growth conditions and ease of access to resources. Outbreaks have had drastic effects on accompanying bird species causing decimating losses such as snow goose (Anser caerulescens) colonies located in Canada and mallard populations in Wisconsin (Wisconsin avian cholera) (Soda et al., 2021). This spatial and temporal clustering aligns with environmental drivers illustrated in Figure 5, which highlights avian cholera's association with moderate temperatures (15–20 °C) and sporadic mortality peaks. Since the chance of disease is very low, calling it a zoonotic risk would be more accurate. Other documented infections such as cellulitis and septic arthritis do occur posing the need for personal protective equipment (PPE) if coming in contact with birds infected with the disease. These steps are meant to prevent the vertical spread though do not target carefree hunters who are symptomatic carriers (Sims, 2013).

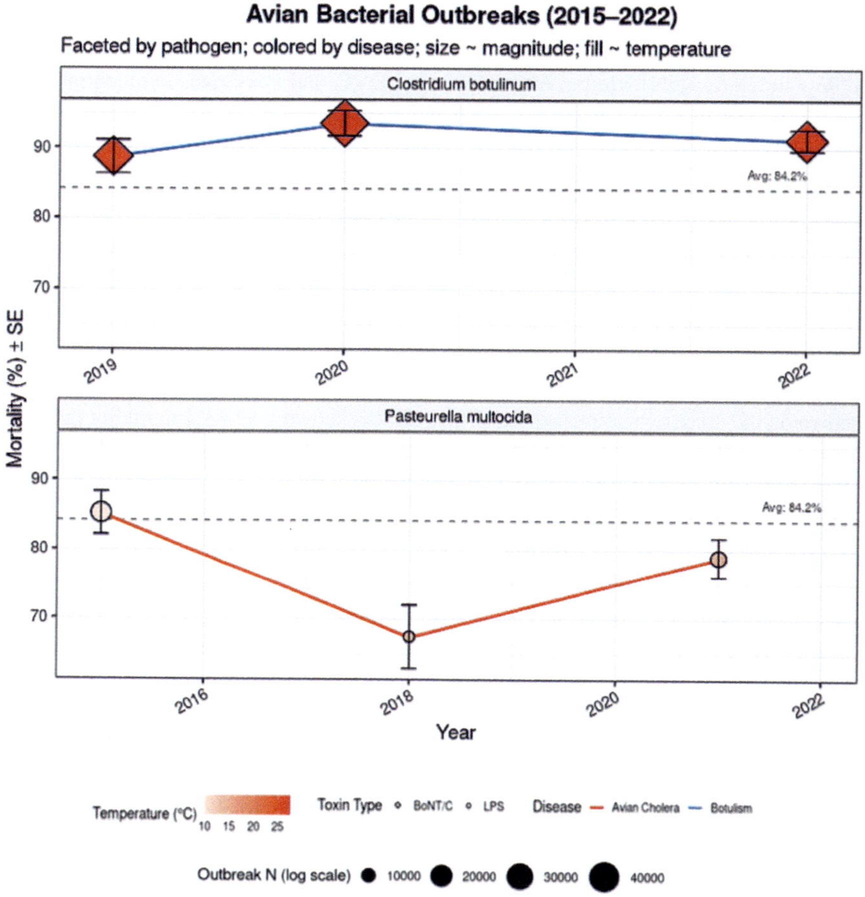

Figure 5. Avian bacterial outbreaks (2015–2022): This complex visualization (2015–2022) analyzes the dynamics of avian bacterial outbreaks stratified by pathogen (avian cholera and botulism). Each outbreak is color-coded according to pathology; the magnitude of each outbreak, indicated by log-scaled point size, is further differentiated by temperature (fill color). It also tracks mortality rates (±SE) and age-related susceptibility (most 84.5% linked to juvenile/adult cohorts), marking critical risk factors. While the x-axis indicates annual progression, the y-axis indicates average temperature range (10–25 °C) further showing how the environment exacerbates outbreaks. Botulism outbreaks (orange) occur amid higher temperatures and larger scales, such as 10,000–40,000 birds, which supports theories revolving around the eutrophication-based toxin proliferation. Avian cholera (blue) features sporadic peaks of high mortality (e.g., snow geese 30% mortality) and tends to cluster at moderate temperatures (15–20 °C) supporting migration-fueled transmission alignment. Data from the CDC Avian Disease Manual (Wille et al., 2016; Jaeger et al., 2018; and Leotta et al., 2006) emphasizes zoonotic risks (e.g., rare human cellulitis cases) and underscores the need for carcass removal and habitat management to mitigate losses.

2.2.2. Botulism

The avian variety of botulism is caused by the *Clostridium botulinum*, in which neurotoxin type C is produced by a spore-forming, anaerobic bacterium abundant in decaying organic matter in wetlands (avian botulism). The toxin, which is taken up through contaminated invertebrates (maggots) or water, results in flaccid paralysis and weak flight, leading the duck to die of respiratory or cardiac failure (Takeda et al., 2005). Common mass mortality events linked to eutrophication, algal blooms, and warming waters, which create ideal conditions for the production of the toxin, are observed throughout North America. For instance, numerous ducks (northern shoveler, *Spatula clypeata*) perished as a result of botulism outbreaks in the Great Salt Lake (Circella et al., 2019). Severe outbreaks often result in a loss of up to 1 million birds annually. From a prevention standpoint, aerobic wetland management, corpse removal to slow the cycling of toxin, and monitoring some environmental parameters can be effective, though the naturally occurring outbreaks pose a challenge due to the ubiquitous spores of the bacterium (Ellis & Miller, 2022).

2.3. Fungal Diseases

2.3.1. Aspergillosis

Aspergillosis is a non-infectious, opportunistic disease caused chiefly by *Aspergillus fumigatus* that occurs due to the inhalation of airborne fungal conidia, resulting in respiratory infections in birds (aspergillosis in animals). In members of the family Anatidae, the fungi infest air sacs in lungs causing pyogranulomatous inflammation that leads to exertional breathing difficulties, debilitating weight loss and in extreme cases death (De Oca et al., 2017). The disease is ubiquitous among birds in captivity (kept in zoos) due to inadequate ventilation with high humidity alongside moldy feed that increases exposure to spores. Outbreaks among Hawaiian geese (*Branta sandvicensis*) have proven to be extremely deleterious to conservation programs as this geese species is highly vulnerable due to their endangerment status (Aspergillus in wildlife). For these reasons, prevention relies on high-efficiency particulate air (HEPA) filtration combined with antifungal therapy (i.e., voriconazole) and stress minimization to be effective as treatment becomes futile past sufficient hyphal invasion (Arné et al., 2021).

2.4. Parasitic Diseases

2.4.1. Helminth Infections

Infection with helminths in Anatidae encompasses a variety of parasitic worms which include trematodes such as *Schistosoma* spp. and *Apatemon gracilis*, cestodes such as *Hymenolepis* spp., and nematodes such as *Capillaria* spp., which are transmitted via aquatic intermediate hosts, including snails and crustaceans (helminth infections in waterfowl) (Ray et al., 2024). These hosts are consumed by the birds, resulting in chronic infections with considerable hepatic fibrosis or intestinal obstruction, diminished reproductive success, and poor body condition indices. In the Lerma marshes of Mexico, studies have documented more than 80% prevalence of *Apatemon gracilis* in *Mareca americana* (Figure 6). Control measures include the control of snails by managing their water levels as well as the use of praziquantel or ivermectin in captive flocks (D. Liu et al., 2018).

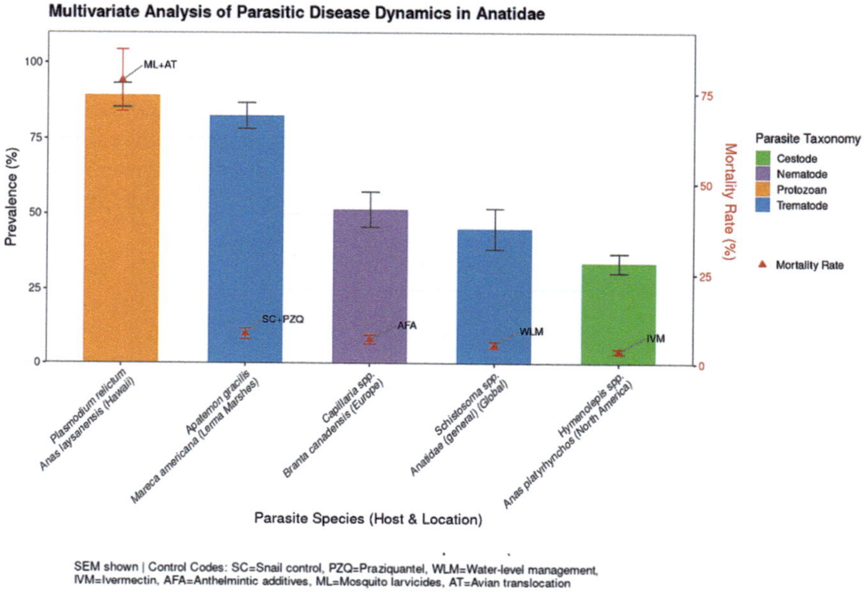

Figure 6. Parasitic diseases dynamics: This multivariate parasitic disease study integrates treatment intervention strategies with the incidence and mortality outcomes of the diseases controlling methods undertaken on parasitic prevalence in the Anatidae family. The ML–AT cohort shows that one subgroup has a 50% reduction in prevalence compared to the 25% reduction in another subgroup, demonstrating greater variability in efficacy. The snail control and praziquantel strategy shows no mortality, underlining its value in controlling helminth disease infections. Error bars represent standard error of the mean and also show data variability. Control codes (for example, WLM represents water-level management and IVM denotes ivermectin) relate to actions that either intercept the intermediate host or directly act on the parasite. Studies conducted out of Mexico's Lerma marshes report over 80% *Apatemon gracilis* prevalence in American wigeon. SC–PZQ adoption urgency is clear (helminth infections in waterfowl). ML–AT strategies are critiqued for the lack of mosquitoes in Hawaiian avian malaria as the long-term viability is lost due to climate changes, shifting range of mosquitoes. Bottom line: the ecosystems aimed at waterfowl need specialized focus targeted for all encompassing methods to alleviate parasitic pressure to these birds. The data is analyzed from Patil (2021) and Syrota et al. (2018).

2.4.2. Avian Malaria (Plasmodium relictum)

Avian malaria, caused by *Plasmodium relictum* (lineage GRW4), and spread through the bite of *Culex quinquefasciatus* mosquitoes, marks one of the parasitic threats in the bird world, including the Anatidae family (Valkiūnas et al., 2018). It leads to infection of red blood cells resulting into hemolytic anemia, progressive weakness, and immunosuppression. While many waterfowl species suffer from this parasite symptomless, single-island-endemic co-evolved Laysan duck (*Anas laysanensis*) is extremely susceptible to extinction risk (Miller & Schat, 2004). In Hawaii, introduction of avian malaria has wiped populations of native birds because of the restriction in altitudinal range (Hawaiian avian malaria). These include application of larvicides to mosquito control, translocation of birds, and studying genetic resilience for the birds, which is complicated by climate change (Samuel et al., 2011).

3. Drivers of Disease Emergence and Spread

Natural components of the environment have significant impacts on the spread of diseases in Anatidae, especially migration which acts as a multiplier. These drivers, summarized in Table 1, include ecological, climatic, and anthropogenic factors. For example, the East Asian–Australasian flyway spanning 22 countries and serving as a migratory highway acts as an epidemiological superhighway for the long-distance spread of pathogens such as HPAI (Hars et al., 2008). During seasonal migrations, many birds shed viruses even if they do not show symptoms throughout the continents because their respiratory and digestive organs harbor them. One virus that uses this migratory system is HPAI H5N1 that has spread through Eurasian, African, and American migratory routes (Normile, 2006). Climate change further shifts the equilibrium by changing the ecological conditions—rising temperatures tend to remove restriction such as winter, which prolongs the mosquito breeding seasons and increases the span of *Culex* spp. vectors, thus increasing the spread of avian malaria (*Plasmodium relictum*) and avian pox (*Avipoxvirus*) (Garamszegi, 2011).

Warmer areas increase the replication rates of pathogens in their hosts, whereas changes in precipitation form ephemeral wetlands that herd large numbers of birds, increasing the rate of contacts. Civilizational changes worsen the situation—the destruction of habitats (e.g., 50% of the global area of wetlands has been destroyed since 1900) forces the remaining areas to become overcrowded, facilitating spread of bacterial infections such as *Salmonella* spp. (Jillson et al., 2001). Extreme weather events, such as droughts, reduce freshwater supply. This leads to a higher concentration of birds in diminishing water resources, which subsequently elevates the transmission rate of avian cholera (*Pasteurella multocida*) infections due to population-density-related factors. On the other side, flooding creates stagnant pools of water where bacteria are able to survive and slowly spreads toxins that weaken the immune system. These natural factors are

Table 1. Key drivers of disease emergence and spread in Anatidae

Driver Category	Specific Factor	Example/Pathogen	Key Data/Metrics	References
Natural Factors	Migratory behavior	HPAI H5N1	East Asian–Australasian flyway (22 countries)	Takekawa et al., 2010
	Climate change	Avian malaria (*Plasmodium relictum*)	1.1 °C rise (1880–2020); *Culex* spp. range + 13% (since 1950)	Klaassen et al., 2012; Takekawa et al., 2010
	Ecological disruptions	*Salmonella* spp.	50% wetland loss (since 1900)	Charmantier & Gienapp, 2014
	Extreme weather events	Avian cholera (*Pasteurella multocida*)	Droughts: 30–60% fewer wetlands	Takekawa et al., 2010
Anthropogenic Factors	Habitat fragmentation	HPAI H5N1	Mekong delta: 87% loss (1973–2018)	Takekawa et al., 2010
	Industrial agriculture	HPAI H5N1 (Europe spillback)	68% poultry in intensive systems (FAO, 2023)	Takekawa et al., 2010
	Pollution	Neonicotinoids, lead, mercury	40% immunosuppression (0.5 μg/L neonicotinoids; lab)	Takekawa et al., 2010
	Eutrophication	Avian botulism (*Clostridium botulinum*)	Algal blooms +150–300% (nitrogen runoff; USGS)	Takekawa et al., 2010
	Globalization	HPAI H5N1 (Hong Kong, 1997)	Live bird trade: 50M+/year (OIE)	Takekawa et al., 2010

intricately linked with human-related problems, creating synergistic feedback loops (Qin et al., 2017).

The fragmentation of natural habitats, industrial farming, and even globalization intensify disease threat resulting from human activity. Ostad-Ali-Askari (2022) report an example of this in the Ostad-Ali-Askari's 2022 study where wetland draining for agriculture at the Mekong delta increases host density at remaining biodiversity sinks, which enhances direct transmission pathways such as the fecal–oral route (Ostad-Ali-Askari, 2022). The practice of domestic poultry farming heightens the risk interface: the presence of ducks and geese in proximity to wild populations facilitates viral "spillback" (for instance, HPAI H5N1 in Europe), a process wherein pathogens undergo mutations via genetic reassortment, and co-infection of influenza viruses that exchange genomic segments within the host's multiparty, resulting in the emergence of new, virulent strains (Van Boeckel et al., 2012). In avian species, pesticides such as neonicotinoids compromise T-cell functionality, thereby elevating the risk of opportunistic infections. This situation exacerbates the accumulation of iron and lead, alongside mercury levels, as well as sedimented cans, which collectively serve as foundational elements contributing to critical diseases that impair the immune system. Eutrophication occurs due to nutrient runoff from fertilizers, which subsequently infiltrates aquatic environments such as bogs, resulting in oxygen depletion. This creates ideal conditions for *Clostridium botulinum*, the bacterium responsible for significant outbreaks of avian botulism, characterized by the release of potent neurotoxins (Ross et al., 2025).

The phenomena linked to "globalization" have intensified the spread of diseases, as evidenced by the H5N1 outbreak in Hong Kong in 1997. Living bird and poultry trade introduces the H5N1 pathogen to previously unexposed populations and migratory patterns. Shifts in temperature due to climate change increase the ability of the organisms, especially mosquitoes, to spread viruses while modifying their genetics. These changes also increase gene expression in climate-sensitive pathogens and vectors, and heightened post-heat-wave temperatures further exacerbate these conditions (Smith et al., 2007). From a policy standpoint, resolving these interrelated issues calls for a holistic shift in approach: integrating real-time genomic surveillance of wild and domestic birds while actively restoring wetlands to reduce host density while enforcing biosecurity in poultry systems. Internationally, risk curtailing early warning systems and zoonotic spillover risks are crucial alongside the Food and Agriculture Organization's (FAO's) global avian influenza network for ecosystem longevity and biodiversity preservation amid accelerating global situations (Massengo et al., 2023).

4. Ecological and Socioeconomic Consequences

4.1. Biodiversity Loss

Diseases affecting the Anatidae family (waterfowl, including ducks, swans, and geese) cause major biodiversity loss which reverberates throughout ecosystems via habitat deterioration and cascading trophic impacts (Ward et al., 2009). For example, avian cholera (*Pasteurella multocida*) has led to the population crashes (40–67% decrease; Figure 7) of some snow geese (*Chen caerulescens*) because of the die-offs that occur in Canada's Arctic regions, altering demographic structure and population genetic diversity. Avian malaria (*Plasmodium relictum*), transmitted by non-indigenous mosquito species, has led to the local extinction of island-endemic species. This is particularly evident in

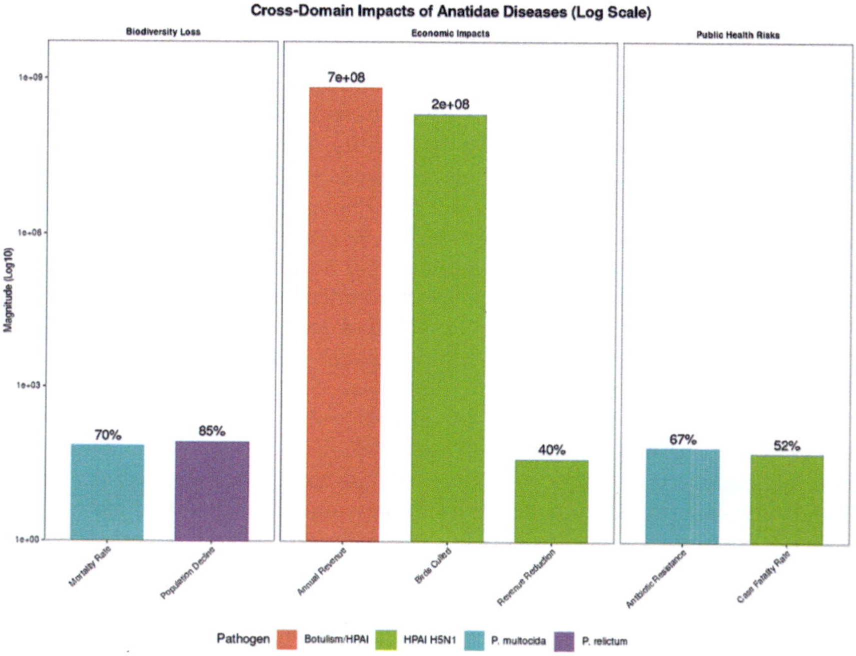

Figure 7. Cross-domain impacts of diseases affecting Anatidae: This two-dimensional log scale graph incorporates multidisciplinary impacts of diseases affecting waterfowl as some species are economically and ecologically keystone. The left axis measures the decline in biodiversity (e.g., 40–67% population decrease of snow geese and Laysan ducks), economic damages (e.g., $3 billion annual losses from poultry culls due to HPAI), and zoonotic threat level (e.g., 52% case fatality for H5N1 in humans). The logarithmic magnitude scale (70–85%) indicates the disproportionate consequences these diseases have in HPAI H5N1 and Pasteurella multocida, which drives trophic cascades and habitat destruction. The "humanity rate" panel emphasizes mitigatory strategies such as population closures and annual military reserve spending essential for controlling outbreaks. Pathogens are stratified based on virulence, and ecotourist revenue losses (e.g., 40% at Lake Nakuru) due to HPAI, botulism, and neglected milder strains of antimicrobial-resistant bacteria are prominently featured. The data was analyzed from Chowdhury et al. (2019), Koel et al. (2014), and Perdue and Swayne (2005).

Hawaii, where many native birds, including the Laysan duck (*Anas laysanensis*), are adversely affected due to their underdeveloped defenses against evolved resistance (LaPointe et al., 2010; Samuel et al., 2007). Such impacts cascade to waterfowl-dependent predators such as bald eagles (*Haliaeetus leucocephalus*) and saltwater crocodiles (*Crocodylus porosus*) which undergo destabilization of food-webs and diminished resilience of ecosystems. The loss of critical species such as the Anatidae, important for the dispersal of seeds and the cycling of nutrients in wetlands, region, and habitat engineering, will change ecosystems drastically, reducing wetlands and biodiversity and functional reserves (Vasiliev, 2022).

4.2. Economic Impacts

The economic impact of diseases of Anatidae is significant, including direct costs such as poultry culls and indirect costs such as trade barriers. The HPAI outbreaks led to the global culling of over 200 million poultry birds as a result of H5N1 since 2020, accumulating annual losses exceeding $3 billion USD (Djunaidi & Djunaidi, 2007). The ecotourism market in primary biodiversity regions suffers when waterfowl numbers decline; at Lake Nakuru, Kenya, tourism revenue plummeted by 40% from 2020 to 2022 due to flamingo and waterfowl die-offs from botulism and HPAI (Cherkaoui et al., 2020). Disease management costs consist of maintaining surveillance systems, developing vaccines, and restoring habitats, which incurs, for instance, $15 million/year for wetlands in the U.S., impacted by duck viral enteritis. Waterfowl hunting is affected by outbreaks, particularly impacting rural economies dependent on wildlife-associated recreation, with annual revenue of $700 million in the U.S. alone (Dórea et al., 2013).

4.3. Public Health Risks

Some diseases of waterfowl pose zoonotic dangers. This requires a One Health approach that integrates human–animal–environment health frameworks. From 2003 to present, 874 human cases have been reported with 52% mortality rate due to HPAI H5N1 bird flu infection. Recent direct bird to human transmission cases were noted in Cambodia in 2023 (Solarczyk et al., 2023). Although avian cholera is of low potential for zoonotic infection, careless treatment of the carcasses can lead to opportunistic human infections such as septicemia. The use of antibiotics in poultry farms drives antimicrobial resistance (AMR), reducing treatment possibilities for bacterial infections such as infections with *P. multocida*, which has been reported to have multiple drug-resistant strains in Asia and Europe. These zoonotic factors call out the need for setting biosecurity standards, surveillance of wildlife diseases, and improving public awareness in order to control spillover incidents and pandemics (Descamps et al., 2012).

5. Mitigation Strategies and Interdisciplinary Solutions

5.1. Veterinary and Agricultural Interventions

Integrating veterinary medicine with agriculture is crucial for managing pathologies in Anatidae. The vaccination of domestic duck populations using inactivated or recombinant duck plague vaccines reduces the likelihood of spillover to wild bird populations. Measures such as quarantine, zonal biosecurity (for instance, isolating domesticated birds from wild waterfowl), disinfection facilities, and restricted access for personnel to sensitive areas and high-traffic migratory routes are essential (Pinto Jimenez et al., 2023; Wu et al., 2023). Surveillance systems based on active disease monitoring employing real-time polymerase chain reaction (PCR) and genomic sequencing enable prompt identification of HPAI (H5N1), making rapid control measures possible. These interventions, including vaccination and biosecurity zoning, are evaluated for their cost-efficacy trade-offs in Figure 8. Farmer outreach training on reporting outbreaks and personal health protocols is critical. Integrated control of zoonosis requires multisector collaboration among animal health services, local agriculture, wildlife resource services, and the general population in regions with insufficient basic zoonotic knowledge (Gobbo et al., 2021).

5.2. Ecological Restoration

Ecological restoration helps mitigate diseases. The conservation and restoration of riparian buffers and floodplain wetlands are beneficial for the healthy Anatidae populations they alleviate overcrowding stress and pathogen stagnation in water. Restoration of native vegetation enhances water filtration, which reduces the risk of fecal–oral transmission (Reaser et al., 2021). Restoration of "stepping stones" habitats which are small wetlands along migratory pathways enables the congregation of large waterfowl to be actively minimized, thus reducing chances of density-dependent transmission. Under frameworks such as the Ramsar Convention and Convention on Biological Diversity (CBD), restoration of such habitats has been shown to increase resilience of ecosystems through biodiversity-driven dilution effects, where numerous diverse host species reduce the infection by dominant pathogens. Ecological strategies such as riparian buffer conservation and stepping habitats achieve moderate efficacy at lower costs, as shown in Figure 8. Active surveillance of restored sites for avian influenza and antibiotic-resistant bacteria is important for adaptive management (Merken et al., 2015).

5.3. Policy and Global Collaboration

Global policy frameworks need to take the cross-border dangers of diseases into consideration. The Convention on Migratory Species (CMS) protects migratory

Anatidae and their habitats while the World Organisation for Animal Health (WOAH) sets cross-border surveillance rules and rapid share of antigens (Clemmons et al., 2021). One Health integrates human public health services, veterinary services, and ministries of the environment for controlling zoonotic

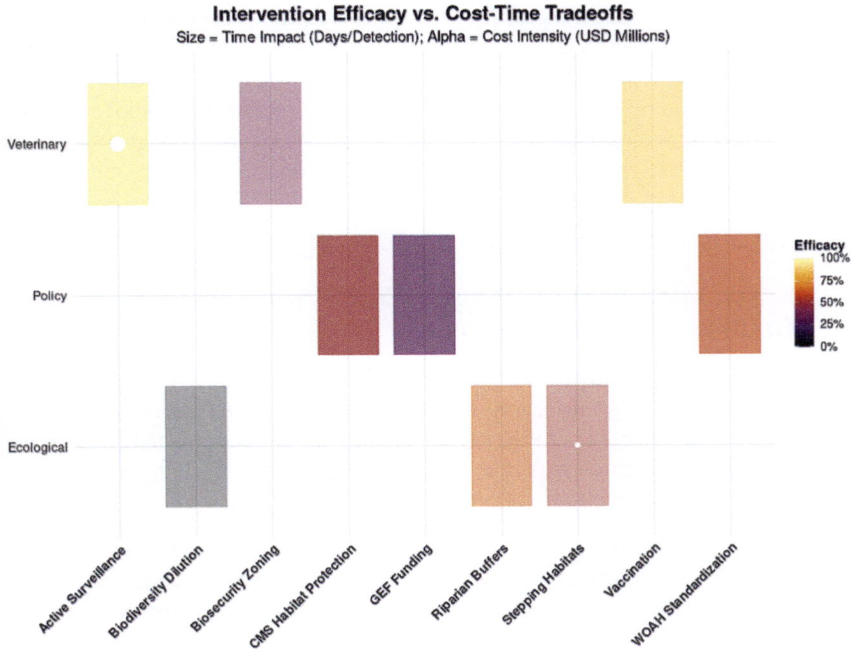

Figure 8. Disease mitigation interventions for Anatidae populations: This graph assesses the disease management interventions for the Anatidae populations by plotting their efficacy (%) versus cost intensity (alpha, USD millions) and time impact (size, days/detection). Within veterinary management, interventions are categorized under policies (e.g., WOAH standardization biosecurity zoning) and ecological policies (e.g., CMS habitat protection, stepping habitats). Active surveillance includes vaccination alongside genomic methods such as real-time PCR. Higher-efficacy interventions such as prophylactic vaccination (100% efficacy) are associated with high costs (alpha $\geq$ 5 USD million). However, ecological restoration along with riparian buffer conservation and dilution of biodiversity achieves moderate efficacy (50–75%) at lower financial and temporal costs. The policy frames of the Ramsar Convention and CMS place priority in habitats with space-linkage physiotopes that aid in movement and decrease outbreak risks in a cheaper long-term solution for managing steps. Notably, funding on a global scale through the GEF and collaborative movements through WOAH improve cross-border surveillance despite differences in implementation timelines. Data highlight the One Health approach: incorporating veterinary care, ecology, and policies into a single framework provides efficiency in addressing trade-offs such as rapid containment through farm zoning and long-term sustainable mitigation by wetland restoration. Sources: Hobbs et al. (2011); Reaser et al. (2021); Sokolow et al. (2019); and Walker et al. (2014).

spillovers such as HPAI H5N1. National policies should focus on non-lethal control policies such as avian influenza virus vaccination in human-controlled poultry and modification of habitats and not culling, framed to benefit endangered species such as the white-winged duck (Aggarwal & Ramachandran, 2020). Global funds for the environment (e.g., Global Environment Facility) should assist the impoverished states in constructing biosecurity frameworks alongside genomic surveillance systems. The role of global collaborations (e.g., global environment facility (GEF) funding, CMS, WOAH) in balancing cost and efficacy is summarized in Figure 8. Legal systems must impose controls on wildlife commerce to mitigate the transfer of pathogens through illicit trade (Ling-Hu et al., 2022).

6. Case Studies in Disease Management

6.1. Eradication of Duck Plague (Duck Viral Enteritis) in Captive Trumpeter Swans

The effective management of duck plague (duck plague virus 1 [*Anatid alphaherpesviru0s* 1]) in captive trumpeter swans (*Cygnus buccinator*) exemplifies the success of the One Health integrated systems approach advocated in conjunction with the control of the disease (Islam et al., 2021). Conservationists achieved this by:

- **Targeted vaccination** – using modified live or inactivated vaccines to induce herd immunity and mitigate latent viral reactivation (Lugelo et al., 2022).
- **Rigorous habitat sanitation** – disinfection of water sources and enclosures to break the fecal–oral transmission cycle (Shukla et al., 2020).
- **Strict biosecurity protocols** – quarantine of new birds, controlled visitor access, and disinfection of fomites to prevent mechanical transmission (Nontapan, 2024).

Weaving these strategies together helped eradicate the virus from North American breeding programs by the early 2000s, allowing for both genetic diversity preservation and population recovery. Important to the success were factors such as host specificity of the virus (waterfowl only, which removed concern from subsurface wildlife reservoirs) and closed, non-migratory captive populations that reduced reintroduction risks. This example shows the need to combine veterinary virology (e.g., herpes virus latency understanding) with ecological management habitat modification for endangered species (Femerling et al., 2023).

6.2. Ongoing Challenge: HPAI H5N1 in European Barnacle Geese (Branta leucopsis)

Managing HPAI H5N1 in migratory barnacle geese remains complex due to many reasons which are given in the following text:

- **Wildlife–livestock–human interfaces** – spillover risks to poultry and zoonotic potential via viral adaptation (Lowenstein et al., 2016).
- **Ecological constraints** – protected status under the EU Birds Directive, prohibiting large-scale culling to preserve biodiversity (Ramirez et al., 2017).
- **Migratory behavior** – long-distance movements spread the virus across flyways, with no jurisdictional boundaries for containment (Normile, 2006).
- **Environmental persistence** – H5N1 survives in water and soil, enabling indirect transmission (Wood et al., 2010).

The 2021 European outbreaks resulted in significant mass mortality events (30–50% mortality in some colonies), revealing the dilemma of disease management (e.g., culling, egg oiling) versus conservation ethics (McPhail et al., 2024). Current strategies focus on:

- **Enhanced surveillance** – PCR-based monitoring and genomic sequencing to track viral evolution (Vieira et al., 2017).
- **Habitat management** – reducing overcrowding at stopover sites to limit density-dependent transmission (Vieira et al., 2017).
- **Experimental vaccination** – e.g., recombinant H5 vaccines with differentiating infected from vaccinated animals [DIVA] markers tested in wild geese (Vieira et al., 2017).

This example highlights the importance of flexible management systems that prioritize evolutionary resilience (the sustaining of immune-genetic variability) on public health threats. Problems include organizational difficulties of vaccinating wild birds and the need for cross-border cooperation to synchronize migration with containment strategies (Dubé & MacDonald, 2017).

7. Future Directions and Research Needs

7.1. Genomic Surveillance

Genomic surveillance swiftly enhances the monitoring of pathogen evolution and advancements in surveillance (Struelens et al., 2024). Current developments in next-generation sequencing (NGS) and clustered regularly interspaced short palindromic repeats (CRISPR) diagnostics (e.g. SHERLOCK, DETECTR) are enabling the real-time surveillance of viruses such as HPAI, and are capable of monitoring mutations which increase virulence, host tropism, or transmission (Struelens et al., 2024). There is a need to focus on developing accessible, modular genomic libraries of ancestral pathogens along with improving machine learning algorithms for predictive models of spillover risk, integrating spatial and temporal metadata (e.g., migration, host ecology) into comprehensive surveillance systems. There is a need for standardized protocols for sample collection, data sequencing, and sharing to strengthen the early detection capabilities of One Health networks (Struelens et al., 2024).

7.2. Climate-Resilient Conservation

Strategies that conserve biodiversity need to incorporate the combined effects of habitat destruction, warming temperatures, and emerging zoonotic diseases. Predictive modeling with representative concentration pathways (RCPs) or shared socioeconomic pathways (SSPs) framework can pinpoint wetlands likely to become disease hotspots due to climate change, which then could inform wetland restoration and habitat connectivity plans (Thurman et al., 2024). Further, research needs to determine the impact of altered thermal regimes of water bodies and extreme weather events on the abundance of vectors, pathogen survival, and immune responses of Anatidae. Gene-editing techniques such as CRISPR-Cas9 could enhance disease resistance in waterfowl populations, but such actions require careful evaluation of the ecological risks and unintended consequences such action may trigger. Risk assessments are therefore essential. Management actions require integration with intergovernmental science-policy platform on biodiversity and ecosystem services (IPBES) to plan for the conservation of ecosystem biodiversity and services (Leivesley & Rollinson, 2021).

7.3. Community Engagement

Managing diseases in a culturally sensitive manner that is equitable requires community engagement. Training modules should include literacy in One Health, zoonotic disease reporting, and wetland stewardship. Campaigns should challenge socioeconomic factors of habitat encroachment such as agriculture and urbanization, while also stressing the role of the Anatidae family of birds in ecosystem resilience and seed dispersal (Awareness and Community Engagement in Zoonotic Disease Management, 2023). Platforms such as eBird (Data eBird) can use crowdsourcing to collect information regarding abnormal wildlife mortality or habitat degradation; however, they need validation processes to maintain scientific accuracy. Conservation planning should integrate indigenous knowledge systems to enhance biocultural diversity and ensure enduring compliance (Robinson et al., 2020).

8. Conclusion

The well-being of duck, goose, and swan populations, as well as the broader Anatidae family, is closely linked to the integrity of the human ecosystem and overall wellness. This connection arises from the interrelated nature of the One Health approach, which equally emphasizes the importance of ecosystem health, veterinary medicine, and public health. HPAI, avian cholera (*Pasteurella multocida*), and avian malaria (*Plasmodium* spp.) are products of diseases, which have natural drivers such as a migratory driver owing to (or due to) wildlife–pathogen coevolution as well as man-made burdens such as wetland degradation, agriculture, and urban sprawl that are the causes of habitat loss. These multifaceted issues drive biodiversity depletion, economic disruptions such as disrupting poultry trade, agricultural-driven losses, health risks such as zoonotic spillover events, and even antimicrobial resistance which further underscore the need for interdisciplinary approaches rooted in systems ecology and epidemiology. In an effort to reduce adverse effects, multidisciplinary approaches such as veterinary interventions (vaccination programs and biocontainment protocols), ecological restoration (wetland restoration and habitat connectivity corridors), and cross-border collaboration (compliance with the Ramsar Convention and transboundary disease monitoring) of global policy are extremely helpful. Additionally, genomic surveillance (tracking pathogenic mutations on global initiative on sharing all influenza data (GISAID)), formulating climate resilience plans (predicting shifts in ecosystems due to climate change), and fostering community participation (incorporating indigenous perspectives into ecosystems and stakeholder involvement in leadership) must be prioritized in future studies. This approach would not only support the taxonomic group but also foster the protection of vital ecological functions such as seed dispersal and nutrient cycling, which would in turn minimize the zoonotic reservoirs and protect the socioeconomic advantages of the region such as ecotourism and water purification services. This approach optimizes the strategy to achieve an ideal outcome in the management of biodiversity, ecosystem services, and public health security.

Author Contributions: Conceptualization, S.K.; resources, S.K., A.S., and A.K.; writing original draft preparation, S.K. and A.S.; writing review and editing, A.K., S.K., and A.S. All authors have read and agreed to the published version of the manuscript.

Funding: This research received no external funding.

Data Availability Statement: This is a book chapter; requests for data can be discussed and data may be provided.

Conflicts of Interest:
The authors declare no conflicts of interest.

References

Aggarwal, D., & Ramachandran, A. (2020). One health approach to address zoonotic diseases. *Indian Journal of Community Medicine*, 45(5), 6. https://doi.org/10.4103/ijcm.IJCM_398_19

Arné, P., Risco-Castillo, V., Jouvion, G., Le Barzic, C., & Guillot, J. (2021). Aspergillosis in Wild Birds. *Journal of Fungi*, 7(3), 241. https://doi.org/10.3390/jof7030241

Awareness and Community Engagement in Zoonotic Disease Management. (2023). *International Journal of Agriculture and Biosciences*, *Zoonosis.* Volume 1, 470–485. https://doi.org/10.47278/book.zoon/2023.036

Bartel, R. D., Sheppard, J. L., Lovas-Kiss, Á., & Green, A. J. (2018). Endozoochory by mallard in New Zealand: What seeds are dispersed and how far? *PeerJ*, 6, e4811. https://doi.org/10.7717/peerj.4811

Beerens, N., Germeraad, E. A., Venema, S., Verheij, E., Pritz-Verschuren, S. B. E., & Gonzales, J. L. (2021). Comparative pathogenicity and environmental transmission of recent highly pathogenic avian influenza H5 viruses. *Emerging Microbes & Infections*, 10(1), 97–108. https://doi.org/10.1080/22221751.2020.1868274

Begum, A., Mukutmoni, M., & Akter, F. (2019). Parasite diversity in mallard: A cross-sectional study on *Anas platyrhynchos* from Munshiganj, Dhaka. *Bangladesh Journal of Zoology*, 47(1), 121–128. https://doi.org/10.3329/bjz.v47i1.42027

Bennett, J., & Whitten, S. (2003). Duck hunting and wetland conservation: Compromise or synergy? *Canadian Journal of Agricultural Economics/Revue Canadienne d'agroeconomie*, 51(2), 161–173. https://doi.org/10.1111/j.1744-7976.2003.tb00171.x

Benzarti, E., Garigliany, M., Hauman, D., Paternostre, J., Linden, A., Franssen, M. et al. (2019). First evidence of fatal Usutu virus natural infections in an *Anatidae*, the common scoter (*Melanitta nigra*). *Vector-Borne and Zoonotic Diseases*, 19(10), 777–780. https://doi.org/10.1089/vbz.2019.2460

Blanchong, J. A., Samuel, M. D., & Mack, G. (2006). Multi-species patterns of avian cholera mortality in Nebraska's rainwater basin. *Journal of Wildlife Diseases*, 42(1), 81–91. https://doi.org/10.7589/0090-3558-42.1.81

Breban, R., Drake, J. M., Stallknecht, D. E., & Rohani, P. (2009). The role of environmental transmission in recurrent avian influenza epidemics. *PLoS Computational Biology*, 5(4), e1000346. https://doi.org/10.1371/journal.pcbi.1000346

CABI. (2019). *Somateria mollissima* (p. 63670) [Dataset]. https://doi.org/10.1079/cabicompendium.63670

Campagnolo, E. R., Banerjee, M., Panigrahy, B., & Jones, R. L. (2001). An outbreak of duck viral enteritis (Duck Plague) in domestic Muscovy Ducks (Cairina moschata domesticus) in Illinois. *Avian Diseases*, 45(2), 522. https://doi.org/10.2307/1592999

Charmantier, A., & Gienapp, P. (2014). Climate change and timing of avian breeding and migration: Evolutionary versus plastic changes. *Evolutionary Applications*, 7(1), 15–28. https://doi.org/10.1111/eva.12126

Cherkaoui, S., Boukherouk, M., Lakhal, T., Aghzar, A., & El Youssfi, L. (2020). Conservation amid COVID-19 pandemic: Ecotourism collapse threatens communities and wildlife in Morocco. *E3S Web of Conferences*, 183, 01003. https://doi.org/10.1051/e3sconf/202018301003

Chowdhury, S., Hossain, M. E., Ghosh, P. K., Ghosh, S., Hossain, M. B., Beard, C. et al. (2019). The pattern of highly pathogenic avian influenza H5N1 outbreaks in South Asia. *Tropical Medicine and Infectious Disease*, 4(4), 138. https://doi.org/10.3390/tropicalmed4040138

Circella, E., Camarda, A., Bano, L., Marzano, G., Lombardi, R., D'Onghia, F., & Greco, G. (2019). Botulism in wild birds and changes in environmental habitat: A relationship to be considered. *Animals*, 9(12), 1034. https://doi.org/10.3390/ani9121034

Clemmons, E. A., Alfson, K. J., & Dutton, J. W. (2021). Transboundary animal diseases, an overview of 17 diseases with potential for global spread and serious consequences. *Animals*, 11(7), 2039. https://doi.org/10.3390/ani11072039

De Oca, V. M., Valdés, S. E., Segundo, C., Gómez, G. G., Ramírez, J., & Cervantes, R. A. (2017). Aspergillosis, a natural infection in poultry: Mycological and molecular characterization and determination of gliotoxin in *Aspergillus fumigatus* isolates. *Avian Diseases*, 61(1), 77–82. https://doi.org/10.1637/11496-092016-Reg

Descamps, S., Jenouvrier, S., Gilchrist, H. G., & Forbes, M. R. (2012). Avian cholera, a threat to the viability of an arctic seabird colony? *PLoS ONE*, 7(2), e29659. https://doi.org/10.1371/journal.pone.0029659

Dhama, K., Kumar, N., Saminathan, M., Tiwari, R., Karthik, K., Kumar, M. A. et al. (2017). Duck virus enteritis (duck plague)—A comprehensive update. *Veterinary Quarterly*, 37(1), 57–80. https://doi.org/10.1080/01652176.2017.1298885

Dima, C., Burlacu, S., & Buzoianu, O. A. C. (2020). Strategic options for the development of ecotourism in the Danube Delta in the context of globalization. *SHS Web of Conferences*, 74, 04005. https://doi.org/10.1051/shsconf/20207404005

Djunaidi, H., & Djunaidi, A. C. M. (2007). The economic impacts of avian influenza on world poultry trade and the U.S. poultry industry: A spatial equilibrium analysis. *Journal of Agricultural and Applied Economics*, 39(2), 313–323. https://doi.org/10.1017/S1074070800023014

Dórea, F. C., Cole, D. J., & Stallknecht, D. E. (2013). Quantitative exposure assessment of waterfowl hunters to avian influenza viruses. *Epidemiology and Infection*, 141(5), 1039–1049. https://doi.org/10.1017/S0950268812001720

Dubé, E., & MacDonald, N. E. (2017). Vaccination resilience: Building and sustaining confidence in and demand for vaccination. *Vaccine*, 35(32), 3907–3909. https://doi.org/10.1016/j.vaccine.2017.06.015

Ellis, M. B., & Miller, C. A. (2022). The effect of a ban on the use of lead ammunition for waterfowl hunting on duck and goose crippling rates in Illinois. *Wildlife Biology*, 2022(2), e01001. https://doi.org/10.1002/wlb3.01001

Femerling, G., Van Oosterhout, C., Feng, S., Bristol, R. M., Zhang, G., Groombridge, J. et al. (2023). Genetic load and adaptive potential of a recovered avian species that narrowly avoided extinction. *Molecular Biology and Evolution*, 40(12), msad256. https://doi.org/10.1093/molbev/msad256

Garamszegi, L. Z. (2011). Climate change increases the risk of malaria in birds: Climate change and malaria prevalence. *Global Change Biology*, 17(5), 1751–1759. https://doi.org/10.1111/j.1365-2486.2010.02346.x

Gilbert, M., Xiao, X., Domenech, J., Lubroth, J., Martin, V., & Slingenbergh, J. (2006). Anatidae migration in the western palearctic and spread of highly pathogenic avian influenza H5N1 virus. *Emerging Infectious Diseases*, 12(11), 1650–1656. https://doi.org/10.3201/eid1211.060223

Gilbert, M., Xiao, X., Pfeiffer, D. U., Epprecht, M., Boles, S., Czarnecki, C. et al. (2008). Mapping H5N1 highly pathogenic avian influenza risk in Southeast Asia. *Proceedings of the National Academy of Sciences*, 105(12), 4769–4774. https://doi.org/10.1073/pnas.0710581105

Gobbo, F., Fornasiero, D., De Marco, M. A., Zecchin, B., Mulatti, P., Delogu, M., & Terregino, C. (2021). Active surveillance for highly pathogenic avian influenza viruses in wintering waterbirds in northeast Italy. *Microorganisms*, 9(11), 2188. https://doi.org/10.3390/microorganisms9112188

Goldstein, D. L., Williams, J. B., & Braun, E. J. (1990). Osmoregulation in the field by salt-marsh savannah sparrows *Passerculus sandwichensis beldingi*. *Physiological Zoology*, 63(4), 669–682. https://doi.org/10.1086/physzool.63.4.30158169

Grado, S. C., Hunt, K. M., Hutt, C. P., Santos, X. T., & Kaminski, R. M. (2011). Economic impacts of waterfowl hunting in Mississippi derived from a state-based mail survey. *Human Dimensions of Wildlife*, 16(2), 100–113. https://doi.org/10.1080/10871209.2011.551175

Gyuranecz, M., Foster, J. T., Dán, Á., Ip, H. S., Egstad, K. F., Parker, P. G. et al. (2013). Worldwide phylogenetic relationship of avian poxviruses. *Journal of Virology*, 87(9), 4938–4951. https://doi.org/10.1128/JVI.03183-12

Hars, J., Ruette, S., Benmergui, M., Fouque, C., Fournier, J.-Y., Legouge, A. et al. (2008). The epidemiology of the highly pathogenic H5N1 avian influenza in mute swan (Cygnus olor) and other Anatidae in the dombes region (France), 2006. *Journal of Wildlife Diseases*, 44(4), 811–823. https://doi.org/10.7589/0090-3558-44.4.811

Hobbs, R. J., Hallett, L. M., Ehrlich, P. R., & Mooney, H. A. (2011). Intervention ecology: Applying ecological science in the twenty-first century. *BioScience*, 61(6), 442–450. https://doi.org/10.1525/bio.2011.61.6.6

Hooper, K. A., & Bloom, J. D. (2013). A mutant influenza virus that uses an N1 neuraminidase as the receptor-binding protein. *Journal of Virology*, 87(23), 12531–12540. https://doi.org/10.1128/JVI.01889-13

Islam, M., Islam, J., Islam, M., Ahamed, T., Islam, M., Khatun, M. et al. (2021). Duck virus enteritis (duck plague) outbreak in an Australian black swan (Cygnus atratus) flock at safari park in Bangladesh: A case report. *Journal of Advanced Veterinary and Animal Research*, 8(4), 1. https://doi.org/10.5455/javar.2021.h545

Ives, A. R., Cardinale, B. J., & Snyder, W. E. (2005). A synthesis of subdisciplines: Predator–prey interactions, and biodiversity and ecosystem functioning. *Ecology Letters*, 8(1), 102–116. https://doi.org/10.1111/j.1461-0248.2004.00698.x

Jaeger, A., Lebarbenchon, C., Bourret, V., Bastien, M., Lagadec, E., Thiebot, J.-B. et al. (2018). Avian cholera outbreaks threaten seabird species on Amsterdam Island. *PLoS One*, 13(5), e0197291. https://doi.org/10.1371/journal.pone.0197291

Jillson, S. J., Dahab, M. F., Woldt, W. E., & Surampalli, R. Y. (2001). Pathogen and pathogen indicator removal characteristics in treatment wetlands systems. *Practice Periodical of Hazardous, Toxic, and Radioactive Waste Management*, 5(3), 153–160. https://doi.org/10.1061/(ASCE)1090-025X(2001)5:3(153)

Klaassen, M., Hoye, B. J., Nolet, B. A., & Buttemer, W. A. (2012). Ecophysiology of avian migration in the face of current global hazards. *Philosophical Transactions of the Royal Society B: Biological Sciences*, 367(1596), 1719–1732. https://doi.org/10.1098/rstb.2012.0008

Klees, S., Schlüter, J.-S., Schellhorn, J., Bertram, H., Kurzweg, A. C., Ramzan, F. et al. (2022). Comparative investigation of gene regulatory processes underlying avian influenza viruses in chicken and duck. *Biology*, 11(2), 219. https://doi.org/10.3390/biology11020219

Knapp, S. (2000). Wet and wild. *Science*, 289(5481), 877–877. https://doi.org/10.1126/science.289.5481.877

Koel, B. F., Van Der Vliet, S., Burke, D. F., Bestebroer, T. M., Bharoto, E. E., Yasa, I. W. W. et al. (2014). Antigenic variation of clade 2.1 H5N1 virus is determined by a few amino acid substitutions immediately adjacent to the receptor binding site. *mBio*, 5(3), e01070–14. https://doi.org/10.1128/mBio.01070-14

LaPointe, D. A., Goff, M. L., & Atkinson, C. T. (2010). Thermal constraints to the sporogonic development and altitudinal distribution of avian malaria *Plasmodium relictum* in Hawaii. *Journal of Parasitology*, 96(2), 318–324. https://doi.org/10.1645/GE-2290.1

Leivesley, J. A., & Rollinson, N. (2021). Maternal provisioning and fluctuating thermal regimes enhance immune response in a reptile with temperature-dependent sex determination. *Journal of Experimental Biology*, 224(5), jeb237016. https://doi.org/10.1242/jeb.237016

Leotta, G. A., Chinen, I., Vigo, G. B., Pecoraro, M., & Rivas, M. (2006). Outbreaks of avian cholera in Hope Bay, Antarctica. *Journal of Wildlife Diseases*, 42(2), 259–270. https://doi.org/10.7589/0090-3558-42.2.259

Ling-Hu, T., Rios-Guzman, E., Lorenzo-Redondo, R., Ozer, E. A., & Hultquist, J. F. (2022). Challenges and opportunities for global genomic surveillance strategies in the COVID-19 era. *Viruses*, 14(11), 2532. https://doi.org/10.3390/v14112532

Liu, D., Zhuo, Z., Tao, J., & Xu, J. (2018). A case report of Apatemon gracilis (Szidat, 1928) infection in domestic geese in mainland China. *Turkish Journal of Veterinary and Animal Sciences*, 42(2). https://doi.org/10.3906/vet-1709-22

Liu, G., Zhou, L., Li, B., & Zhang, L. (2014). The complete mitochondrial genome of Aix galericulata and Tadorna ferruginea: Bearings on their phylogenetic position in the Anseriformes. *PLoS One*, 9(11), e109701. https://doi.org/10.1371/journal.pone.0109701

Livezey, B. C. (1996). A phylogenetic reassessment of the tadornine–anatine divergence (Aves: Anseriformes: Anatidae). *Annals of the Carnegie Museum*, 65(1), 27–88. https://doi.org/10.5962/p.215133

Livezey, B. C. (1997). A phylogenetic classification of waterfowl (Aves: Anseriformes), including selected fossil species. *Annals of the Carnegie Museum*, 66(4), 457–496. https://doi.org/10.5962/p.215141

Lowenstein, C., Waters, W. F., Roess, A., Leibler, J. H., & Graham, J. P. (2016). Animal husbandry practices and perceptions of zoonotic infectious disease risks among livestock keepers in a rural Parish of Quito, Ecuador. *The American Society of Tropical Medicine and Hygiene*, 95(6), 1450–1458. https://doi.org/10.4269/ajtmh.16-0485

Lugelo, A., Hampson, K., Ferguson, E. A., Czupryna, A., Bigambo, M., Duamor, C. T. et al. (2022). Development of dog vaccination strategies to maintain herd immunity against rabies. *Viruses*, 14(4), 830. https://doi.org/10.3390/v14040830

Massengo, N. R. B., Tinto, B., & Simonin, Y. (2023). One health approach to arbovirus control in Africa: Interests, challenges, and difficulties. *Microorganisms*, 11(6), 1496. https://doi.org/10.3390/microorganisms11061496

Maurer, A. J. (2003). POULTRY | Ducks and Geese. *In: Encyclopedia of Food Sciences and Nutrition* (pp. 4686–4689). Elsevier. https://doi.org/10.1016/B0-12-227055-X/00953-6

McPhail, G. M., Collins, S. M., Burt, T. V., Careen, N. G., Doiron, P. B., Avery-Gomm, S. et al. (2024). Geographic, ecological, and temporal patterns of seabird mortality during the 2022 HPAI H5N1 outbreak on the island of Newfoundland. *Zoology*. https://doi.org/10.1101/2024.01.17.575746

Merken, R., Deboelpaep, E., Teunen, J., Saura, S., & Koedam, N. (2015). Wetland suitability and connectivity for trans-Saharan Migratory Waterbirds. *PLoS One*, 10(8), e0135445. https://doi.org/10.1371/journal.pone.0135445

Miller, M. M., & Schat, K. A. (2004). Chicken infectious anemia virus: An example of the ultimate host–parasite relationship. *Avian Diseases*, 48(4), 734–745. https://doi.org/10.1637/7271-090304R

Nontapan, N. (2024). Role of biosecurity measures in controlling infectious diseases in poultry farms in Thailand. *Journal of Animal Health*, 4(2), 51–62. https://doi.org/10.47604/jah.2768

Normile, D. (2006). Evidence points to migratory birds in H5N1 spread. *Science*, 311(5765), 1225–1225. https://doi.org/10.1126/science.311.5765.1225

Ostad-Ali-Askari, K. (2022). Review of the effects of the anthropogenic on the wetland environment. *Applied Water Science*, 12(12), 260. https://doi.org/10.1007/s13201-022-01767-4

Parker, G. R., Petrie, M. J., & Sears, D. T. (1992). Waterfowl distribution relative to wetland acidity. *The Journal of Wildlife Management*, 56(2), 268. https://doi.org/10.2307/3808822

Patil, S. D. (2021). Prevalence and histopathological studies on intestinal cestode parasite of avian host Gallus Domesticus (L) from Nashik Region, Maharashtra State, India. *International Journal for Research in Applied Science and Engineering Technology*, 9(11), 206–212. https://doi.org/10.22214/ijraset.2021.38778

Pavia, M. (2013). The Anatidae and Scolopacidae (Aves: Anseriformes, Charadriiformes) from the late Neogene of Gargano, Italy. *Geobios*, 46(1–2), 43–48. https://doi.org/10.1016/j.geobios.2012.10.013

Pedersen, K., Dietz, H. H., Jørgensen, J. C., Christensen, T. K., Bregnballe, T., & Andersen, T. H. (2003). *Pasteurella multocida* from outbreaks of avian cholera in wild and captive birds in Denmark. *Journal of Wildlife Diseases*, 39(4), 808–816. https://doi.org/10.7589/0090-3558-39.4.808

Perdue, M. L., & Swayne, D. E. (2005). Public health risk from avian influenza viruses. *Avian Diseases*, 49(3), 317–327. https://doi.org/10.1637/7390-060305R.1

Pinto Jimenez, C. E., Keestra, S., Tandon, P., Cumming, O., Pickering, A. J., Moodley, A. et al. (2023). Biosecurity and water, sanitation, and hygiene (WASH) interventions in animal agricultural settings for reducing infection burden, antibiotic use, and antibiotic resistance: A One Health systematic review. *The Lancet Planetary Health*, 7(5), e418–e434. https://doi.org/10.1016/S2542-5196(23)00049-9

Qin, H., Xiao, J., Li, J., Gao, X., & Wang, H. (2017). Climate variability and avian cholera transmission in Guangxi, China. *Revista Brasileira de Ciência Avícola*, 19(2), 211–220. https://doi.org/10.1590/1806-9061-2016-0411

Ramesh, S., Soundararajan, C., Selvaraj, J., & Subapriya, S. (2018). Avian Pox in a blue rock pigeon—A case report. *International Journal of Current Microbiology and Applied Sciences*, 7(12), 3690–3692. https://doi.org/10.20546/ijcmas.2018.712.418

Ramirez, I., Tarzia, M., Dias, M. P., Burfield, I. J., Ramos, J. A., Garthe, S. et al. (2017). How well is the EU protecting its seabirds? Progress in implementing the Birds Directive at sea. *Marine Policy*, 81, 179–184. https://doi.org/10.1016/j.marpol.2017.03.034

Ray, M., Barton, D., & Shamsi, S. (2024). Current knowledge of helminth parasites in Australian Anatidae species. *Journal of Natural History*, 58(45–48), 2249–2269. https://doi.org/10.1080/00222933.2024.2420950

Reaser, J. K., Witt, A., Tabor, G. M., Hudson, P. J., & Plowright, R. K. (2021). Ecological countermeasures for preventing zoonotic disease outbreaks: When ecological restoration is a human health imperative. *Restoration Ecology*, 29(4), e13357. https://doi.org/10.1111/rec.13357

Robinson, O. J., Ruiz-Gutierrez, Viviana, Reynolds, M. D., Golet, G. H., Strimas-Mackey, M., & Fink, D. (2020). Integrating citizen science data with expert surveys increases accuracy and spatial extent of species distribution models. *Diversity and Distributions*, 26(8), 976–986. https://doi.org/10.1111/ddi.13068

Ross, T. A., Zhang, J., Wille, M., Asimakopoulos, A. G., Jaspers, V. L. B., & Klaassen, M. (2025). Interplay between pollution and avian influenza virus in shorebirds and waterfowl. *EcoHealth*. https://doi.org/10.1007/s10393-025-01707-z

Samuel, M. D., Botzler, R. G., & Wobeser, G. A. (2007). Avian cholera. *In:* N. J. Thomas,

D. B. Hunter, & C. T. Atkinson (Eds.), *Infectious Diseases of Wild Birds* (1st ed., pp. 239–269). Wiley. https://doi.org/10.1002/9780470344668.ch12

Samuel, M. D., Hobbelen, P. H. F., DeCastro, F., Ahumada, J. A., LaPointe, D. A., Atkinson, C. T. et al. (2011). The dynamics, transmission, and population impacts of avian malaria in native Hawaiian birds: A modeling approach. *Ecological Applications*, 21(8), 2960–2973. https://doi.org/10.1890/10-1311.1

Samuel, M. D., Shadduck, D. J., Goldberg, D. R., & Johnson, W. P. (2005). Avian cholera in waterfowl: The role of lesser snow and ross's geese as disease carriers in the Playa Lakes Region. *Journal of Wildlife Diseases*, 41(1), 48–57. https://doi.org/10.7589/0090-3558-41.1.48

Santos, X. T., Grado, S. C., & Hunt, K. M. (2016). Assessing the economic impact of waterfowl hunting in Mississippi: Effects of changes in industry sector data using the IMPLAN model. *International Journal of Social Economics*, 43(1), 86–103. https://doi.org/10.1108/IJSE-04-2014-0068

Shukla, J. B., Naresh, R., Verma, S. R., & Agarwal, M. (2020). Modeling the effect of sanitation in a human habitat to control the spread of bacterial diseases. *Modeling Earth Systems and Environment*, 6(1), 39–49. https://doi.org/10.1007/s40808-019-00653-4

Si, Y., Wang, T., Skidmore, A. K., De Boer, W. F., Li, L., & Prins, H. H. T. (2010). Environmental factors influencing the spread of the highly pathogenic avian influenza H5N1 virus in wild birds in Europe. *Ecology and Society*, 15(3), art26. https://doi.org/10.5751/ES-03622-150326

Sims, L. D. (2013). Intervention strategies to reduce the risk of zoonotic infection with avian influenza viruses: Scientific basis, challenges and knowledge gaps. *Influenza and Other Respiratory Viruses*, 7(s2), 15–25. https://doi.org/10.1111/irv.12076

Smith, K. F., Sax, D. F., Gaines, S. D., Guernier, V., & Guégan, J.-F. (2007). Globalization of human infectious disease. *Ecology*, 88(8), 1903–1910. https://doi.org/10.1890/06-1052.1

Soda, K., Tomioka, Y., Hidaka, C., Matsushita, M., Usui, T., & Yamaguchi, T. (2021). Susceptibility of common family anatidae bird species to Clade 2.3.4.4e H5N6 high pathogenicity avian influenza virus: An experimental infection study. *In:* Review. https://doi.org/10.21203/rs.3.rs-1169623/v1

Sokolow, S. H., Nova, N., Pepin, K. M., Peel, A. J., Pulliam, J. R. C., Manlove, K. et al. (2019). Ecological interventions to prevent and manage zoonotic pathogen spillover. *Philosophical Transactions of the Royal Society B: Biological Sciences*, 374(1782), 20180342. https://doi.org/10.1098/rstb.2018.0342

Solarczyk, P., Wojtkowiak-Giera, A., & Heddergott, M. (2023). Migrating anatidae as sources of environmental contamination with zoonotic giardia, cryptosporidium, cyclospora and microsporidia. *Pathogens*, 12(3), 487. https://doi.org/10.3390/pathogens12030487

Struelens, M. J., Ludden, C., Werner, G., Sintchenko, V., Jokelainen, P., & Ip, M. (2024a). Real-time genomic surveillance for enhanced control of infectious diseases and antimicrobial resistance. *Frontiers in Science*, 2, 1298248. https://doi.org/10.3389/fsci.2024.1298248

Sun, Z., Pan, T., Hu, C., Sun, L., Ding, H., Wang, H. et al. (2017). Rapid and recent diversification patterns in Anseriformes birds: Inferred from molecular phylogeny and diversification analyses. *PLoS One*, 12(9), e0184529. https://doi.org/10.1371/journal.pone.0184529

Syrota, Ya. Yu., Greben, O. B., Poluda, A. M., Maleha, O. M., Lisitsyna, O. I., & Kornyushin, V. V. (2018). Helminths of the Mallard, *Anas Platyrhynchos* (Aves, Anatidae), in Ukraine: Analysis of the diversity in mixed forest zone and the black sea region. *Vestnik Zoologii*, 52(4), 267–278. https://doi.org/10.2478/vzoo-2018-0028

Takeda, M., Tsukamoto, K., Kohda, T., Matsui, M., Mukamoto, M., & Kozaki, S. (2005). Characterization of the neurotoxin produced by isolates associated with avian botulism. *Avian Diseases*, 49(3), 376–381. https://doi.org/10.1637/7347-022305R1.1

Takekawa, J. Y., Newman, S. H., Xiao, X., Prosser, D. J., Spragens, K. A., Palm, E. C. et al. (2010). Migration of waterfowl in the East Asian Flyway and spatial relationship to HPAI H5N1 outbreaks. *Avian Diseases*, 54(s1), 466–476. https://doi.org/10.1637/8914-043009-Reg.1

Thurman, L. L., Alger, K., LeDee, O., Thompson, L. M., Hofmeister, E., Hudson, J. M. et al. (2024). Disease-smart climate adaptation for wildlife management and conservation. *Frontiers in Ecology and the Environment*, 22(4), e2716. https://doi.org/10.1002/fee.2716

Trott, K. A., Giannitti, F., Rimoldi, G., Hill, A., Woods, L., Barr, B. et al. (2014). Fatty liver hemorrhagic syndrome in the backyard chicken: A retrospective histopathologic case series. *Veterinary Pathology*, 51(4), 787–795. https://doi.org/10.1177/0300985813503569

Valkiūnas, G., Ilgūnas, M., Bukauskaitė, D., Fragner, K., Weissenböck, H., Atkinson, C. T. et al. (2018). Characterization of *Plasmodium relictum*, a cosmopolitan agent of avian malaria. *Malaria Journal*, 17(1), 184. https://doi.org/10.1186/s12936-018-2325-2

Van Boeckel, T. P., Thanapongtharm, W., Robinson, T., Biradar, C. M., Xiao, X., & Gilbert, M. (2012). Improving risk models for avian influenza: The role of intensive poultry farming and flooded land during the 2004 Thailand epidemic. *PLoS One*, 7(11), e49528. https://doi.org/10.1371/journal.pone.0049528

Van Leeuwen, C. H. A., Van Der Velde, G., Van Groenendael, J. M., & Klaassen, M. (2012). Gut travellers: Internal dispersal of aquatic organisms by waterfowl. *Journal of Biogeography*, 39(11), 2031–2040. https://doi.org/10.1111/jbi.12004

Van Riper, C., Van Riper, S. G., & Hansen, W. R. (2002). Epizootiology and effect of avian pox on Hawaiian forest Birds. *The Auk*, 119(4), 929–942. https://doi.org/10.1093/auk/119.4.929

Vasiliev, D. (2022). The role of biodiversity in ecosystem resilience. *IOP Conference Series: Earth and Environmental Science*, 1072(1), 012012. https://doi.org/10.1088/1755-1315/1072/1/012012

Vieira, A. R., Salzer, J. S., Traxler, R. M., Hendricks, K. A., Kadzik, M. E., Marston, C. K. et al. (2017). Enhancing surveillance and diagnostics in anthrax-endemic countries. *Emerging Infectious Diseases*, 23(13). https://doi.org/10.3201/eid2313.170431

Vrtiska, M. P., Gammonley, J. H., Naylor, L. W., & Raedeke, A. H. (2013). Economic and conservation ramifications from the decline of waterfowl hunters. *Wildlife Society Bulletin*, 37(2), 380–388. https://doi.org/10.1002/wsb.245

Walker, L. R., Hölzel, N., Marrs, R., Del Moral, R., & Prach, K. (2014). Optimization of intervention levels in ecological restoration. *Applied Vegetation Science*, 17(2), 187–192. https://doi.org/10.1111/avsc.12082

Ward, M. P., Maftei, D. N., Apostu, C. L., & Suru, A. R. (2009). Association between outbreaks of highly pathogenic avian influenza subtype H5N1 and migratory waterfowl (Family *Anatidae*) populations. *Zoonoses and Public Health*, 56(1), 1–9. https://doi.org/10.1111/j.1863-2378.2008.01150.x

Webby, R. J., & Uyeki, T. M. (2024). An update on highly pathogenic avian influenza A(H5N1) virus, Clade 2.3.4.4b. *The Journal of Infectious Diseases*, 230(3), 533–542. https://doi.org/10.1093/infdis/jiae379

Wille, M., McBurney, S., Robertson, G. J., Wilhelm, S. I., Blehert, D. S., Soos, C. et al. (2016). A pelagic outbreak of avian cholera in north American gulls: Scavenging as a primary mechanism for transmission? *Journal of Wildlife Diseases*, 52(4), 793. https://doi.org/10.7589/2015-12-342

Wood, J. P., Choi, Y. W., Chappie, D. J., Rogers, J. V., & Kaye, J. Z. (2010). Environmental persistence of a highly pathogenic avian influenza (H5N1) virus. *Environmental Science & Technology*, 44(19), 7515–7520. https://doi.org/10.1021/es1016153

Wu, Y., Liu, L., Zhang, M., Zhan, H., Wang, C., Wang, M. et al. (2023). A recombinant duck plague virus containing the ICP27 deletion marker provides robust protection in ducks. *Microbiology Spectrum*, 11(4), e00983–23. https://doi.org/10.1128/spectrum.00983-23

Xu, Y., Gong, P., Wielstra, B., & Si, Y. (2016). Southward autumn migration of waterfowl facilitates cross-continental transmission of the highly pathogenic avian influenza H5N1 virus. *Scientific Reports*, 6(1), 30262. https://doi.org/10.1038/srep30262

Yamaguchi, N., Hupp, J. W., Higuchi, H., Flint, P. L., & Pearce, J. M. (2010). Satellite-tracking of northern Pintail *Anas acuta* during outbreaks of the H5N1 virus in Japan: Implications for virus spread. *Ibis*, 152(2), 262–271. https://doi.org/10.1111/j.1474-919X.2010.01010.x

CHAPTER

11

An Assessment of the Threats to Anatidae Birds on a Global Level and Potential Conservation Strategies

Vaithianathan Kannan

Arulagam, Asokapuram, Coimbatore, Tamil Nadu, India

1. Introduction

The abundance and distribution of birds have been employed in monitoring the health of the ecosystems and have thus served as indicators of environmental change (Mekonen, 2017). One of them is the Anatidae family which consists of ducks, geese, and swans that have their own unique ecological and cultural roles. Anatidae consists of more than 170 species which are distributed across all the continents except Antarctica and are found under a variety of conditions, including tundra wetlands and mangroves in the tropics and inland lakes and coastal estuaries. They are key ecological agents, as grazers, seed dispersers, prey, predators, and are indicators of the welfare of wetlands (Green, 1996). In addition to their ecological importance, Anatidae birds have been ingrained in human lifestyles and cultures and are found to be esteemed as food, leisure, and esthetics. However, they are facing more risks and the danger to survival in the world despite the fact that they are flexible. The knowledge of these threats, and of the conservation methods designed to combat them, can give not only a clear picture of the situation of the Anatidae, but also that of the whole issue of the sustainability of the biodiversity of wetlands in a rapidly changing environment.

Being one of the most ecologically significant and widespread families of waterbirds (Kear, 2005), Anatidae are the core of global biodiversity and ecosystem services (Hester & Harrison, 2010) as they need wetlands and migration routes. Nevertheless, they are also exposed to mounting demands by depending on

*Corresponding author: kannan.vaithianathan@gmail.com

DOI: 10.1201/9781003669784-11

them (Wetlands International, 2021). The most widespread threat in the world is the destruction and loss of wetlands as a result of reclamation, intensive farming, urbanization, and shifting hydrology (Davidson, 2014; Finlayson et al., 2017). Changing rainfall, aiding drought, and rising sea level are some of the stressors aggravated by climate change and threatening inland and shoreline wetlands (IPCC, 2022; Galbraith et al., 2002). The other threat is the very perilous highly pathogenic avian influenza (HPAI) that has caused mass fatalities among wild populations and has augmented the biosecurity dangers at the wildlife, poultry, and human interface (Globig et al., 2020; Wille & Barr, 2022). Other threats are unsustainable hunting, lead poisoning by ammunition, agrochemical and industry pollution, oil spills, invasion of other species, and hybridization with feral waterfowl, which reduce the strength of the population and genetic purity (Green & Pain, 2019; Mateo & Kanstrup, 2019; Kirby et al., 2008; BirdLife International, 2022; BirdLife International, 2023).

A successful conservation should be multi-scaled, collaborative, and adaptive. Protection and restoration of important wetlands and migratory bottlenecks in specific areas (Stroud et al., 2016); the ban on lead ammunition and sustainable hunting policies (Kanstrup et al., 2018); the improvement of surveillance and preparedness to HPAI (Munster et al., 2007); and dealing with pollution and invasive species (Millennium Ecosystem Assessment, 2005) take the top of the list. The relevant policy framework is possible with the help of the international frameworks, such as the Agreement on the Conservation of African–Eurasian Migratory Waterbirds (AEWA) or the Ramsar Convention, but its efficacy relies on the effectiveness of the national implementation and cross-border coordination (Ramsar Convention Secretariat, 2013; AEWA, 2018). Community-based stewardship, wetland design, and adaptive management strategies linking conservation goals and local livelihoods are also important to be taken into account (Reynolds et al., 2017). The last century has seen unprecedented landscape transformation, where the waterbirds are one of the most affected, especially Anatidae. Since 1970, their main habitats, wetlands, have lost approximately 35% of the area through drainage and pollution, development of infrastructure, and climate change (Convention on Wetlands, 2025). This loss eliminates breeding areas, migration resting points, and wintering areas. Anatidae are particularly susceptible to habitat fragmentation and hunting along flyways, as well as disruption of key stopover resources by climate (Duan et al., 2025).

These issues are intensified by human activities. Some areas continue with unsustainable hunting and egg harvesting, and agricultural, industrial effluents, and invasive species disturb the ecology of the wetlands. The reduction of safe habitats is caused by competition over water resources, due to increased farming and aquaculture. This is further strained by disease outbreaks such as the avian influenza, which threatens populations of the wild and brings conflicts against the poultry industries and the government health departments. This is a combination of forces that highlights how vague the Anatidae position in the Anthropocene is. Vulnerability is affected by the biological characteristics of species. Although most Anatidae possess a high reproductive capacity and versatile diets, other more specialized species that are restricted to fragmented habitats, including Baer's pochard (*Aythya baeri*) and the white-winged wood duck (*Asarcornis scutulata*), are endangered (Wu et al., 2022). Conversely, generalists such as the mallard (*Anas platyrhynchos*) survive in the man-made conditions. Conservation should thus be species-specific, differentiating resilient widespread populations and those populations needing immediate action.

The Anatidae conservation is also rather complex socio-economically and culturally. These birds also supply the local economies through hunting and trade, bird watching, ecotourism, and restoring wetlands in different communities. Sensible and comprehensive methods are needed for human needs and ecological sustainability. Conservation of Anatidae needs international collaboration, unification of policies, and flyway-scale manoeuvres; the same efforts being demonstrated by the Ramsar Convention, AEWA, Convention on Migratory Species (CMS), Important Bird Areas (IBA) and other multilateral agreements due to the long migrations of the species. The gaps may be filled with more general landscape management, habitat restoration, domestic stewardship, and advanced technology, such as satellite tracking, gene tracking, and citizen science. More resiliency is enhanced by policy reform, enforcement of anti-poaching, and sustainable wetland management. A general challenge that is not predictable is the issue of climate change. Temperature changes because of changes in breeding periods, the wetland hydrology, and food supply. There are already beginning to be mismatches between the time of migration and the occurrence of resources, and sea-level rise is threatening the very survival of wetlands in the coastal environment. The solution to this should focus on local measures and global climate pledges, and the conservation of Anatidae provides a prism through which one can see the bigger picture of biodiversity, human welfare, and environmental health. Finally, it is not only about saving the lives of ducks, geese, and swans as such but also about wetlands, which ensure the sustainability of water security, climate stability, and human well-being. It is an ethical responsibility and what makes their preservation a practical necessity in the greater struggle of biodiversity loss and climate change. With a combination of pre-existing stressors in the ecosystem and society, the state of Anatidae makes the global intertwined conservation response more urgent, innovative, and adaptive.

2. Wetland Loss, Degradation, and Hydrological Change

Wetlands represent the environmental heart of the Anatidae population, offering essential breeding, feeding, and staging environments for their survival and movement (Milton & Finlayson, 2018). Nevertheless, these systems are dying and depleting at an alarming rate, and this poses the greatest danger to waterfowl all over the world (Davidson, 2014). Sixty-four percent of the wetlands that exist globally have been overtaken by agricultural development, urbanization, industrialization, and infrastructure developments in the past century (Ramsar Convention Secretariat, 2018). These losses in the case of the Anatidae that depend on intercontinental wetlands end the migration routes and access to good stopover sites (Stroud et al., 2016). Hydrological changes have contributed to this downward trend. Floodplains and marshes that are vital to the populations of waterfowl are altered during flood events by water controls, dams, and water diversions (Kingsford et al., 2016). Asian Mekong River dams have changed the movement of sediments and seasonal flooding, which has swept away the quality of the habitats of migratory ducks and geese (Ziv et al., 2012). Clearing potholes of prairies to farmlands in North America has removed one of the most productive breeding sites of ducks with far-reaching ramifications of population regulation (Anteau & Afton, 2004).

Climate change is building up these pressures as well. Changes in precipitation patterns are causing more frequent and severe droughts in wetlands in both Africa and Central Asia (IPCC, 2022). The rise of water level is a threat to coastal marshes and estuaries that drive wetlands to the inland areas where artificial structures are prone to inhibiting migration causing a so-called coastal squeeze (Nicholls & Hines, 2007). The decrease in snowmelts and increasing temperatures in arctic breeding grounds are changing tundra habitats and changing the food availability and timing without matching the time of migrating geese (Jónsson & Afton, 2015; Fox & Abraham, 2017). It has an extra layer of stress from pollution and nutrient loading. The eutrophication, algal blooms, and depletion of oxygen in wetlands are a result of runoff of nitrogen and phosphorus in agricultural practices (Millennium Ecosystem Assessment, 2005; Balaji-Prasath et al., 2023). The situation abuses the quality of habitat and reduces the valuable food materials that consist of the aquatic vegetation and invertebrates (Carpenter, 2005). The destruction of wetlands undermines waterfowl habitats and ecosystem services of the world as a whole due to their ability to filter water, absorb carbon, and the ecosystem in general (Zedler & Kercher, 2005).

The loss of any wetland, which is essential in the case of migratory Anatidae, can disturb the entire flyway networks (Boere & Stroud, 2006). These effects are manifested in the East Asian–Australasian flyway, where the disappearance of wetlands in China, Korea, and Southeast Asia has resulted in precipitous populations of the different duck and goose species (Murray et al., 2015; Murray & Fuller, 2012). These complex threats require dynamic conservation work. Some components of Europe and North America have been found to be effective with the restoration of wetlands, including reintroduction of natural hydrological regimes (Ma et al., 2010). Preservation of the still existing wetlands is required and mandatory through the use of such tools as the Ramsar Convention; however, national laws and regulations should be strong to help in preservation lately (Ramsar Convention Secretariat, 2018). Floodplain reconnection and managed coastal realignment methods can be used to strengthen climate adaptation strategies to counter the developing pressures (Erwin, 2009). When protecting and rehabilitating wetlands, the quality of the habitat and ecological relationships, which are the cornerstone of the global migration of the species, should not be disturbed in the case of Anatidae (Stroud et al., 2016).

3. Highly Pathogenic Avian Influenza (HPAI/H5) and Other Diseases

There is an increasing danger with the emerging infectious diseases against the wild bird species and it is the HPAI, and the H5Nx variants, in this case, that are causing the panic around the world. Anatidae have great migration distances as well as concentrations in wetlands that are hosts and carriers of influenza viruses of birds (Munster et al., 2007). The recent HPAI outbreaks were accompanied by the high mortality rate of wild birds, resulting in the death of swans, geese, and ducks in vast numbers in Europe and Asia (Globig et al., 2020). These mortality events not only lead to the reduction of populations but also lead to changes in the migration and breeding patterns, which create further pressure on already threatened species (Wille & Barr, 2022).

The wetlands increase the spread of HPAI because the higher the concentration of waterfowl in the wetlands, the higher the chances of spreading the virus (Hesterberg et al., 2009). Migration also increases the issue as the infected anatids can infect other continents and other places and poultry systems with the viruses (Kilpatrick et al., 2006). Particularly, domesticated poultry in Asia comes into contact with wildlife, and in this regard, it is especially important that intensive poultry production is loosely coupled with major migration routes, which preconditions the cross-species infection and a possible zoonotic spillover (Takekawa et al., 2010). Other than the avian flu, other illnesses such as duck viral enteritis, botulism, and cholera have led to some cases of massive yet localized death outbreaks (Friend et al., 1999). These disease outbreaks can become more frequent and serious because of the ecological changes, such as higher temperatures and nutrient pollution, which will change the duration of the pathogen and the host vulnerability (Altizer et al., 2013). The responses must be effective by means of coordinated global monitoring. The early warning systems, such as the Global Influenza Surveillance and Response System (GISRS), have increased surveillance of migratory birds (Gilbert et al., 2006). However, scarcity of resources and inconsistency in the availability of information hinder the rapid response at flyway levels (Keawcharoen et al., 2008). Vaccination of poultry, improvement of biosecurity, and isolation of domestic flocks of waterfowl habitats with wild waterfowl are the primary preventative strategies (Swayne & Sims, 2021). In order to save, disease risk consideration in habitat management, artificial congregation reduction through artificial feeding, and heightened monitoring at a global scale would be significant in capitalizing upon the adverse impacts of the HPAI on the population of Anatidae (Wille & Barr, 2022).

4. Hunting (Overexploitation) and Crippling/Lead Poisoning

Hunting of Anatidae has been a cultural, recreational, and subsistence practice. However, uncontrolled hunting is also a major threat to their survival in all parts of the globe (Madsen et al., 2015). Poor hunting laws have led to the decrease of several populations of wild duck and goose (Wetlands International, 2021). Indicatively, poaching and capturing to obtain food and trade remain common in some parts of Asia and Africa, which significantly impacts migratory waterbirds (Brochet et al., 2016). It is exacerbated by the mode of hunting that leads to crippling rather than straight harvesting. It has been postulated that one bird that is found, perishes, or becomes more vulnerable to predation and disease by another (Noer et al., 1995). This hidden death is difficult to quantify yet might significantly underestimate the real impact of hunting on the population of Anatidae (Madsen & Guillemain, 2013).

Another major problem is lead poisoning caused by disposal of ammunition (Mateo et al., 2007). Waterfowl ingest lead shot in the sediments through foraging or ingest it through grit, mistaking it for food (Mateo & Kanstrup, 2019). Such consumption can cause severe intoxication, reduced survival, low reproductive competence, and the risk of becoming more susceptible to diseases (Pain, 2019). In the past century, millions of waterfowl died throughout the world as a result of lead poisoning (Green & Pain, 2019). Several countries have taken steps to address this issue. In 1991, the United States banned the use of lead shot to hunt waterfowl, and with this the exposure rates of lead to the ducks drastically reduced in terms of the resulting deaths (Anderson et al., 2000). Recently, the European Union has implemented restrictions on the use of lead ammunition in wetlands as part of the Registration, Evaluation, Authorisation and Restriction of Chemicals (REACH) regulation, which suggests the introduction of a positive trend in the process of the global abolition of lead ammunition (Kanstrup et al., 2018). However, the degree of conformance and implementation is very different in different regions (Stroud & Pain, 2011).

Scientifically monitored population requires operating sustainable hunting models to reconcile between cultural values and conservation efforts (Madsen et al., 2015). Some international accords, such as the AEWA, offer harmonization of hunting policies, but national-level practices are vastly different (AEWA, 2018). To save the ducks population, or more generally the Anatidae family, from risk of extinction, not only the quotas of hunting must be introduced, but also the prohibition of the use of lead cartridges for hunting should be implemented, the losses of birds which are injured must be compensated by imposing fines on hunters, and the use of nontoxic pellets (substitutes for lead pellets) should be promoted (Mateo & Kanstrup, 2019).

5. Pollution and Toxicants (Oil Spills, Agrochemicals, Heavy Metals, Microplastics)

Pollution is a wide and diverse menace to Anatidae, and it directly affects their survival, reproduction, and quality of habitat. One of the most pronounced and destructive types of waterfowl pollution is oil spills. The oil smears feathers, weakening insulation and buoyancy to cause hypothermia, drowning, and the inability to fly (Camphuysen, 2003). The birds that strive to cleanse the oil off their feathers consume the hydrocarbons that may result in internal organ destruction and death (Jenssen, 1994). Major oil spillages, including the Deepwater Horizon accident in the Gulf of Mexico, caused the death of thousands of waterbirds, including ducks and geese, and this indicates the magnitude of exposure (Haney et al., 2014).

Agrochemicals are also dangerous. Direct poisoning may arise as a result of ingestion of pesticides such as organophosphates and carbamates by waterfowl feeding on agricultural landscapes (Mineau & Whiteside, 2013). Herbicides change the composition of the aquatic vegetation and lower the food supply of the dabbling ducks, whereas the insecticides decrease the invertebrate population which is essential for breeding females and ducklings (Taylor et al., 2006). The result of fertilizer runoff is eutrophication which causes harmful algal blooms, resulting in oxygen depletion and consequent decreases in the biodiversity of wetlands (Carpenter, 2005; Balaji-Prasath et al., 2021). In the case of Anatidae, the effects of these changes are lesser food availability and reduced breeding success (Zedler & Kercher, 2005). The heavy metals such as mercury, cadmium, and arsenic in wetlands are bioaccumulated by food chains, thus subjecting waterfowl to toxicity (Burger & Gochfeld, 2004). The most notable effect comes from mercury, which disrupts reproduction and the nervous system as high concentrations have reportedly been found in the ducks that live in polluted wetlands (Evers et al., 2005). Sub-lethal effects, such as a decrease in hatching success and changes in behavior, are contributing to population-level losses (Scheuhammer, 1987). The microplastics, which are created as a result of the irresponsible dumping of plastic waste and are later deposited into the bodies of water, are a comparatively new but global menace to various species such as Anatidae. Microplastics are a developing threat to Anatidae due to ingestion and entanglement that results in internal trauma and low nutrition and fitness. Also, chemical exposure to plastic additives and pollutants has been observed in freshwater and marine waterbirds.

Besides direct physical harm, microplastics may alter gut microbiomes, serve as reservoirs of long-lasting chemicals and pathogens, and increase other pressures on the Anatidae (gut loss, disturbance, hunting), and the physiological and reproductive functions have been shown to be affected in wild birds (Fackelmann et al., 2023; Bilal et al., 2023; Kühn & van Franeker, 2020). The effects of pollution on Anatidae do not occur independently but build up pressures that interact with other hazards. The case in point, wetlands already degraded by agrochemical runoff may at the same time experience a decrease in resistance to oil spills or epidemics (Millennium Ecosystem Assessment, 2005). It leads to the need to have unified strategies, such as tightening the control over agrochemicals, rehabilitation of the polluted wetlands, and the establishment of the mechanisms of timely response to oil spills (Haney et al., 2014). A shift to sustainable agricultural activities and enhancement of global conventions on harmful chemicals will also decrease risks to the population of Anatidae in the long-term (Mineau & Whiteside, 2013).

6. Habitat Fragmentation, Disturbance, and Loss of Stopover Sites

The life of Anatidae depends on the existence of a chain of suitable living environments on their migration routes. However, these ecological networks are being disturbed and disrupted by habitat fragmentation which threatens both migratory and resident populations (Boere & Stroud, 2006). Fragmentation occurs as a result of infrastructure and agricultural development or urban sprawl. This fragmentation separates once-linked wetlands, isolating populations and reducing genetic exchange (Fahrig, 2003). One of the most important impacts of fragmentation is the extinction of stopover sites. The migratory anatids often have specific wetlands as stopover points, and there they replenish fat reserves before continuing their long trip (Newton, 2006). Disruption of the migration strategies by the loss or degradation of only one of the stopover sites can lead to a decline in survival and reproductive performance (Alerstam et al., 2003). The East Asian–Australasian flyway is one of the examples of this threat, with the major loss of intertidal habitats used by ducks and geese as a result of massive coastal reclamation, particularly around the Yellow Sea (Murray & Fuller, 2012).

The impacts of habitat changes are enhanced by the disruptions caused by various human activities, for example, leisure activities. Recreational activities at wetlands such as boating, fishing, and visitation may disrupt feeding and roosting behaviors of anatids, and thus birds may expend additional energy to avoid humans (Klein et al., 1995). Even in other habitats, a routine disturbance due to machines and hunting may reduce the fitness of habitats, although it may remain physically intact (Gill et al., 2001). Fragmentations are increased by climate change. Altered water patterns and distribution of plants reduce the predictability of access to a stopover site and migratory birds have to adjust their routes or schedules with uncertain outcomes (Jónsson & Afton, 2015). In the case of long-distance migrants, these inconsistencies may lead to severe consequences as new habitats may not be available in the pathways of migration (Runge et al., 2015). The conservation should be directed to the identification, protection, and restoration of the essential stopover sites on the magnitude of migratory patterns (Stroud et al., 2016). It is extremely essential to cooperate globally because migratory species cross a lot of political boundaries. National implementation is often ineffective;

however, such institutions as the Ramsar Convention and AEWA have underlined the importance of the networks of sites (Ramsar Convention Secretariat, 2013). The practical solutions include conservation of uninhabited coastal lands, control of human activities in critical wetlands, and establishment of habitat corridors that connect broken environments (Fahrig, 2003; Newton, 2006). Protecting stopover habitats does not only entail the preservation of the environment but also the continuity of ecology that is fundamental to the successful movement of Anatidae on the planet.

7. Conservation Imperatives

Addressing these threats is a matter that needs concerted efforts to conserve. It is important to protect big interconnected wetland areas. Programs such as the Ramsar Convention on Wetlands and flyway-based programs such as the AEWA place special emphasis on international cooperation in the protection of important habitats. Fragmentation can be reduced by revitalizing degraded wetlands (and maintaining hydrological connections). The disturbances during critical breeding and migration periods can be minimized by controlling human activity through zoning, seasonal limits, as well as educational efforts. In the case of the stopover sites, great importance should be put on the protection of key staging wetlands along major migratory pathways. By tracking the migration routes with satellites and observing them over a long period of time, it is possible to pinpoint the bottlenecks in the migration pathways and implement specific conservation efforts. It is also vital to incorporate the preservation of wetlands into the overall land-use planning that would require agricultural development and urbanization to take into account the ecology of the waterfowl. The schemes to fund ecosystem services, ecotourism initiatives, and conservation works by the communities can encourage the local stakeholders to protect and conserve wetlands which also act as stopover habitats. As has been mentioned before, habitat fragmentation, disturbance, and loss of stopover sites can be listed among the most important and complicated threats to Anatidae across the world. These stresses not only endanger the survival of particular species but also the ecological and functional integrity of whole flyways as the presence or absence of these habitats disrupts the resting and breeding sites of species, as well as damaging migration paths. Effective conservation cannot be achieved in isolation, instead it is a global effort that should be able to balance the needs of humans and the conservation of migrating waterfowl. The sustainability of the Anatidae is dependent on our ability to sustain the regularity, quality, and safety of their habitat throughout the annual cycle.

8. Data and Monitoring Gaps and Weak Governance at the Flyway Scale

The safeguarding of the Anatidae necessitates a cross-border perspective of the geese, as these migratory birds use interdependent breeding, wintering, and stopover areas along flyways. Despite decades of conservation awareness, there has not been a significant change as there are still gaps in data collection and ineffective governance structures. These loopholes undermine coordination of efforts at the international level, threat evaluation, and threat quantification capacities in line with biodiversity objectives. Standardized and trusted data are extremely important, yet the majority of the species in particular those located in Asia, Africa, and South America are less studied, hence leading to biased conservation priority that overlooks endangered populations (Guerrero-Moreno et al., 2025). It is also possible that the population estimates of most species of Anatidae are uncertain and they are obsolete, attributed to the inconsistent or incomplete monitoring (Guerrero-Moreno et al., 2025). Rather, many nations lack the resources, expertise, and political stability to conduct a synchronized waterbird census. In the case where the practice of monitoring is implemented, there is usually unequal coverage and the smaller but ecologically important wetlands are left out. Such scanty information does not provide the opportunity to identify any worsening over time and hides the whole outcomes of the threats such as the disappearance of habitats, pollution, and disease outbreaks. With technological progress, however, even in migration routes in some portions of the globe, much of the picture remains elusive, especially along the strained fly-paths, such as the East Asian–Australasian, because the conservation efforts are not coordinated with the actual ecological needs of the organisms (Iswandaru et al., 2022).

Poor governance adds to these data problems. The Anatidae migration spans the borders of nations that have varying priorities on politics, conservation, and economic pressure, resulting in disjointed policies and poor cross-border coordination. Despite focusing on collaboration, frameworks such as the Ramsar Convention, CMS, and AEWA are in place and seem to work unevenly. The majority of the wetlands that are placed under protection are not under proper local management; conservation goals are rarely integrated into the larger sectors such as agriculture or urban development. Also, insufficient funding, inadequate institution capacity, and accountability undermine long-term conservation commitments and enforcement is of a voluntary nature. The scientific and institutional bases must be improved in order to fill the gap between the data and governance. Via more surveillance, including International Waterbird Census, remote sensing, citizen science, and advanced tracking, the quotas and coverage can be improved but only on the condition of being standardized across regions. On the governance level, it is important that the international agreements are implemented at the national level, using strong institutions, better accountability mechanisms, and implementing the agreements. The idea of wetlands conservation should be encompassed in broader sustainable development agendas and this will give this move more power. Lastly, migratory Anatidae will not survive without coordinating science and policy on entire flyways—by locating means to govern on data to assist conservation, and by locating means to enact data to support coordinated international action (Convention on Wetlands, 2025).

9. Priority Actions: Practical, High-Impact, and Implementable

The complexity of the challenges and successful conservation can be managed through practical, impactful, and actionable measures that directly tackle primary threats. The subsequent priority actions emphasize strategies that can provide tangible advantages for Anatidae and wetland ecosystems globally.

9.1. Protect and Restore Critical Wetlands

Safeguarding wetlands remains the cornerstone of Anatidae conservation. Many populations depend on a network of breeding grounds, staging areas, and wintering sites. Protecting large, intact wetlands through legal designation (e.g., Ramsar Sites, UNESCO Biosphere Reserves) ensures the preservation of core habitats. Restoration of degraded wetlands by re-establishing natural hydrology, removing invasive plants, and improving water quality can quickly enhance habitat suitability for ducks, geese, and swans. Practical steps include mapping and prioritizing wetlands most important for Anatidae along major flyways and focusing investments there. For example, rehabilitating prairie potholes in North America, floodplains in Europe, and tidal flats in East Asia might protect essential habitats for millions of birds. Collaborating with local communities in these initiatives guarantees lasting sustainability (Lobat et al., 2021).

9.2. Reduce Hunting Pressure and Eliminate Lead Ammunition

Unregulated hunting and the continued use of lead-based ammunition remain significant threats. Practical solutions include strengthening enforcement of hunting regulations, implementing science-based harvest limits, and engaging hunting communities in co-management schemes. A particularly high-impact action is the global phase-out of lead shot. Nontoxic alternatives are already available, affordable, and effective. Several regions, including the European Union, have begun implementing bans on lead ammunition in wetlands. Expanding and enforcing such policies worldwide would reduce poisoning, benefiting not only Anatidae but also raptors, mammals, and human health (Convention on Wetlands, 2025).

9.3. Strengthen Monitoring and Data Sharing Across Flyways

Effective conservation depends on knowing where birds are, how populations are changing, and what pressures they face. Expanding and standardizing waterbird monitoring through programs such as the International Waterbird Census are both feasible and impactful. Investments in satellite telemetry, citizen science platforms, and remote sensing can fill gaps in migration knowledge, especially in underrepresented regions such as Africa and South America. Creating centralized data platforms accessible to governments, NGOs, and researchers ensures that conservation decisions are informed by the best available evidence. Such transparency also builds accountability and robust data at the flyway level (Eren et al. 2024).

9.4. Improve Governance and Flyway-Level Coordination

Migratory Anatidae require cooperation among nations that share populations. Strengthening flyway-scale frameworks such as the AEWA and the East Asian–Australasian Flyway Partnership ensures coordinated action.

Practical steps include:

- Establishing regional working groups focused specifically on Anatidae.
- Mainstreaming waterfowl conservation into national wetland and climate policies.
- Mobilizing financial resources for flyway initiatives, particularly in developing and underdeveloped countries.

These actions are achievable through modest investments and political will, creating a platform for collective accountability.

9.5. Manage Disease Risks Proactively

HPAI/H5 has emerged as a global threat to both wild Anatidae and poultry. Mitigation requires proactive measures rather than reactive crisis responses. High-impact actions include early-warning surveillance programs at key wetlands and their environs, rapid response networks linking veterinary and conservation agencies, and strict biosecurity measures in poultry operations near migratory bird habitats. Training local field staff to collect samples and report mortalities ensures timely detection. Importantly, fostering cooperation between the wildlife and agriculture sectors builds trust and minimizes unnecessary scapegoating of wild birds (Lewandrowski & Ingram, 1999).

9.6. Reduce Pollution and Promote Robust Waste Management and Sustainable Agriculture

Practical measures to address pollution can yield immediate benefits. These include:

- Rapid response protocols for oil spills in sensitive waterfowl habitats.
- Strict enforcement of industrial and municipal discharge standards near wetlands.
- Incentives for farmers to adopt pesticide-free buffer zones around wetlands.

Sustainable agricultural practices, such as integrated pest management, reduced fertilizer use, and the maintenance of traditional rice paddies, can simultaneously support Anatidae and human livelihoods. Demonstration projects in Asia and Africa have shown that farmers can benefit economically while conserving waterfowl habitat.

9.7. Engage Communities and Build Awareness

Local communities play a vital role in the conservation of Anatidae. Numerous wetlands intersect with human communities, where waterfowl offer sustenance, leisure, and cultural significance. Strengthening communities via participatory governance, ecotourism projects, and benefit-sharing programs promotes stewardship. Campaigns aimed at schools, farmers, hunters, and policymakers foster public backing. For instance, emphasizing the importance of wetlands in managing floods and providing clean water connects the conservation of Anatidae to wider human well-being. Cost-effective yet powerful awareness-raising initiatives, such as signage in wetlands, local bird festivals, and media outreach, are practical solutions (Rutter et al., 2022).

9.8. Incorporate Climate Resilience into Conservation Strategies

Climate change is changing wetland water dynamics, altering migration schedules, and influencing food supplies (Meng et al., 2016). Practical measures involve recognizing and safeguarding climate refuges and wetlands anticipated to stay sturdy amid evolving circumstances. Improving links between wetlands allows birds to modify their migration paths due to changing circumstances.

At the local level, adaptive water management (such as regulated flooding or water storage) can protect wetlands from drought or rising sea levels (Ivajnšič & Kaligarič, 2014). These actions can be put into practice using current knowledge and technologies, needing only incorporation into wetland management strategies. Preserving Anatidae in particular and other waterbirds worldwide necessitates immediate yet feasible measures. Safeguarding and rehabilitating wetlands, banning lead ammunition, enhancing oversight, improving leadership, controlling diseases, decreasing pollution, involving communities, and preparing for climate change are actionable measures that can yield significant outcomes. The effectiveness of these actions lies in their feasibility: the tools, technologies, and successful models are already available. To scale them up, a robust political commitment, significant financial backing, and efficient cross-border cooperation are necessary. Through decisive action, the conservation community can ensure the future of Anatidae and, in turn, protect the well-being of wetlands that support both wildlife and humans (Pullin et al., 2013, Waliczky et al., 2019).

10. Metrics to Track Success in Anatidae Conservation

The protection of Anatidae ducks, geese, and swans necessitates extensive, collaborative efforts spanning continents and ecosystems (Green, 1996, Nourani et al., 2017). Considering the various pressures these birds encounter, from habitat loss to hunting and illness, it is insufficient to create strategies; their success must also be evaluated and guaranteed. Success metrics allow conservationists, policymakers, and local communities to assess the impact of interventions, improve their strategies, and encourage accountability. Strong metrics also aid in obtaining funding, inform adaptive management, and preserve international collaboration along flyways. For this, reliant on interlinked wetland systems, metrics should include ecological, social, and governance aspects (Turner et al., 2000).

10.1. Population Status and Trends

The core of successful conservation depends on the capacity to maintain or boost Anatidae populations. Metrics on population status offer a clear measure of species well-being and the success of interventions.

- **Population estimates:** Frequent waterbird surveys at breeding, wintering, and stopover locations provide insights into population size. Success is indicated by stable or rising figures, especially for endangered species such as the Baer's pochard (*Aythya baeri*) or white-headed duck (*Oxyura leucocephala*).
- **Breeding success:** Nest counts, clutch size, and fledgling survival rates measure reproductive output. Positive trends suggest that habitat quality and predator management are effective (Lennox et al., 2025).
- **Demographic metrics:** Age proportions (youths compared to adults) offer an understanding of recruitment levels and sustainability over time. A healthy population structure is a key metric of resilience.

Population-based metrics are essential but must be interpreted in the context of migration (Cayuela et al., 2018). An increase in numbers at one site may, at times, mask declines elsewhere along the flyway, underscoring the importance of coordinated monitoring.

10.2. Habitat Extent and Quality

Anatidae conservation depends on wetlands, grasslands, and freshwater systems that provide food, shelter, and breeding grounds. Habitat metrics track whether these ecosystems are being maintained or restored.

- **Wetland extent:** Remote sensing and GIS analyses can measure changes in wetland area over time. Success is indicated by slowing or reversing the rate of wetland loss, especially in biodiversity hotspots such as prairie potholes or Asian floodplains (Rebelo et al., 2009).
- **Habitat quality indicators:** Water quality (nutrient levels, salinity, pollution, etc.), vegetation diversity, and invertebrate abundance provide insights into the capacity of wetlands to support waterfowl.
- **Connectivity:** For migratory species, metrics should capture habitat connectivity. The number and condition of stopover sites secured along flyways are a powerful measure of conservation progress.

Restoration success can also be tracked by the speed and scale at which restored habitats are recolonized by Anatidae.

10.3. Threat Reduction

Conservation efforts must also measure whether key threats are being monitored and mitigated. Practical threat-reduction metrics include:

- **Hunting and lead poisoning:** Monitoring the proportion of birds with embedded shot or elevated blood-lead levels indicates progress toward eliminating toxic ammunition (De Francisco et al., 2003). Compliance with hunting regulations can be tracked through law enforcement records and community surveys.
- **Disease prevalence:** Outbreak frequency and mortality rates from HPAI or botulism provide critical metrics for assessing disease prevalence. A reduction in mass mortality events signals improved disease surveillance and management.
- **Pollution levels:** Declines in pesticide residues, heavy metal concentrations, microplastics, and oil-spill frequency within key wetlands reflect effective environmental regulation (Pettigrove et al., 2023).

By linking actions to measurable threat reduction, conservationists can demonstrate clear outcomes beyond species counts.

10.4. Governance and Policy Effectiveness

Because Anatidae cross political boundaries, governance is a cornerstone of conservation success. Metrics here capture the strength of institutions and the implementation of commitments.

- **Site protection status:** The proportion of critical Anatidae sites formally designated as protected (e.g., Ramsar, important bird areas (IBAs), national reserves) indicates progress.
- **Management effectiveness:** The existence of management plans, their level of implementation, and adequacy of staffing/funding are practical governance indicators.
- **Policy adoption:** Tracking the number of countries that have phased out lead ammunition, integrated wetlands into climate strategies, or joined flyway agreements shows international momentum.
- **Compliance and reporting:** Rates of timely reporting under international frameworks (e.g., AEWA, CMS) can be used as accountability metrics.

Governance metrics emphasize that conservation is not only ecological but also political.

10.5. Community and Stakeholder Engagement

Long-term success depends on the support of people who live and work around wetlands. Social metrics measure the extent to which communities are engaged and benefit from conservation efforts.

- **Awareness and attitudes:** Surveys of local communities, farmers, and hunters reveal whether knowledge and attitudes toward Anatidae and wetlands are improving.
- **Participation:** The number of community members involved in wetland management committees, citizen science initiatives, and ecotourism programs can serve as measurable indicators.
- **Livelihood benefits:** Metrics such as income from sustainable tourism, payment-for-ecosystem-services schemes, or reduced conflict with agriculture demonstrate that conservation aligns with human well-being.

These indicators ensure conservation is socially inclusive and resilient.

10.6. Climate Adaptation and Resilience

Given the rapid pace of climate change, success must also be measured in terms of adaptation capacity.

- **Climate refugia secured:** The number of wetlands identified and protected as long-term refuges under climate change scenarios is a forward-looking metric.
- **Adaptive management practices:** Evidence of dynamic water regulation (e.g., managed flooding, drought buffers) at sites reflects readiness for climate variability.
- **Migration shifts:** Monitoring changes in timing, routes, and duration of migration provides a measure of how Anatidae adapt and whether conservation networks remain adequate.

These metrics connect conservation efforts with worldwide climate-resilience objectives.

10.7. Integration into Broader Development Agendas

Ultimately, the success of Anatidae cannot be separated from wider sustainability frameworks. Measurements that show the incorporation of waterfowl and wetland preservation into national and international plans offer leverage for sustainable backing.

- **Sustainable Development Goals (SDGs):** Advancements in SDG-6 (Clean Water), SDG-13 (Climate Action), and SDG-15 (Life on Land) can be correlated with conservation results for Anatidae.
- **Ecosystem service valuation:** Quantifying flood control, carbon storage, or fisheries benefits from wetlands conserved for Anatidae provides metrics linking biodiversity with human welfare.

This integration ensures enduring political and financial support.

Metrics for assessing success in Anatidae conservation should extend beyond merely tallying birds. They should record the condition of populations, the quality and connectivity of habitats, the level of threats, the robustness of governance, community involvement, and the ability for climate adaptation. Collectively, these indicators form a complex depiction of conservation success. Crucially, metrics must be practical, standardized, and scalable across flyways, facilitating comparison and cooperation among nations. Incorporating precise metrics into conservation strategies allows the global community to transition from hopeful pledges to observable results, guaranteeing that Anatidae and their associated wetlands flourish for future generations (Cimon-Morin & Poulin, 2018).

Evaluating the threats and conservation approaches for Anatidae worldwide serves as a critical alert that waterbirds, especially ducks, geese, and swans, encounter a combination of pressures jeopardizing their existence. From loss of habitat to pollution, from hunting to disease, the combined effects of these pressures are gradually diminishing the resilience of populations worldwide. The key conclusions from this extensive assessment highlight both the severity of the challenges and the real possibilities for reversing declines, provided that coordinated, science-driven, and inclusive conservation efforts are implemented.

A diverse range of threats necessitates cohesive responses on individual factors contributing to the decline that can be tackled separately. The loss and degradation of wetlands weaken the essential basis for Anatidae survival, while pollution, toxins, and diseases increase risks by directly impacting health, reproduction, and mortality. Unregulated hunting increases pressure, particularly in areas where subsistence and commercial harvesting are inadequately overseen. Compounding all of this are global forces such as climate change and weak governance at the flyway level, which fragment responses and limit the effectiveness of conservation efforts. A final key message is the interdependency of threats: an Anatidae population weakened by habitat loss is more vulnerable to disease outbreaks; a population stressed by pollution may suffer higher mortality when exposed to hunting. Conservation strategies must therefore reflect this interplay, adopting a systems-level approach (White et al., 2010; Kumar et al., 2023).

11. Conservation of Anatidae is Inseparable from Wetland Protection

The centrality of wetlands cannot be overstated that their loss and degradation translate directly into population declines. Protecting and restoring wetlands is therefore the single most important conservation strategy, with ripple effects benefitting not only birds but also human societies reliant on wetland ecosystem services. This requires integrating wetland values into national land-use planning, agriculture, and water management policies. Investments in wetland conservation are both ecological and socio-economic, enhancing water security, carbon sequestration, and disaster risk reduction (Bassi et al., 2014; Prasanya et al., 2024).

12. Flyways Demand Flyway-Level Governance

Anatidae are migratory birds, moving across thousands of kilometres and multiple jurisdictions each year. Their survival depends on the integrity of the entire flyway, making international cooperation indispensable. Weak governance and gaps in data monitoring at this scale hinder effective protection. A final takeaway is that global treaties and flyway initiatives, such as the Ramsar Convention and the African–Eurasian Waterbird Agreement, must be strengthened through improved compliance, more effective resource allocation, and enhanced community-level engagement. The conservation of Anatidae is a litmus test for our ability to manage migratory species collectively across political borders (Green, 1996; Cao et al., 2020).

13. Science and Monitoring are Cornerstones of Conservation

Effective conservation is only as strong as the evidence—yet, monitoring gaps remain severe, especially in parts of Asia, Africa, and South America. Populations are declining in regions where even baseline abundance, distribution, and demographic data are missing. A priority message is the urgent need to invest in standardized, globally connected, and technologically enhanced monitoring systems. This means embracing satellite telemetry, genetic tools, and artificial intelligence for survey analysis, and citizen science networks. Only with robust data can managers detect trends, forecast risks, and evaluate the success of interventions (Green, 1996).

14. Addressing Hunting and Poisoning Requires Pragmatism and Local Engagement

Hunting of Anatidae is a cultural tradition, a source of livelihood, and a recreational activity in many parts of the world. Outright bans are often impractical, risking alienation of communities and illegal harvests. Instead, sustainable harvest frameworks, education, and strong enforcement are needed. Such changes require collaboration with hunters, manufacturers, and regulators, ensuring that conservation does not conflict with livelihoods but instead fosters long-term sustainability (Smalley & Reeves, 2022).

15. Health and Disease are Shared Risks

The spread of HPAI H5 underscores the interconnectedness of wildlife, domestic animals, and humans. Anatidae conservation cannot ignore the One Health perspective, which emphasizes the overlap between ecosystem health and human well-being. Reducing disease risks involves improved surveillance, better farm biosecurity, and responsible wetland management. Protecting Anatidae is therefore also an investment in global public health security (Green, 1996; Shi et al., 2024).

16. Conservation is Achievable, but Urgency is the Key

A key conclusion is that Anatidae conservation is not only necessary but also achievable. There are numerous examples of successful wetland restoration projects that have revived bird populations, policy reforms that have phased out lead shot, and community-based conservation efforts that have secured local buy-in (Gyurácz et al., 2011). These instances show that when science, governance, and community involvement come together, populations can bounce back. However, time is limited. The ongoing decline of ecosystems, along with the intensifying effects of climate change, indicates that chances for intervention are diminishing (Ward et al., 2010; Quesnelle et al., 2013; Tozer et al., 2018).

17. Metrics Matter: Success Must be Tracked and Demonstrated

Clear, quantifiable metrics such as population trends, wetland area, sustainable harvesting, and pollution decrease are vital for showcasing advancement. Without them, conservation may turn into hopeful words instead of tangible transformation. Transparency is equally vital: openly sharing successes and failures fosters accountability, allows for strategy adjustments, and maintains public trust (Nourani et al., 2017, Convention on Wetlands, 2025).

18. The Broader Significance: Anatidae as Ambassadors of Wetlands

Ultimately, Anatidae represent a broader narrative. They are prominent, engaging, and frequently meaningful species that can act as representatives for wetland preservation. Safeguarding them fosters wider backing for biodiversity, climate action, and human welfare. They remind us that conservation involves protecting ecosystems and processes rather than just individual species that support life on Earth (Convention on Wetlands, 2025).

19. Conclusion

The facts that wetlands are being lost, degraded, and their hydrology is being altered are the indicators that conservation and restoration are needed. In a bid to safeguard the remaining wetlands, we need sound legislation, ecologically balanced land-use strategies, and effective implementation of these strategies to combat such activities as encroachment and pollution. Re-wetting of lost parts, removal of invasive plants, and reinstating the lost hydrological processes have shown promising results in other areas of the world. It is also important to recognize wetlands as natural infrastructure that serves very critical services to human beings, a view that may affect investment and policy concentration. As climate change intensifies, wetlands should no longer be viewed as waste sites, which are constantly flooded, but rather as a very important defense against floods, droughts, and the output of carbon. Their protection harmonizes the ecological health and human power. The future of wetlands is a gauge of how man can manage to balance the pursuit of development with the need to take good care of the environment, and the choices undertaken in this century will determine whether these important ecological treasures are available to future generations.

The fate of the Anatidae birds is an indicator of the state of our wetlands, our control, and our capacity to control common resources in the world. The fact that they still exist is not just a conservation problem, but also an indicator of how well the world works together, the ability to develop policies based on evidence, and the willingness of the community to live in harmony with the environment. The point is quite obvious: the preservation of Anatidae is necessary and possible. It requires the improvement of habitat preservation, strengthening flyway control, reducing hunting, dealing with pollution, eliminating disease threats, and bridging gaps in the data. Obtaining success will guarantee the future of ducks, geese, and swans, as well as ensuring our ability to maintain the well-being of the planet. Habitat and migration pathways are the basis of the conservation of Anatidae, complicated even more by intrusions, diseases, and poisons. Conserving and rehabilitating wetlands, working internationally, eradicating unnecessary toxic hazards, improving disease surveillance, and working with local people will all have the most significant positive outcomes. There is a need to act early and with coordinated action at important locations as small bottlenecks may make the difference in the survival of whole populations.

Acknowledgments

I express my deepest gratitude to Dr. P. A. Azeez for his invaluable guidance, inspiration, and encouragement. I am grateful for his patience, insightful advice, and unwavering support throughout this journey. I thank my colleagues Dr. Balaji-Prasath Barathan and Sri Jehu Pandian Nadar from Mumbai for their assistance in reviewing the manuscript.

Data Availability Statement: Data are available from the authors on request.

Conflict of Interest: None.

Ethics Approval: Not applicable.

Declarations/Informed Consent Statement: Not applicable.

References

AEWA. (2018). Agreement on the Conservation of African–Eurasian Migratory Waterbirds: Strategic Plan 2019–2027. UNEP/AEWA Secretariat.

Alerstam, T., Hedenström, A., & Åkesson, S. (2003). Long-distance migration: Evolution and determinants. *Oikos*, 103(2), 247–260.

Altizer, S., Ostfeld, R. S., Johnson, P. T., Kutz, S., & Harvell, C. D. (2013). Climate change and infectious diseases: From evidence to a predictive framework. *Science*, 341(6145), 514–519.

Anderson, W. L., Havera, S. P., & Zercher, B. W. (2000). Ingestion of lead and nontoxic shotgun pellets by ducks in the Mississippi Flyway. *Journal of Wildlife Management*, 64(3), 848–857.

Anteau, M. J., & Afton, A. D. (2004). Nutrient reserves of lesser scaup (Aythya affinis) during spring migration in the Mississippi Flyway: A test of the spring condition hypothesis. *The Auk*, 121(3), 917–929.

Balaji-Prasath, B., Chen, W., Su, Y. et al. (2023). The effects of nutrient loading from different sources on eutrophication in a large shallow lake in Southeast China. *Environ. Geochem. Health.*, 45, 7603–7620. https://doi.org/10.1007/s10653-023-01641-5.

Balaji-Prasath, B., Lin, Z.-R., Su, Y.-P., She, C.-X., Lin, H., Zhang, C.-W. et al. (2021). Adsorption-release characteristics of phosphorus and the community of phosphorus—Accumulating organisms of sediments in a shallow lake. *Sustainability*, 13, 11501. https://doi.org/10.3390/su132011501

Bassi, N., Kumar, M. D., Sharma, A., & Pardha-Saradhi, P. (2014). Status of wetlands in India: A review of extent, ecosystem benefits, threats and management strategies. *Journal of Hydrology: Regional Studies*, 2, 1–19.

Bilal, M., Yaqub, A., Hassan, H. U., Akhtar, S., Rafiq, N., Ali Shah, M. I. et al. (2023). Microplastic quantification in aquatic birds: Biomonitoring the environmental health of the Panjkora river freshwater ecosystem in Pakistan. *Toxics*, 11(12), 972. doi: 10.3390/toxics11120972.

BirdLife International. (2022). *State of the World's Birds 2022: Insights and Solutions for the Biodiversity Crisis*. BirdLife International.

BirdLife International. (2023). *State of the World's Birds: Indicators for our Changing World*. BirdLife International. https://www.birdlife.org

Boere, G. C., & Stroud, D. A. (2006). The flyway concept: What it is and what it isn't. *In:* G. C. Boere, C. A. Galbraith, & D. A. Stroud (Eds.), *Waterbirds Around the World* (pp. 40–47). The Stationery Office.

Brochet, A. L., Van den Bossche, W., Jbour, S., Ndang'ang'a, P. K., Jones, V. R., Abdou, W. A. L. I., ... & Butchart, S. H. (2016). Preliminary assessment of the scope and scale of illegal killing and taking of birds in the Mediterranean. *Bird Conservation International*, 26(1), 1–28.

Burger, J., & Gochfeld, M. (2004). Marine birds as sentinels of environmental pollution. *EcoHealth*, 1(3), 263–274.

Camphuysen, C. J. (2003). Characteristics of Atlantic puffins (*Fratercula arctica*) wrecked in the Netherlands. *Atlantic Seabirds*, 5(1), 21–30.

Cao, L., Deng, X., Meng, F., & Fox, A. D. (2020). Defining flyways, discerning population trends and assessing conservation challenges of key Far East Asian Anatidae species: An introduction. *Wildfowl*, 6, 1–12.

Carpenter, S. R. (2005). Eutrophication of aquatic ecosystems: Bistability and soil phosphorus. *Proceedings of the National Academy of Sciences*, 102(29), 10002–10005.

Cayuela, H., Rougemont, Q., Prunier, J. G., Moore, J. S., Clobert, J., Besnard, A. et al. (2018). Demographic and genetic approaches to study dispersal in wild animal populations: A methodological review. *Molecular Ecology*, 27(20), 3976–4010.

Cimon-Morin, J., & Poulin, M. (2018). Setting conservation priorities in cities: Approaches, targets and planning units adapted to wetland biodiversity and ecosystem services. *Landscape Ecology*, 33(11), 1975–1995.

Convention on Wetlands. (2025). *Global Wetland Outlook 2025: Valuing, Conserving, Restoring and Financing Wetlands*. Gland, Switzerland: Secretariat of the Convention on Wetlands. DOI: 10.69556/GWO-2025-eng.

Davidson, N. C. (2014). How much wetland has the world lost? Long-term and recent trends in global wetland area. *Marine and Freshwater Research*, 65(10), 934–941. https://doi.org/10.1071/MF14173

De Francisco, N., Ruiz Troya, J. D., & Agüera, E. I. (2003). Lead and lead toxicity in domestic and free-living birds. *Avian Pathology*, 32(1), 3–13.

Duan Y., Wang, S., Cao R., Feng, J., Ge J., & Wang, T. (2025). Tracking habitat use of migratory birds in a human-dominated stopover site using deep learning and acoustics indices. *Ecosphere*, 16(2), e70493. Doi: org/10.1002/ecs2.70493.

Eren, S., Beaulieu, A., Piersma, T., & Crockford, N. (2024). Flyways beyond migratory pathways: The case of waterbird conservation. *Conservation and Society*, 22(2), 74–85.

Erwin, K. L. (2009). Wetlands and global climate change: The role of wetland restoration in a changing world. *Wetlands Ecology and Management*, 17(1), 71–84. https://doi.org/10.1007/s11273-008-9119-1

Evers, D. C., Burgess, N. M., Champoux, L., Hoskins, B., Major, A., Goodale, W. M. et al. (2005). Patterns and interpretation of mercury exposure in freshwater avian communities in northeastern North America. *Ecotoxicology*, 14(1), 193–221.

Fackelmann, G., Pham, C. K., Rodríguez, Y. et al. (2023). Current levels of microplastic pollution impact wild seabird gut microbiomes. *Nat. Ecol. Evol.*, 7, 698–706. https://doi.org/10.1038/s41559-023-02013-z

Fahrig, L. (2003). Effects of habitat fragmentation on biodiversity. *Annual Review of Ecology, Evolution, and Systematics*, 34(1), 487–515.

Finlayson, C. M., Everard, M., Irvine, K., McInnes, R. J., Middleton, B. A., van Dam, A. A. et al. (2017). *The Wetland Book: I. Structure and Function, Management, and Methods*. Springer.

Fox, A. D., & Abraham, K. F. (2017). Why geese benefit from the transition from natural vegetation to agriculture. *Ambio*, 46(S2), 188–197. https://doi.org/10.1007/s13280-016-0879-1

Friend, M., Franson, J. C., & Ciganovich, E. A. (Eds.). (1999). *Field Manual of Wildlife Diseases: General Field Procedures and Diseases of Birds*. US Geological Survey (USGS).

Galbraith, H., Jones, R., Park, R., Clough, J., Herrod-Julius, S., Harrington, B. et al. (2002). Global climate change and sea level rise: Potential losses of intertidal habitat for shorebirds. *Waterbirds*, 25(2), 173–183.

Gilbert, M., Chaitaweesub, P., Parakamawongsa, T., Premashthira, S., Tiensin, T., Kalpravidh, W. et al. (2006). Free-grazing ducks and highly pathogenic avian influenza, Thailand. *Emerging Infectious Diseases*, 12(2), 227–234.

Gill, J. A., Norris, K., Potts, P. M., Gunnarsson, T. G., Atkinson, P. W., & Sutherland, W. J. (2001). The buffer effect and large-scale population regulation in migratory birds. *Nature*, 412(6845), 436–438.

Globig, A., Staubach, C., Beer, M., Köppen, U., Fiedler, W., Nieburg, M. et al. (2020). Epidemiological and molecular analysis of an outbreak of highly pathogenic avian influenza H5N8 in Germany, 2016. *Transboundary and Emerging Diseases*, 67(3), 1012–1025. https://doi.org/10.1111/tbed.13409

Green, A. J. (1996). Analyses of globally threatened Anatidae in relation to threats, distribution, migration patterns, and habitat use. *Conservation Biology*, 10(5), 1435–1445.

Green, R. E., & Pain, D. J. (2019). Risks to human health from ammunition-derived lead in Europe. *Ambio*, 48(9), 954–968. https://doi.org/10.1007/s13280-019-01194-x

Gyurácz, J., Bánhidi, P., & Csuka, A. (2011). Successful restoration of water level and surface area restored migrant bird populations in a Hungarian wetland. *Biologia*, 66(6), 1177–1182.

Haney, J. C., Geiger, H. J., & Short, J. W. (2014). Bird mortality from the Deepwater Horizon oil spill. II. Carcass sampling and exposure probability in the coastal Gulf of Mexico. *Marine Ecology Progress Series*, 513, 239–252.

Hester, R. E., & Harrison, R. M. (Eds.). (2010). *Ecosystem Services* (Vol. 15). Royal Society of Chemistry.

Hesterberg, U., Harris, K., Stroud, D., Guberti, V., Busani, L., Pittman, M. et al. (2009). Avian influenza surveillance in wild birds in the European Union in 2006. *Influenza and Other Respiratory Viruses*, 3(1), 1–14.

IPCC. (2022). *Climate Change 2022: Impacts, Adaptation, and Vulnerability*. Contribution of Working Group II to the Sixth Assessment Report of the Intergovernmental Panel on Climate Change. Cambridge University Press.

Iswandaru, D., Nugraha, G., Iswanto, A. D. D., Fitriana, Y., & Webliana, K. (2022). Between hopes and threats: New migratory birds records on the Sawala Mandapa education and training forest, Indonesia. *Forest and Society*, 6(1), 469–488.

Ivajnšič, D., & Kaligarič, M. (2014). How to preserve coastal wetlands, threatened by climate change-driven rises in sea level. *Environmental Management*, 54(4), 671–684.

Jenssen, B. M. (1994). Review article: Effects of oil pollution, chemically treated oil, and cleaning on the thermal balance of birds. *Environmental Pollution*, 86(2), 207–215. https://doi.org/10.1016/0269-7491(94)90134-2

Jetz, W., Thomas, G. H., Joy, J. B., Hartmann, K., & Mooers, A. O. (2012). The global diversity of birds in space and time. *Nature*, 491(7424), 444–448. https://doi.org/10.1038/nature11631

Jónsson, J. E., & Afton, A. D. (2015). Are wintering areas shifting north? Learning from Lesser Snow Geese banded in Southwest Louisiana. *Southeastern Naturalist*, 14(2), 293–307.

Kanstrup, N., Swift, J., Stroud, D. A., & Lewis, M. (2018). Hunting with lead ammunition is not sustainable: European perspectives. *Ambio*, 47(8), 846–857. https://doi.org/10.1007/s13280-018-1042-y

Kear, J. (Ed.). (2005). *Ducks, Geese and Swans*. Oxford University Press.

Keawcharoen, J., van Riel, D., van Amerongen, G., Bestebroer, T., Beyer, W. E., van Lavieren, R. et al. (2008). Wild ducks as long-distance vectors of highly pathogenic avian influenza virus (H5N1). *Emerging Infectious Diseases*, 14(4), 600–607. https://doi.org/10.3201/eid1404.071016

Kilpatrick, A. M., Chmura, A. A., Gibbons, D. W., Fleischer, R. C., Marra, P. P., & Daszak, P. (2006). Predicting the global spread of H5N1 avian influenza. *Proceedings of the National Academy of Sciences of the United States of America*, 103(51), 19368–19373. https://doi.org/10.1073/pnas.0609227103

Kingsford, R. T., Bino, G., & Porter, J. L. (2016). Continental impacts of water development on waterbirds, contrasting two Australian river basins: Global implications for sustainable water use. *Global Change Biology*, 22(2), 673–688. https://doi.org/10.1111/gcb.13176

Kirby, J. S., Stattersfield, A. J., Butchart, S. H. M., Evans, M. I., Grimmett, R. F. A., Jones, V. R. et al. (2008). Key conservation issues for migratory land and waterbird species on the world's major flyways. *Bird Conservation International*, 18(S1), S49–S73. https://doi.org/10.1017/S0959270908000439

Klein, M. L., Humphrey, S. R., & Percival, H. F. (1995). Effects of ecotourism on distribution of waterbirds in a wildlife refuge. *Conservation Biology*, 9(6), 1454–1465.

Kühn, S., & van Franeker, J. A. (2020). Quantitative overview of marine debris ingested by marine megafauna. *Mar Pollut Bull.*, 151, 110858. doi: 10.1016/j.marpolbul.2019.110858. Epub 2020 Jan 29. PMID: 32056640.

Kumar, R., Horwitz, P., & Finlayson, C. M. (2023). Wetlands as social–ecological systems: Bridging nature and society. *In: Ramsar Wetlands* (pp. 525–553). Elsevier.

Lennox, R. J., Kambestad, M., Berhe, S., Birnie-Gauvin, K., Cooke, S. J., Dahlmo, L. S. et al. (2025). The role of habitat in predator–prey dynamics with applications to restoration. *RestorEcol*, 33, e14354. https://doi.org/10.1111/rec.14354

Lewandrowski, J., & Ingram, K. (1999). Policy considerations for increasing compatibilities between agriculture and wildlife. *Natural Resources Journal*, 229–269.

Lobat, Z., Saeed, A., Jafari, H. R., & Masoud, B. K. (2021). Sustainable wetland management through bridging the communication gap between conservation projects and local communities. *Environment, Development and Sustainability*, 23(7), 11098–11119.

Ma, Z., Cai, Y., Li, B., & Chen, J. (2010). Managing wetland habitats for waterbirds: An international perspective. *Wetlands*, 30(1), 15–27. https://doi.org/10.1007/s13157-009-0001-6

Madsen, J., & Guillemain, M. (2013). Incidence of embedded shotgun pellets and inferred hunting kill amongst Russian/Baltic barnacle geese Branta leucopsis. *European Journal of Wildlife Research*, 59(1), 77–80. https://doi.org/10.1007/s10344-012-0675

Madsen, J., Bunnefeld, N., Nagy, S., Griffin, C., Defos du Rau, P., Mondain-Monval, J.-Y. et al. (2015). *Guidelines on Sustainable Harvest of Migratory Waterbirds*. AEWA Conservation Guidelines No. 5. African–Eurasian Waterbird Agreement Secretariat.

Mekonen, S. (2017). Birds as biodiversity and environmental indicator. *Indicator*, 7(21), 28–34.

Mateo, R., & Kanstrup, N. (2019). Regulations on lead ammunition adopted in Europe and evidence of compliance. *Ambio*, 48(9), 989–998. https://doi.org/10.1007/s13280-019-01191-0

Mateo, R., Green, A. J., Lefranc, H., Baos, R., & Figuerola, J. (2007). Lead poisoning in wild birds from spent ammunition: A review and update. *Ambio*, 36(6), 461–469. https://doi.org/10.1579/0044-7447(2007)36[461:LPIWBF]2.0.CO;2

Guerrero-Moreno, M. A., da Silva, E. C., Oliveira, F. A., Nascimento, A. C. L., Michelan, T. S., Dias-Silva, K. et al. (2025). Use of aquatic organisms as flagship species in selecting priority areas for conservation. *Water Biology and Security*. https://doi.org/10.1016/j.watbs.2025.100509.

Meng, L., Roulet, N., Zhuang, Q., Christensen, T. R., & Frolking, S. (2016). Focus on the impact of climate change on wetland ecosystems and carbon dynamics. *Environmental Research Letters*, 11(10), 100201.

Millennium Ecosystem Assessment. (2005). *Ecosystems and Human Well-Being: Wetlands and Water Synthesis*. World Resources Institute.

Milton, G. R., & Finlayson, C. M. (2018). Diversity of freshwater ecosystems and global distributions. *Freshwater Ecology and Conservation: Approaches and Techniques*, 3–19.

Mineau, P., & Whiteside, M. (2013). Pesticide acute toxicity is a better correlate of U.S. grassland bird declines than agricultural intensification. *PLoS One*, 8(2), e57457. https://doi.org/10.1371/journal.pone.0057457

Munster, V. J., Veen, J., Olsen, B., Vogel, R., Osterhaus, A. D., & Fouchier, R. A. (2007). Towards improved influenza A virus surveillance in migrating birds. *Vaccine*, 25(16), 3172–3177. https://doi.org/10.1016/j.vaccine.2007.01.069

Murray, N. J., & Fuller, R. A. (2012). Coordinated effort to maintain East Asian–Australasian flyway. *Oryx*, 46(4), 479–480. https://doi.org/10.1017/S0030605312001135

Murray, N. J., Ma, Z. J., & Fuller, R. A. (2015). Tidal flats of the Yellow Sea: A review of ecosystem status and anthropogenic threats. *Austral Ecology*, 40(4), 472–481. https://doi.org/10.1111/aec.12211

Newton, W. E. (2006). Can conditions experienced during migration limit the population levels of birds? *Journal of Ornithology*, 147, 146–166.

Nicholls, D. G., & Hines, J. E. (2007). Effects of lead shot ingestion on survival of wildfowl in the United States. *Journal of Wildlife Management*, 71(5), 1575–1582. https://doi.org/10.2193/2006-206

Noer, H., Madsen, J., & Clausen, P. (1995). Lead shot ingestion and lead poisoning in waterfowl in Denmark. *Environmental Pollution*, 88(1), 1–7. https://doi.org/10.1016/0269-7491(94)00072-7

Nourani, E., Kaboli, M., Farhoodinia, M., & Collen, B. (2017). National assessment of threatened species using sparse data: IUCN Red List classification of Anatidae in Iran. *Animal Conservation*, 20(1), 42–50.

Pain, D. J. (2019). Effects of lead from ammunition on birds and other wildlife. *Science of the Total Environment*, 676, 1–6. https://doi.org/10.1016/j.scitotenv.2019.04.054

Pettigrove, V., Hassell, K., Kellar, C., Long, S., Macmahon, D., Myers, J. et al. (2023). Catchment sourcing urban pesticide pollution using constructed wetlands in Melbourne, Australia. *Science of the Total Environment*, 863, 160556.

Prasanya, J., Kanmani, S., & Senthil Kumar, P. (2024). A review of the wetland's restoration mechanisms and its economic and social benefits. *Water Practice & Technology*, 19(11), 4355–4377.

Pullin, A. S., Sutherland, W., Gardner, T., Kapos, V., & Fa, J. E. (2013). Conservation priorities: Identifying need, taking action and evaluating success. *Key Topics in Conservation Biology*, 2, 1–22.

Quesnelle, P. E., Fahrig, L., & Lindsay, K. E. (2013). Effects of habitat loss, habitat configuration and matrix composition on declining wetland species. *Biological Conservation*, 160, 200–208.

Ramsar Convention Secretariat. (2013). *The Ramsar Convention Manual: A Guide to the Convention on Wetlands* (6th ed.). Ramsar Convention Secretariat.

Ramsar Convention Secretariat. (2018). *Global Wetland Outlook: State of the World's Wetlands and their Services to People 2018*. Ramsar Convention Secretariat.

Rebelo, L. M., Finlayson, C. M., & Nagabhatla, N. (2009). Remote sensing and GIS for wetland inventory, mapping and change analysis. *Journal of Environmental Management*, 90(7), 2144–2153.

Reynolds, R. E., Shaffer, T. L., Renner, R. W., Newton, W. E., Batt, B. D. J., & Anderson, M. G. (2017). Conservation of wetlands and waterfowl in the prairie pothole region of North America. *Wildfowl*, 67, 1–23.

Runge, C. A., Watson, J. E. M., Butchart, S. H. M., Hanson, J. O., Possingham, H. P., & Fuller, R. A. (2015). Protected areas and global conservation of migratory birds. *Science*, 350(6265), 1255–1258. https://doi.org/10.1126/science.aac9180

Rutter, J. D., Dayer, A. A., & Raedeke, A. H. (2022). Ecological awareness, connection to wetlands, and wildlife recreation as drivers of wetland conservation involvement. *Wetlands*, 42, 18. https://doi.org/10.1007/s13157-021-01522-6

Scheuhammer, A. M. (1987). The chronic toxicity of aluminium, cadmium, mercury, and lead in birds: A review. *Environmental Pollution*, 46(4), 263–295.

Shi, J., Meng, L., Xia, S., Liu, S., & Zhou, L. (2024). Habitat suitability and determinants for anatidae in multi-watershed composite wetlands in Anhui, China. *Animals*, 14(7), 1010.

Smalley, A. L., & Reeves, H. M. (2022). *The Market in Birds: Commercial Hunting, Conservation, and the Origins of Wildlife Consumerism, 1850–1920*. JHU Press.

Stroud, D. A., & Pain, D. J. (2011). *Implementing the AEWA Action Plan for Reducing, and Eventually Eliminating the Use of Lead Shot in Wetlands*. Agreement on the Conservation of African–Eurasian Migratory Waterbirds (AEWA) Technical Series No. 6. Bonn, Germany: UNEP/AEWA Secretariat.

Stroud, D. A., Baker, A., Blanco, D. E., Davidson, N. C., Delany, S., Ganter, B. et al. (2016). *The Conservation Status of Migratory Waterbirds and Wetlands: A Compilation of Analyses of Relevant International Treaties, Conventions, Agreements and Other Instruments*. UNEP-WCMC.

Swayne, D. E., & Sims, L. D. (2021). Avian influenza. *Veterinary Vaccines: Principles and Applications*, 229–251.

Takekawa, J. Y., Newman, S. H., Xiao, X., Prosser, D. J., Spragens, K. A., Palm, E. C. et al. (2010). Migration of waterfowl in the East Asian flyway and spatial relationship to HPAI H5N1 outbreaks. *Avian Diseases*, 54(s1), 466–476.

Taylor, R. L., Maxwell, B. D., & Boik, R. J. (2006). Indirect effects of herbicides on bird food resources and beneficial arthropods. *Agriculture, Ecosystems & Environment*, 116(3–4), 157–164. https://doi.org/10.1016/j.agee.2006.01.012s

Tozer, D. C., Steele, O., & Gloutney, M. (2018). Multispecies benefits of wetland conservation for marsh birds, frogs, and species at risk. *Journal of Environmental Management*, 212, 160–168.

Turner, R. K., Van Den Bergh, J. C., Söderqvist, T., Barendregt, A., Van Der Straaten, J., Maltby, E. et al. (2000). Ecological–economic analysis of wetlands: Scientific integration for management and policy. *Ecological Economics*, 35(1), 7–23.

Waliczky, Z., Fishpool, L. D., Butchart, S. H., Thomas, D., Heath, M. F., Hazin, C. et al. (2019). Important bird and biodiversity areas (IBAs): Their impact on conservation policy, advocacy and action. *Bird Conservation International*, 29(2), 199–215.

Ward, M. P., Semel, B., & Herkert, J. R. (2010). Identifying the ecological causes of long-term declines of wetland-dependent birds in an urbanizing landscape. *Biodiversity and Conservation*, 19(11), 3287–3300.

Wetlands International. (2021). *Waterbird Population Estimates*, 6th edition. Wetlands International. https://wpe.wetlands.org

White, P. C., Godbold, J. A., Solan, M., Wiegand, J., & Holt, A. R. (2010). Ecosystem services and policy: A review of coastal wetland ecosystem services and an efficiency-based framework for implementing the ecosystem approach. *Ecosystem Services.* Royal Society of Chemistry, Cambridge, 29–51.

Wille, M., & Barr, I. G. (2022). Resurgence of avian influenza virus. *Science*, 376(6592), 459–460. https://doi.org/10.1126/science.abo1232

Wu, L., Wang, Y., Mo, X., Wei, Q., Ma, C., Wang, H. et al. (2022). Shifted to the south, shifted to the north, but no expansion: Potential suitable habitat distribution shift and conservation gap of the critically endangered Baer's Pochard (Aythya baeri). *Remote Sensing*, 14(9), 2171. https://doi.org/10.3390/rs14092171

Zedler, J. B., & Kercher, S. (2005). Wetland resources: Status, trends, ecosystem services, and restorability. *Annual Review of Environment and Resources*, 30(1), 39–74. https://doi.org/10.1146/annurev.energy.30.050504.144248

Ziv, G., Baran, E., Nam, S., Rodríguez-Iturbe, I., & Levin, S. A. (2012). Trading-off fish biodiversity, food security, and hydropower in the Mekong River Basin. *Proceedings of the National Academy of Sciences of the United States of America* (PNAS), 109(15), 5609–5614. https://doi.org/10.1073/pnas.1201423109

Index

A

Adaptive radiation, 210, 211
Adoption, 115, 116
Aerodynamic drafting, 159
Aerodynamic efficiency, 165
AEWA, 307, 308, 314
Aggression, 115
Alloparenting, 94, 115, 116
Anatidae, 1-5, 36-38, 41, 43, 94-98, 130-134, 201-210, 249-256, 306-309
Anatidae family diseases, 277, 279, 281, 283
Anatomical specializations, 166
Anatomy, 36-38, 44, 49, 50, 57
Anseriformes, 158, 162
Anthropogenic drivers, 278, 287, 288
Aquatic habitat, 165
Avian cholera (*Pasteurella multocida*), 277-279, 281, 283
Avian influenza containment, 293, 295
Avian malaria (*Plasmodium relictum*), 86-288, 290, 296

B

Behavior, 94, 95, 97–100
Bill morphology, 79, 80
Bill structure, 98-100
Biodiversity, 161, 162, 171, 174, 178, 179, 180
Biodiversity indicator, 69, 72, 73
Biodiversity loss, 289, 290
Bioindicator, 225
Biosecurity measures, 281, 289, 291, 292, 293, 294
Bird–Human coexistence, 255, 265, 266

C

Celestial navigation, 171, 172
Climate, 118
Climate change, 13, 19, 21, 22, 162, 163, 170, 172, 173, 176
Climate change impact, 287, 289, 296
Cluster-N, 171
Community-based conservation, 261
Conservation, 64, 66, 68, 70, 72-74, 225, 226, 228, 230, 232
Conservation biology, 19, 30
Conservation challenges, 180
Conservation genomics, 213, 217
Corticosterone, 166,
Courtship displays, 110
Critically endangered, 158, 169
Cultural symbolism, 256

D

Demographic trends, 233
Dietary guilds, 137
Disease transmission dynamics, 278

DNA barcoding, 201-203, 205, 207
Dominance hierarchies, 94, 113, 115
Duck plague (Duck viral enteritis), 279, 280, 281, 282, 283, 291, 292
Ducks, 36-41, 225, 227, 232, 234
Duct, 102, 103, 106–110, 114

E

Echelon formation, 166
Ecological niches, 62-64, 70, 74
Ecological restoration, 292, 293
Ecosystem, 94, 95, 112, 118
Ecosystem resilience, 291, 292, 296
Ecosystem services, 29
Endangered, 228-230, 235-237, 239, 240
Ethology, 116
Evolutionary adaptations, 95, 96
Extinct, 225, 226, 228-230

F

Feeding ecology, 133
Feeding guilds, 75
Flyways, 234, 235, 242
Folklore and birds, 253, 255, 256
Food web, 160, 161
Foraging, 160-162, 174, 175, 177
Foraging strategies 63, 64, 75-79, 136, 139, 142, 144
Fragmented habitat, 165, 176
Freshwater, 225, 229, 230
Fungal aspergillosis, 285, 297

G

Geese, 36-40, 42, 94, 97, 99, 102, 104
Genomic surveillance, 289, 294, 295
Ghrelin, 166, 192
Global status, 225, 238, 242
Goose, 228-232
Gregarious, 159

H

Habitat, 95, 97, 98, 101–103, 106, 112–115, 117
Habitat degradation, 308
Habitat destruction, 162, 174
Habitat fragmentation, 288
Habitat preferences, 63, 64, 67, 68
Highly pathogenic avian influenza (HPAI H5N1), 279, 280, 282, 283, 287, 288, 291, 294
Hybridization, 201, 202, 204-206
Hyperphagia, 166, 167

I

Illegal poaching, 174
Invasive predators, 164, 184
IUCN Red List, 230, 238, 241

M

Magnetoreceptive cryptochrome 171
Microplastic pollution, 175
Migration 8-13, 158-165
Migratory flyways, 161
Migratory waterfowl, 276-278, 280, 281
Monogamous, 159
Morphological adaptations, 5, 131, 143

N

Navigation, 159, 166, 171, 172
Neuroethology, 116
Neuroplasticity, 171
Northern hemisphere, 182

O

One health framework, 291, 293, 294, 295, 296,

P

Parasitic helminth infections, 279, 285, 286
Pathogen ecology, 277-279, 282, 283
Photoperiod, 166, 173
Phylogeography, 203, 210
Physiology, 36, 41, 44-47
Pollution, 160, 172, 174-176

Population trends, 227, 229, 233, 234, 240, 242
Public health risks, 291

R

Ramsar convention, 307-309
Ramsar sites, 176, 179, 180, 182
Reproduction, 102, 108-110, 116
Reproductive biology, 13, 14
Rete ophthalmicum, 166

S

Seasonal migration, 160, 161, 183
Seed dispersion, 161
Social learning, 116–118, 165
Social structure, 95, 97, 99, 101
Southern hemisphere, 160
Survival, 98, 106–117
Swan, 94, 97, 99–101, 225, 228, 229, 231

T

Temperature, 162, 167, 173, 183
Thermoregulatory organ, 167

U

Unsustainable hunting, 174

V

Vector-borne diseases, 281, 283, 287, 289, 296
V-formation flight, 159
Viral reassortment, 289
Visceral fat, 166
Vocalization, 95, 104, 106–109

W

Waterbirds, 228, 235, 236, 240
Waterfowl, 1-3, 6, 10, 11
Waterfowl ecotourism, 276, 277, 291
Wetlands, 3, 10-13, 15, 130, 131, 140-142, 147, 159, 161-163, 169, 170, 171, 225, 229, 231-234
Wetland conservation, 276, 277, 293, 296
Wildlife–livestock interface, 295
Wing strokes, 165

Z

Zoonotic risk, 264
Zoonotic transmission, 279, 283, 284, 289

About the Editors

Ashish Kumar Arya is a Senior Researcher at the Department of Environmental Science at Graphic Era Deemed to be University, Dehradun, India. His expertise includes avian biology, wetland ecology, wildlife, and biodiversity conservation, with a strong focus on diversity and conservation of water birds, as well as the breeding biology of birds.

Archana Bachheti is a Professor at Graphic Era University in Dehradun, India. With an extensive experience of close to 20 years, she has served in different capacities within academia in India and has also provided expertise internationally. Her teaching portfolio includes subjects like Ecology and Environment, Environmental Science, Freshwater Ecology, Disaster Management, and Bryophytes and Pteridophytes. Dr. Archana's research interests cover a broad and interdisciplinary field of plant ecology. Her focus areas include eco restoration, green chemistry, especially the synthesis of nanomaterials, and exploring the medicinal properties of plants.

Vinaya Kumar Sethi is a Faculty at Environmental Science at Uttarakhand Sanskrit University, India. His research area includes breeding behaviour and acoustic communication in birds as well as wildlife and biodiversity conservation.

For Product Safety Concerns and Information please contact our EU representative GPSR@taylorandfrancis.com Taylor & Francis Verlag GmbH, Kaufingerstraße 24, 80331 München, Germany